Biology: A modern int
GCSE edition

B. S. Beckett

The author is a graduate of Hull University where he obtained a B.Sc. degree in Zoology and Botany in 1963. After that he taught in schools in Leeds and Hull, and from 1968–71 he was Head of Biology at King George V School in Hong Kong. While he was at this last post he carried out investigations into the use and evaluation of discovery methods in science teaching, which led to an M.A. (Ed.) degree. When he returned to England he continued his research on discovery and formal teaching methods, for which he was awarded a B.Phil. degree by Hull University.

Since then the author has divided his time between writing, illustrating, and teaching.

B.S. Beckett

Biology

A modern introduction

GCSE edition

Oxford University Press

Oxford University Press, Walton Street, Oxford OX2 6DP

Oxford New York Toronto
Delhi Bombay Calcutta Madras Karachi
Petaling Jaya Singapore Hong Kong Tokyo
Nairobi Dar es Salaam Cape Town
Melbourne Auckland

and associated companies in
Beirut Ibadan Berlin Nicosia

Oxford is a trade mark of Oxford University Press

© *Oxford University Press 1976, 1982, 1986*
First published 1976
GCSE edition 1986
Reprinted 1986
ISBN 0 19 914260 2

Acknowledgements
Illustrations are by B S Beckett.

The publisher would like to thank the
following for permission to reproduce
photographs: AC Allison, p 67; Anatomical Institute, Bern, p 100; Ardea London, p 131 centre; Ardea/John E Swedburg, p 199; Ardea/Weaver, p 290 right; British Kidney Patient Association, p 116; British Museum (Natural History), p 212; Camera Press, p 226; J Allan Cash, p 283, p 285; Central Electricity Generating Board, p 228; Bruce Coleman, p 167; Cow and Gate Limited, p 193; Julian Cremona, p 131 right, p 132; Friends of the Earth/C Rose, p 273 top; Philip Harris Biological Ltd, p 119; DM Kendall, p 168; Keystone Press, p 211; JH Kugler, p 11; KR Lewis, p 21; DB Moffat, p 115; Network Photographers/Martin Mayer, p 34 bottom right; Network/Mike Goldwater, p 267; Oxford Scientific Films/Steve Dalton front cover; OSF/Bob Fredrick back cover; OSF/Doug Allan p 131 left; OSF/Dr JAL Cooke p 273 bottom; Oxfam, p 27, p 34 bottom left and right, p 256; Petit Format/Nestle/Science Photo Library, p 191; Popperfoto, p 195; JJ Pritchard, p 132; Department of Histopathology, John Radcliffe Hospital, p 146, p 241 right; Radio Times Hulton Picture Library, p 31; Journal of Pathological Bacteriology 97, p 48; Department of Medical Illustration, St Bartholomew's Hospital, London, p 241 left; Science Photo Library/Dr Jeremy Burgess, p 15 bottom left; SPL, p 101; SPL/Sinclair Stammers, p 207; SPL/CNRI, p 236; Sport and General Press Agency, p 91; John Watney Photo Library, p 75; C James Webb, p 158; Welsh Development Agency, p 290 left; RHJ Williams, p 283

The photographs on the front and back of this book are of the Purple Emperor butterfly.

Notice to readers:
The author welcomes constructive criticism of his books, suggested improvements, and general correspondence related to topics covered in the text. Correspondence should be addressed to B.S. Beckett, care of the Oxford University Press.

Printed in Great Britain by Butler & Tanner Ltd, Frome and London

To the teacher

Biology is an established and successful textbook for secondary schools. For the GCSE examinations it has been thoroughly revised to provide a comprehensive coverage of all the five syllabuses.

The presentation is more visual: new drawings and diagrams have been added, and certain existing ones have been improved. Many new photographs are included, some specially taken by the author to illustrate syllabus topics.

There is a greater emphasis on the display and interpretation of data, in the form of tables, graphs, and diagrams.

The book is more relevant to everyday life, in that it includes more of the social, environmental, and technological aspects of biology.

The organization and methodology of the GCSE edition remain unaltered. Plant and animal biology are still treated together, with special emphasis on the concepts and principles common to both.

Chapters are arranged sequentially so that some topics such as cells and photosynthesis can be developed gradually until their biological significance is fully appreciated.

There are two different types of illustration. Some are an attempt to represent subjects as they appear in nature, and from these students can see how specimens appear when viewed under ideal conditions. Other illustrations are diagrammatic representations of the realistic drawings, or summaries of biological principles described in the text. Important plant and animal structures are drawn in both ways so that, by comparison, the student can see how to reduce complex information to a simple diagram. Almost all illustrations have been drawn by the author and are carefully integrated into the text and referred to wherever they will aid understanding.

Chapters now include a much larger selection of tests. There are a number of examination-type questions in which students are asked to recall facts, label diagrams, and interpret experimental results, tables, and graphs. The factual recall tests which have proved very popular in previous editions have also been included and up-dated.

The practical investigations included in most chapters have been revised and added to. Students are shown how to verify what they have learned and explore for themselves the processes of scientific method by designing their own experiments and experiencing the excitement of discovery. Students are encouraged to display and interpret their experimental data in the form of graphs and tables.

Of course, no book at this level, however modern, can be any more than an introduction. This one will continue to serve its purpose if it helps teachers to interest their students in the vast, rich, and exciting world which is the subject of *Biology*.

To the student

Scientists approach their subject in the following way. First they *study* existing knowledge that others have already discovered. Second, they *learn* this body of knowledge including all its technical terms. Third, they make sure they *understand* what they have learned. Fourth, they may repeat some experiments to *verify* the results. Fifth, and most important, they devise new experiments to *investigate* new subjects and perhaps create new knowledge.

This book is written in such a way that you can follow roughly the same five steps yourself. This is possible because most chapters consist of the following parts.

The text

The text of each chapter introduces the existing knowledge on a biological topic. These accounts are up to date, fully illustrated, simply worded, and contain explanations of the more important technical terms.

Practical work

Most chapters contain exercises which show you how to verify the accuracy of factual statements contained in the text. You can do this by repeating some of the experiments from which these facts were discovered.

In addition, most chapters also contain opportunities for you to make investigations into knowledge which is not in the chapter, but which follows naturally from it. Sometimes these investigations state problems for you to solve and instructions on how to solve them; but in other cases there is no more than a problem, a list of necessary apparatus, and a few clues on how to proceed.

Work on an investigation of this type is obviously not the same as the work of a scientist who is solving a problem and creating new knowledge. But it is like the work of a scientist in some ways: it gives you a chance to use knowledge gained from a chapter to discover knowledge which is new *to you*. At the same time it allows you to experience what it is like to solve a problem, making use of controlled experiments which, in some cases, you have devised yourself.

Questions

Questions are provided to test your ability to recall facts, and your ability to understand these facts. Some questions are similar to those which you will meet in examination papers and will give you experience in attempting these tests.

Summary and factual recall tests

In addition, each chapter ends with a brief summary of the basic facts contained in the text. You can take part in constructing these summaries by supplying missing words and phrases, and sometimes by summarizing ideas in your own words. You should try to supply these missing parts from memory before referring to the chapter.

This book introduces biology and the way biologists work. This involves more than memorizing facts. It involves *understanding* the facts by reading them, summarizing them, checking them, and investigating beyond them.

Contents

Contents

1

The characteristics and variety of living things

1.1 The characteristics of life

Biology is the study of living things, but it has not yet produced a definition of life. It has, however, produced a list of the characteristics of life. It has done this by studying the differences between living and non-living things. It is not difficult, for instance, to see the differences between a man and waxwork dummy. The man walks, eats, sees, etc., whereas the dummy does not. However, characteristics such as these cannot be applied to all living things. Most plants do not move about from place to place, eat, or perform breathing movements, but this does not mean they are not alive. Furthermore, some objects such as plant seeds show no signs of life whatsoever until they begin to grow. Lotus seeds are an interesting example: some were once grown after being stored for 160 years.

When listing the characteristics of life, therefore, only those which are common to all living things must be included. It must also be remembered that under certain circumstances all these characteristics can be suspended for a period, as in seeds, and reappear later.

It is now generally accepted that a thing is alive if it exhibits, or is capable of exhibiting, the seven characteristics listed below. From a biological standpoint, these are the features which distinguish living from non-living things:

1. *Movement*
Living things move in a directed and controlled way. In other words they move of their own accord, whereas non-living things move only if pushed or pulled by something else. Animals usually move their whole bodies and often have special organs which do this, such as fins, wings, and legs. These are called **locomotory organs** because they move the animal from place to place. Plants also move in a directed and controlled manner but generally only parts of their bodies move. Consequently, their movements

are not locomotory. Leaves which turn towards the light, and roots which grow down into the soil are examples of plant movements. Plant movements are generally very slow and not always obvious to the casual observer.

2. *Sensitivity*
Living things are sensitive to their environment. This means that they detect and respond to events in the world around them. Simple organisms such as *Amoeba* (page 8) have limited sensitivity, while higher animals such as humans are more sensitive and can react to small changes in light, sound, touch, smell, taste, temperature, etc. Humans have highly developed sense organs and a complex nervous system through which responses are co-ordinated. Plants, on the other hand, have no sense organs, but the way they move shows that certain regions of their bodies are sensitive to light, gravity, water, and various chemicals.

3. *Feeding*
Living things feed. Food is the material from which organisms obtain energy for movement, and the raw materials necessary for growth and repair of the body. The scientific study of food and the different ways in which organisms feed is called **nutrition**. There are two types of nutrition:

a) *Autotrophic nutrition* Autotrophic organisms make their own food. Green plants, for example, manufacture sugar and starch from carbon dioxide and water, using the energy of sunlight to drive the necessary chemical reactions. This process is called **photosynthesis**.

b) *Heterotrophic nutrition* Heterotrophic organisms obtain food from the bodies of other organisms. This is done in various ways. **Carnivores,** such as cats and foxes, eat the flesh of animals. **Herbivores,** such as cattle, horses, and rabbits, eat plants.

Omnivores, such as humans, eat both plants and animals. **Parasites**, such as fleas, mosquitoes, and tapeworms, live on or in another living organism called the **host** from which they obtain food. **Saprotrophs** (sometimes called saprophytes), which include many types of fungi and bacteria, obtain their food in liquid form from the decaying remains of dead organisms.

4. Respiration

Living things respire. Respiration is a complex sequence of chemical reactions which results in the release of energy from food. These reactions are vital to life because they provide the power for the numerous chemical and physical processes within living organisms. Most organisms respire using oxygen which they absorb from the surrounding air or water. The oxygen reacts with food substances in the body of the organism, releasing energy and producing carbon dioxide gas and water as waste substances. Most animals obtain oxygen by means of respiratory organs, such as gills and lungs. These animals carry out breathing movements to ensure that waste carbon dioxide is removed from the body. Plant respiration is less obvious, but it can be detected using methods described in chapter 8.

5. Excretion

Living things excrete. Excretion is the removal from the body of waste products which result from normal life processes. Waste products such as carbon dioxide must be excreted. If they accumulate in the body they cause poisoning which slows down vital chemical reactions. Excretion must not be confused with **egestion**, which is the removal from the body of substances that have passed partly, or completely undigested though the digestive system. Most animals have special excretory organs, such as kidneys.

6. Reproduction

Living things are able to reproduce. Unless reproduction occurs populations of organisms will diminish and eventually disappear as their members die from old age, disease, accidents, attacks from other organisms, etc. It is a fundamental law of nature that living things can only be produced by other living things, i.e. every living organism owes its existence to the reproductive activities of other living organisms. This fact has not always been accepted. At one time it was believed that life could develop spontaneously. People believed, for example, that mould formed out of decaying bread, that cockroaches were formed out of crumbs and dust on a bakery floor, and that rotting sacks of grain turned into rats and mice. These beliefs

contributed to the theory that living things can arise spontaneously out of non-living material This theory of spontaneous generation has now been totally rejected in its original form.

7. Growth

Living things grow. Most animals grow until they reach maturity. Plants usually continue to increase in size throughout their life span.

1.2 Metabolism and enzymes

Metabolism is a word used to describe all the chemical changes within an organism which are necessary for life. It is one of the most useful technical words in biology because it summarizes the most vital process of life. Apart from its use in this general sense, metabolism also refers to specific processes within part of an organism, such as **muscle metabolism**. In addition, it is used to refer to the part played by a particular substance within an organism's life processes, such as **protein metabolism**.

The word **metabolite** refers to substances which undergo various changes during metabolism. For example, carbon dioxide and water are metabolites used in the process of photosynthesis.

There are two different types of metabolism:

Catabolism

Catabolism is a process in which complex substances are **broken down** into simpler ones, resulting in the release of energy. During respiration, for example, glucose sugar is broken down into carbon dioxide gas and water, releasing energy for life.

Anabolism

Anabolism is a process which uses energy released during catabolism to **build up** (synthesize) complex substances from simpler ones. Examples of anabolism are photosynthesis, and all the processes of growth and repair in the bodies of organisms.

Both catabolic (breaking down) and anabolic (building up) processes occur in living things. In growing organisms anabolism proceeds at a faster rate than catabolism, whereas in most healthy adult animals the two are more or less balanced. Neither process stops until the organism dies. Furthermore, new experimental techniques have shown that the bodies of living organisms are constantly changing. Hardly any body substance remains for long in a stable condition. The idea that the adult human body is a relatively permanent structure is an illusion.

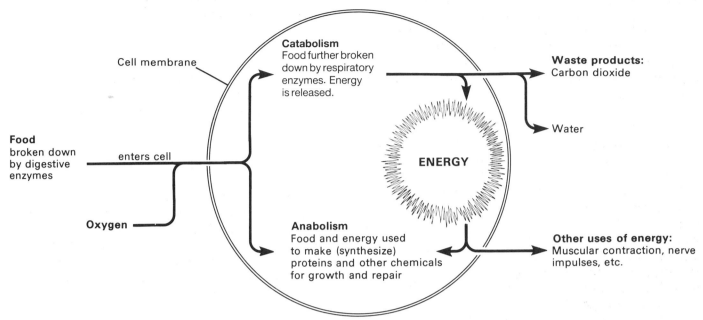

Fig. 1.1 Diagram of metabolism: all the chemical changes which are necessary for life

Almost every part, even the solid parts such as bones, are constantly and simultaneously being broken down and remade. The raw materials for these changes come from food, and the energy needed comes from respiration.

Enzymes

Metabolism would occur very slowly or not at all if it were not for chemicals called **enzymes**. Enzymes speed up chemical changes inside organisms without themselves being used up in the change. The word **catalyst** is used to describe chemicals which do this.

Each type of enzyme acts as a catalyst for only *one* type of chemical change. For examples, there are enzymes which join glucose molecules together to make starch – an example of anabolism. During respiration other enzymes break down glucose molecules to release energy (Fig. 1.1). Section 4.3 describes enzymes which break down food into simple, soluble substances during digestion.

1.3 Identifying organisms

It is often necessary to identify various specimens during the course of a biological investigation. This task is quite easy when it involves telling the difference between major groups. For instance, there is

little difficulty in distinguishing between beetles and butterflies. In detailed studies, however, there remains the problem of finding the name of an organism. Having found a beetle, for example, how is it to be identified from among the 300 000 different species in the world? First, it is almost certain that the beetle's country of origin will be known, which reduces the range of possibilities a little. Second, there is the possibility of making an identification by looking through books with accurate illustrations of living things from various countries, but in the case of organisms which look very much alike it is difficult to know where to begin. Devices called **keys** make this task easier.

Table 1 shows one of the simplest types of key. It is made up of brief descriptions arranged in numbered pairs. To use this type of key you begin at the first pair of descriptions and decide which one fits the organism to be identified. The key either names the organism or gives the number of the next pair of descriptions which must be consulted. This procedure is continued until an identification is made.

Keys designed to identify animals or plants down to species level have the same structure as Table 1. But they can be much more difficult to use because they include many specialist technical terms, and often use features which require a hand lens, or even a microscope to observe.

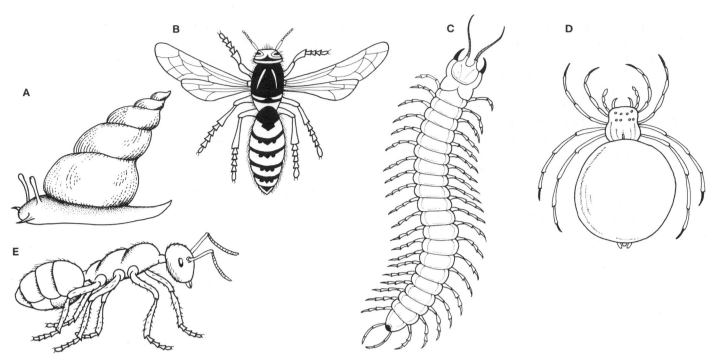

Fig. 1.2 Identify these animals using Table 3

Table 1 Key to animals with backbones

1	Hair present	Mammals
	Hair absent	Go to 2
2	Feathers present	Birds
	Feathers absent	Go to 3
3	Breathe with lungs	Go to 4
	Breathe with gills	Go to 5
4	Dry scaly skin	Reptiles
	Moist scaleless skin	Amphibians
5	Fins and cartilage skeleton	Chondrichthyes (cartilage fish)
	Fins and bony skeleton	Osteichthyes (bony fish)

How to make a key

Imagine you have to make a key which can be used to identify the animals illustrated in Figure 1.2. First, you would list the main features of these animals, looking for those which make each one different from all the others, as shown in Table 2. Next, you would arrange these features together in pairs so that they make a key, as shown in Table 3. Use this key in the following way.

Look at animal A in Figure 1.2 and read the first pair of descriptions in Table 3. Decide which description fits this animal. Opposite this description you will find either the name of this animal or the number of the next group of descriptions. Continue in this way until you reach the name of the animal. Identify animals B, C, D, and E.

Table 2 Features of animals illustrated in Figure 1.2

Features	A	B	C	D	E
Shell	Present	Absent	Absent	Absent	Absent
Wings	Absent	Present	Absent	Absent	Absent
Limbs	Absent	Six	Many	Eight	Six

Table 3 Key to the animals illustrated in Figure 1.2

1	Shell present	Snail
	Shell absent	Go to 2
2	Wings present	Wasp
	Wings absent	Go to 3
3	Eight legs	Spider
	Many legs	Centipede
	Six legs	Ant

1.4 Classification: sorting living things into groups

Using a key is one way of identifying organisms. More than 1 500 000 different kinds of living thing have been found so far, and more are being discovered all the time. The task of sorting is mammoth, keys alone will not suffice. One of the methods biologists use is **classification**.

How organisms are classified

Living things can be sorted into groups on the basis of shared features. In other words, different creatures can be grouped together if they have something in common.

It is important, however, to identify as many shared features as possible before deciding which creatures to group together. For example, all animals that fly *could* form a group, but this would mean including animals as different as sparrows, bats, and butterflies. Biologically, though, birds have more in common with fish than with butterflies.

Modern classification systems are based on a careful study of all the main features of organisms, including body shapes, different types of limbs and skeletons, the arrangement of internal organs, and many other characteristics.

Biological classification

Organisms are first divided into very large groups called **kingdoms**. Most organisms can be placed in either the **plant kingdom** or the **animal kingdom**. Some simple organisms which cannot easily be fitted into either of these groups can be put together into the **protista kingdom**. These large groups are then gradually broken down into smaller and smaller groups, containing fewer and fewer types of organisms, but with more and more features in common. Kingdoms are sub-divided into smaller groups called **phyla** (singular **phylum**). Each phylum or division is divided into **classes**, the classes are divided into **orders**, and the orders are divided into **families**. Each family is divided into **genera** (singular **genus**), and genera are divided into **species**.

Species

A species is a group of organisms so alike that they can mate together and produce young. Humans, cats, dogs, and sunflowers are examples of species. Usually, members of one species cannot breed with members of another. Different types of dog can mate and produce puppies, but dogs and cats cannot breed with each other.

Naming organisms

Every species of organism known to science, living and extinct, has been given a double scientific name: one name for its genus (the generic name) and one for its species (the specific name). This scientific name is always written in Latin, which may seem strange and unnecessarily complicated. But it does mean that all scientists, whatever part of the world they come from, can use the same name and be sure they are talking about the same organism. For example, the scientific name for humans is *Homo sapiens*. This is usually written *H. sapiens* for short. The generic name is always written with a capital letter and the specific name with a small letter.

The following pages show an illustrated outline classification of some of the main phyla. It must be remembered that this is only one of several ways in which living things can be classified.

Kingdom Protista

Protists are organisms whose bodies are quite simple in structure, when compared with animals and plants. Many protists consist of only one cell and are too small to see without a microscope. Others consist of many cells (they are multicellular) and have groups of cells specialized for different jobs.

Protozoa

Protozoa are microscopic protists which consist of one cell. Some feed on other protists, and some are parasites of larger organisms causing diseases such as malaria and dysentery. Other protozoa have chlorophyll and feed by photosynthesis.

Rhizopods are protozoa which move and feed by pushing out parts of their bodies to form temporary arms called pseudopodia. *Amoeba* is an example.

Ciliates are covered with microscopic hairs called cilia which they use for feeding and movement. *Paramecium* is an example.

Flagellates move by lashing a whip-like hair called a flagellum. Many have chlorophyll. *Euglena* is an example.

Fungi

Moulds, mushrooms, and yeasts are fungi. Most are multicellular, yeast is an exception. Multicellular fungi are usually made up of fine threads called hyphae, which are collectively known as the mycelium of the fungus. In moulds, the mycelium is spread out in a thin network. Mucor, the pin mould, is an example.

In mushrooms and toadstools, the mycelium is bundled into a compact structure. Most moulds, mushrooms, and toadstools are saprophytes, feeding on dead organisms and stored food. Some fungi cause disease, like ringworm in animals and mildew in plants. Yeasts feed by changing sugars into alcohol, a process called fermentation.

Algae

Algae are simple plant-like organisms. They contain chlorophyll and live by photosynthesis. Green algae live in the sea, in fresh water, and in damp places on land. They occur as single cells, hollow balls of cells, and fine threads. Spirogyra is an example. Brown algae live in the sea and often reach several metres in length. Bladderwrack is an example. Diatoms are microscopic, single celled algae whose cell walls are made of silica.

Plant kingdom

Plants are multicellular organisms which contain chlorophyll and make food by photosynthesis (i.e. they are autotrophic).

Mosses and liverworts

These are plants without true roots, stems, or leaves, but they often possess structures which resemble these parts of higher plants. They all need water for reproduction, because male plants produce sperms which must swim to a female to fertilize it. After fertilization, the female plant grows a capsule. The capsule bursts, releasing spores which grow into new plants.

Mosses have structures which resemble roots, stems, and leaves. Polytrichum and sphagnum are examples.

Some liverworts are flat, and leaf-like in shape. Others have a 'stem' with leaf-like structures arranged on either side. Pellia and marchantia are examples.

Ferns, horsetails, and club mosses

These plants have **true** roots, stems, and leaves, but no flowers. They produce spores which grow into plants resembling liverworts. These tiny plants produce sperms and eggs. After fertilization, the egg grows into a new fern.

Seed Plants

These are plants which reproduce by making seeds.

Cone-bearing plants (gymnosperms) produce seeds inside cones. Firs and pines are examples.

Flowering plants (angiosperms) have flowers containing reproductive organs which produce seeds. *Monocotyledons* are flowering plants whose seeds grow into seedlings with only one seed leaf (cotyledon). Their leaves have parallel veins. Grasses, tulips, and lilies are examples. *Dicotyledons* are the largest group of flowering plants. Their seedlings have two seed leaves, and their leaves have a branching network of veins. Oak trees, and buttercups are examples.

Animal kingdom

Animals are multicellular organisms which feed by eating other organisms (i.e. they are heterotrophic).

Coelenterates

Coelenterates are found mainly in the sea. They have a hollow, sac-like body with a single opening, the mouth, at one end which is often surrounded by tentacles. The tentacles and body are armed with sting cells which are used to paralyse their prey before eating. Sea anemonies and jellyfish are examples.

True worms (Annelids)

Annelids, or true worms, have bodies consisting of many similar parts called segments. Each segment is marked off from the next by a ring around the body. Some live in the soil (e.g. earthworms), some in the sea under sand or mud (e.g. sandworms), and some are parasites (e.g. leeches).

Arthropods

There are more species of arthropod than of any other type of animal. Arthropods have a segmented body enclosed in a tough outer skin, the cuticle. This forms an external skeleton, or exoskeleton, which protects and supports the body. The cuticle is thin and flexible where the body and legs bend. Most arthropods have antennae sensitive to touch, temperature, sound, taste, and smell. Some have compound eyes made up of thousands of visual units. *Crustaceans* are arthropods with two pairs of antennae. Crabs and lobsters are examples, they have a hard cuticle. Woodlice and water fleas have a thin cuticle.

Arachnids have four pairs of legs and no antennae. Spiders, scorpions, and harvestmen are examples.

Insects have three pairs of legs. The body is divided into a head (with eyes and antennae), a thorax (with legs and sometimes wings), and an abdomen (containing digestive and other organs). Over 700 000 types of insect are known.

Myriapods

Myriapods have a long body consisting of many segments, each of which bears one or two pairs of legs. Centipedes and millipedes are examples.

Molluscs

Molluscs have a soft body which is usually enclosed in one or two shells. Limpets and snails have one shell, and crawl about on a large slimy 'foot'. Slugs, cuttlefish, and octopus have shells inside the body. The octopus and cuttlefish also have long tentacles armed with suckers to catch their prey.

Echinoderms

Echinoderms live in the sea. They have a spiny skin and a body which is usually divided into five portions, such as the 'arms' of a *starfish*. They move about on tiny tube feet with suckers at the ends. Sea urchins and sea cucumbers are examples.

Vertebrates

These are animals with a backbone, or vertebral column.

Fish have a streamlined shape, are covered with overlapping scales, breathe through gills, and move using fins. Fish such as sharks have a skeleton of gristle (cartilage), but most have a skeleton of bone. Most bony fish are weightless in water because they are buoyed up by an air-filled swim bladder inside their bodies.

Amphibia have a moist, scaleless skin. They can live under water – breathing through their skin, or on land – breathing with lungs. They lay their eggs in water, and these hatch into larvae (tadpoles) with gills and a tail.

Reptiles have a dry scaly skin, breathe with lungs, and lay eggs with a tough leathery shell.

Birds have a constant warm body temperature. Their skin is covered with feathers, and their front limbs are wings. Birds which can fly have a streamlined shape, light hollow bones, huge lungs, and very powerful flight muscles. Birds feed using a beak with no teeth.

Mammals also have a constant warm body temperature. They have a hairy skin, and females suckle their young on milk from mammary glands (breasts). *Monotremes* are egg-laying mammals. The duck-billed platypus is an example.

Marsupials are pouched mammals. Kangaroos and koalas are examples. *Primates* are the largest group of mammals. Humans and apes are examples. They complete their development attached to their mother's womb by an organ called a placenta, which supplies the baby with food and oxygen.

Protista kingdom

Protozoa

Amoeba
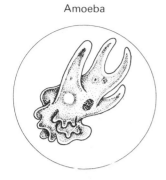

Foraminifera
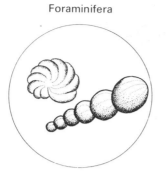

Vorticella
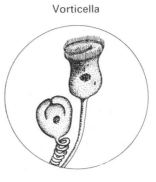

Euglena
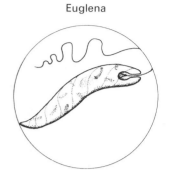

Mycophyta (fungi)

Yeast
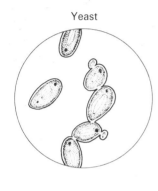

Mushroom

Mucor
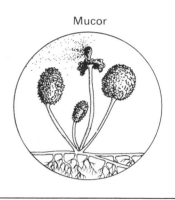

Chlorophyta (green algae)

Spirogyra
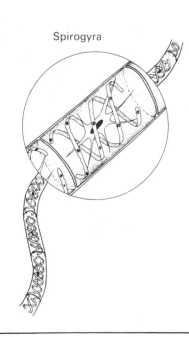

Phaeophyta (brown algae)

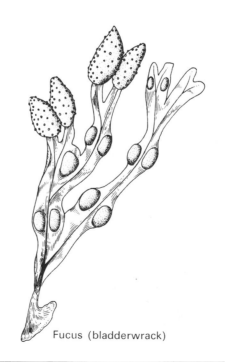
Fucus (bladderwrack)

Bacilliarophyta

Diatoms
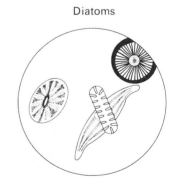

Plant kingdom

Bryophyta (mosses and liverworts)

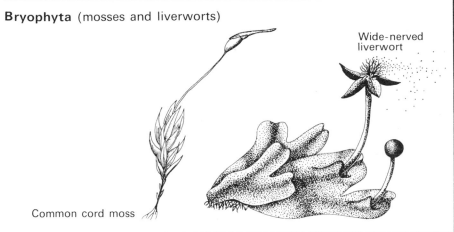

Wide-nerved liverwort

Common cord moss

Pteridophyta (ferns)

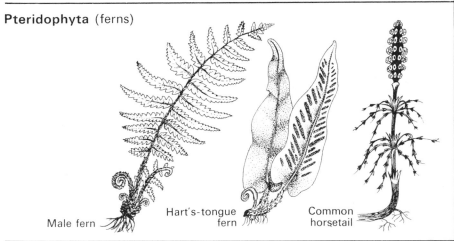

Male fern

Hart's-tongue fern

Common horsetail

Gymnospermae (pines, firs, and spruces)

Scots pine (branch with female cone)

Angiospermae (flowering plants)

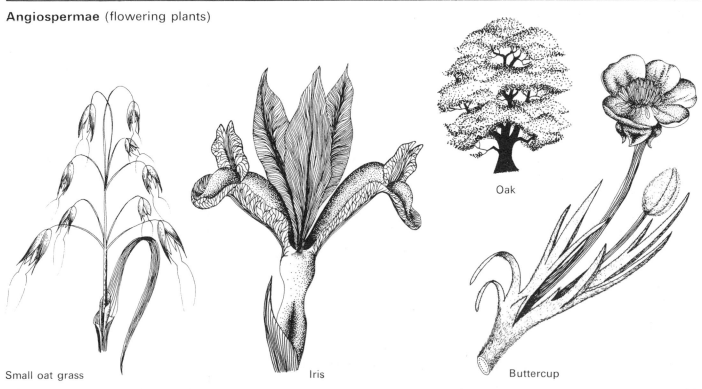

Small oat grass

Iris

Oak

Buttercup

Animal kingdom

Coelenterata

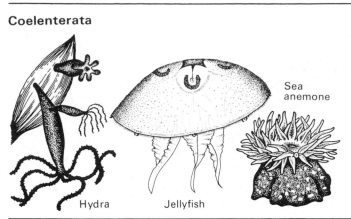

Sea anemone

Hydra   Jellyfish

Annelida

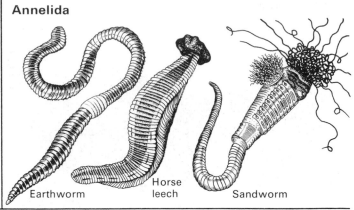

Earthworm   Horse leech   Sandworm

Arthropoda

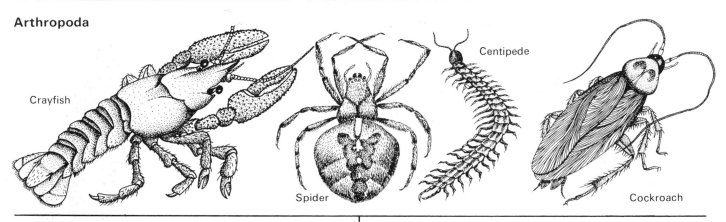

Centipede

Crayfish   Spider   Cockroach

Mollusca

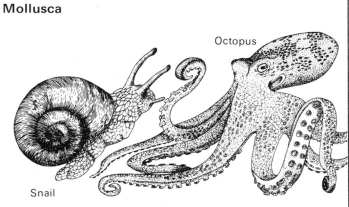

Octopus

Snail

Echinodermata

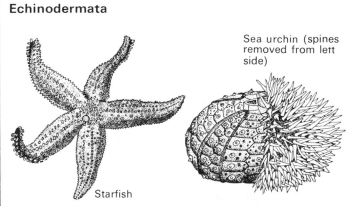

Sea urchin (spines removed from left side)

Starfish

Chordata

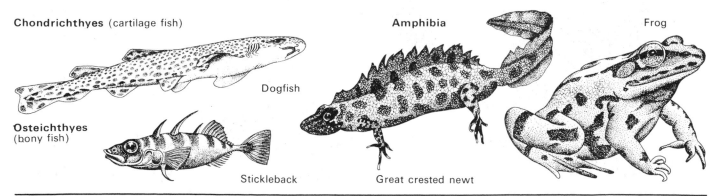

Chondrichthyes (cartilage fish)

Dogfish

Osteichthyes (bony fish)

Stickleback

Amphibia   Frog

Great crested newt

Chordata (continued)

Reptilia

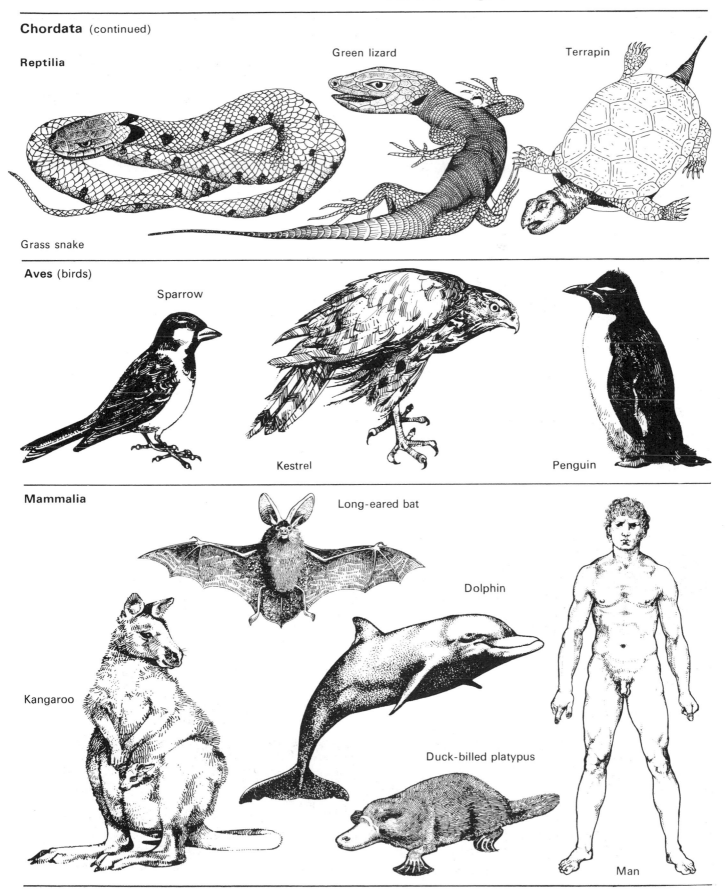

Green lizard

Terrapin

Grass snake

Aves (birds)

Sparrow

Kestrel

Penguin

Mammalia

Long-eared bat

Dolphin

Kangaroo

Duck-billed platypus

Man

Questions

1. Study the following leaves carefully.

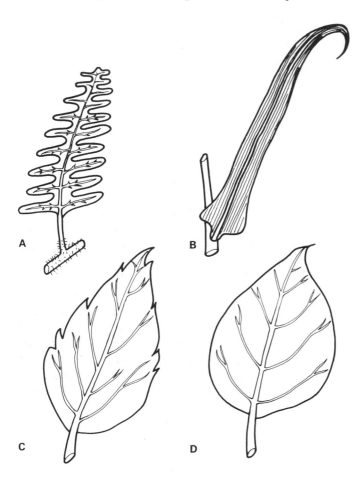

a) With the help of your teacher, copy and complete the table below, using one of the two words after the colon in column one (see example given in line one).

b) Use the information in this table to construct a key, like the one on page 4, which could be used to identify the leaves. Insert the letters instead of the names.

	Features	A	B	C	D
1	leaf blade: divided or undivided	divided	undivided	undivided	undivided
2	leaf margin: smooth or toothed				
3	leaf type: simple or compound				
4	arrangement of veins: net-veined or parallel-veined				

2. Match the structures in the first column with the organisms which possess them in the second:

antennae	snail
flagella	mould
fins	beetle
flowers	*Tilapia*
tube feet	*Euglena*
spores	dandelion
coiled shell	starfish
cones	pine tree

Summary and factual recall test

Examples of locomotory organs are (1–name three). Plant movements are not locomotory because (2). Examples of plant movements are (3–name two).

Higher animals are sensitive to (4–name five things to which animals are sensitive), and they have complex nervous systems which (5) their responses. Plants are sensitive to (6–name three things to which plants are sensitive).

Food provides (7) for movement, and raw materials necessary for (8) and (9) of the body. (10) nutrition is typical of green plants, in which (11) and (12) are absorbed and made into food using the energy of (13). This process is called (14). The main types of heterotrophic organisms are (15–name five).

During respiration (16) is absorbed and reacts with food in the body releasing (17), and producing (18) gas as waste. Examples of respiratory organs are (19–name two).

(20) is the removal of waste substances produced by the body, and must not be confused with (21) which is the removal of undigested substances from the intestine.

Living organisms arise only from (22). This is the opposite of the now disproved theory of (23) generation which proposed that (24).

One difference between animal and plant growth is that (25).

Metabolism is the word which stands for (26). An example of catabolism is (27), and (28) is an example of anabolism. In a healthy adult animal catabolism and anabolism are more or less (29).

Enzymes (30) chemical changes. The word (31) describes chemicals which do this.

The scientific name for a human being is (32). The first of these names denotes the (33) and the second the (34). Scientific names are useful because (35).

A species is a group of organisms which can mate and produce (36) young.

2

Cells

In 1665 an English naturalist called Robert Hooke made a chance observation while using a microscope which he had designed himself. When examining a thin slice of cork, a substance which comes from the bark of a tree, he saw that it looked 'much like a honeycomb' consisting of 'a great many little boxes'. Hooke called these boxes **cells**, from the Latin for 'a little room'.

This seemingly trivial incident is important because it was the first time anyone had noticed that living things are not necessarily made up of continuous material, but sometimes appear to consist of separate units. Furthermore, Hooke's use of the word 'cell' to describe these units has survived to this day and has become a fundamental part of the language of biology.

Two of the most interesting ways of describing cells are to call them 'the units of life', or 'the building-blocks of which living things are made'. These descriptions are useful because they emphasize that cells are the structural units of life. In simple terms cells are like the bricks which make up a wall. But bricks are dead, identical in shape, and quite large; cells are living, of many different shapes, and microscopic in size.

The human body is made up of several million million cells. These are invisible except under high magnification because they measure on average between 0.005 mm and 0.02 mm in diameter. If it were possible to increase the size of a man to two hundred times his volume his cells would still be only the size of a pin-head. But if his size could be increased a million times his cells would then be the size of a cricket ball and it might be possible to see a little of their structure with the naked eye.

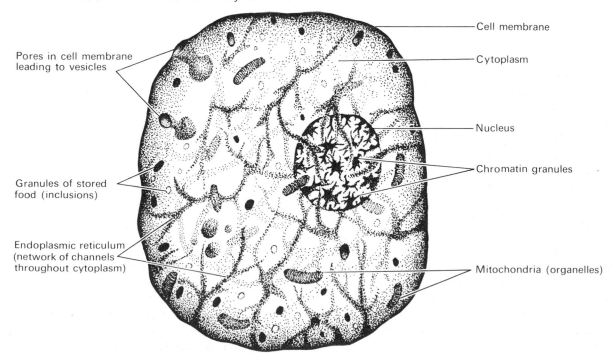

Fig. 2.1 An unspecialized animal cell magnified approximately one million times

2.1 Cell structure

One of the most astonishing things about cells is that they nearly always have the same basic structure, no matter what their function is or what organism they are found in. The single cell which forms the body of an amoeba, a brain cell of a frog, and a leaf cell of a buttercup all have certain features in common. All cells contain a round or oval object called a **nucleus**, surrounded by a jelly-like substance called **cytoplasm**, both of which are enclosed within a very thin skin known as the **cell membrane**.

Cell membrane

The cell membrane is 0.00001 mm thick and forms the outer boundary of the cell. It is here that all exchanges take place between a cell and its surrounding environment. In a way which is not yet fully understood this membrane allows certain chemicals to pass in and out of the cell, but prevents the passage of others. Hence, cell membranes are said to be **semi-permeable**, or, to be more accurate, **selectively permeable**.

Cytoplasm

The term cytoplasm refers to all the living substances of a cell except the nucleus. Cytoplasm is a jelly-like material containing a large number of important substances. These include the many different enzymes concerned with metabolism, oil droplets, glycogen granules in animal cells, starch grains in plant cells, and crystals of excretory substances.

The cytoplasm also contains tiny living structures called **organelles**. All cells contain round or sausage-shaped organelles called **mitochondria**. These are a cell's power plants: they contain enzymes which release energy during respiration. Plant cells have organelles called **chloroplasts**. These trap the sunlight energy needed for photosynthesis.

The nucleus

At least one nucleus is found in the cells of all organisms. The nucleus of a cell contains rod-shaped objects called **chromosomes**. These are only visible when a cell is about to divide into two. Chromosomes contain a complex chemical called **deoxyribonucleic acid**, or **DNA**. DNA controls the development of the features that an organism inherits from its parents. In other words it contains the 'instructions', in chemical code, for making an organism. These hereditary instructions are 'obeyed' as a fertilized egg cell grows into an adult organism.

2.2 Plant and animal cells compared

Having described features common to both plant and animal cells it is now necessary to discuss those which are found only in one group or the other.

Cellulose

A tough and fairly rigid layer of a substance called cellulose completely surrounds the cell membrane in plant cells. The cellulose forms the **cell wall** (Fig. 2.2B). The presence of this wall around every cell in a plant gives a great deal of support to the plant.

Plant cells are not isolated from each other by their cellulose cell walls. Because cellulose is permeable to all fluids and is perforated at intervals by tiny holes through which cells are interconnected by fine cytoplasmic threads. The presence of a cell wall makes plant cells clearly visible as distinct units when viewed under a microscope.

Almost all types of animal cell are naked: that is, there is no wall of any kind outside the cell membrane. Animal cells have a less distinct outline when seen under a microscope.

A Diagram of an animal cell

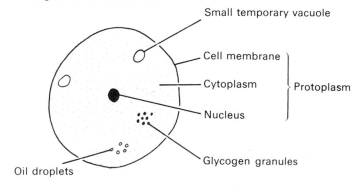

B Diagram of a plant cell

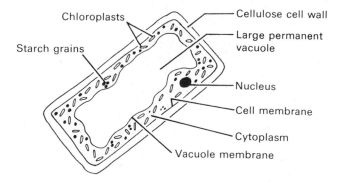

Fig. 2.2 Animal and plant cells compared

Vacuoles

A vacuole is a space filled with fluid in the cytoplasm of a cell. Vacuoles are lined with a semipermeable membrane similar to the cell membrane. Mature plant cells usually have a large permanent vacuole at their centre (Fig. 2.2B). Small temporary vacuoles, or **vesicles**, are often found in animal cells (Fig. 2.2A), and are particularly common in freshwater protists such as *Paramecium* (Fig. 1.00).

Chlorophyll

Chlorophyll is a green substance which absorbs light for use as a source of energy in the chemical reactions of photosynthesis. Chlorophyll is present in most plants, and is usually though not exclusively found in the cells of leaves. It is always contained within disc-shaped organelles called **chloroplasts**. Animal cells never contain chlorophyll.

2.3 **Cells, tissues, and organs**

Multicellular organisms (those made of many cells) usually consist of many different parts. These parts consist of cells with special features that enable them to perform a particular task efficiently. In other words, the cells are **specialized**. Groups of specialized cells are called **tissues**. Muscle tissue, for example, is made of specialized cells which can contract and move the body. Nervous tissue consists of cells which can carry nerve impulses. Photosynthetic tissue in plants consists of cells filled with chloroplasts. These absorb sunlight energy which is used in other parts of the cell to drive the chemical reactions of photosynthesis.

An **organ** is made up of several different tissues, each of which contributes to the functions of the organ as a whole. The heart and liver are animal organs, and leaves and roots are plant organs.

Several organs working in conjunction form an **organ system**. The circulatory system, composed of the heart, blood, and blood vessels, is an example.

Division of labour

The way in which various parts of multicellular organisms are specialized for one particular function is an example of division of labour. This means that the *labour* involved in maintaining the organism's life processes is *divided* among specialized parts. The advantage of division of labour is that it is highly efficient. Just as a man trained to do one particular job is likely to be better at it than a jack-of-all-trades, so a specialized body tissue performs its function more efficiently than a group of unspecialized cells with many different functions.

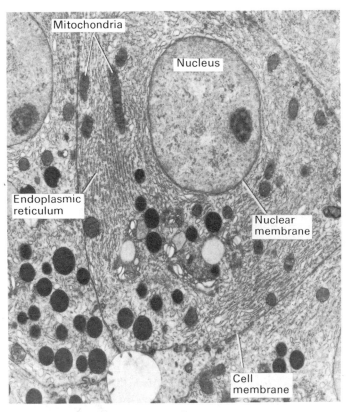

Electron micrograph of an animal cell

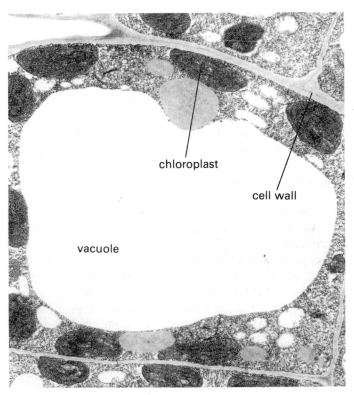

A plant cell

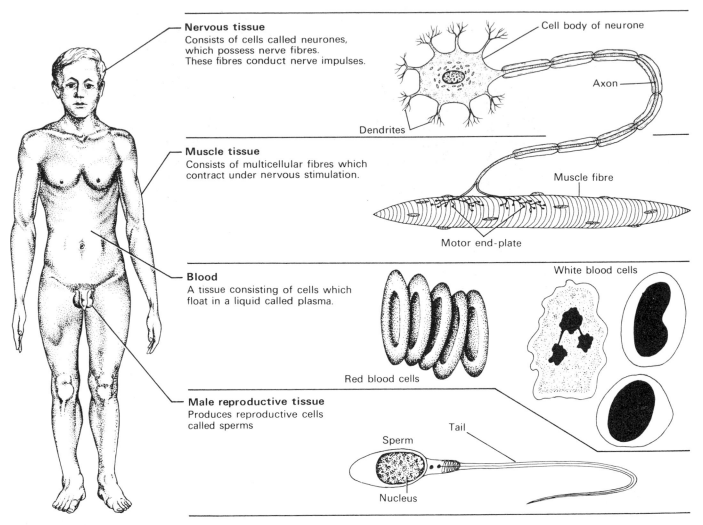

Fig. 2.3 A selection of human cells and tissues

2.4 Cell size

The majority of cells in the human body are between 0.005 mm and 0.02 mm in diameter. A human hair is about 0.1 mm thick.

All cells absorb food and oxygen and remove waste products through their cell membranes, and so they cannot grow so large that the surface area of the membrane is insufficient to support the volume of the cell. For this reason, when cells reach a size at which their surface area is just sufficient to support their volume – a point known as **optimum size** – they either stop growing or divide. Therefore, the upper limit of cell growth is determined by the ratio of volume to surface area.

The smallest cells are bacteria. Three thousand million could live in a chemical solution in an ordinary test tube. The largest cells are eggs. Ostrich eggs can be 20 cm long.

2.5 Cell division and growth

In certain circumstances a cell, usually referred to as the **parent cell**, divides into two cells which are called **daughter cells**. The nature, occurrence, and location in the body of cell division differ according to the species and age of an organism.

Growth

When a cell divides in a multicellular organism the daughter cells remain attached to each other and grow, usually without much delay, to the size of the original parent cell. As a result, the organism grows in size. **Growth can be defined as an increase in the over-all size of an organism, and an increase in the number of its cells.**

By far the greatest amount of cell division and growth takes place as a multicellular organism develops from a single fertilized egg cell called the

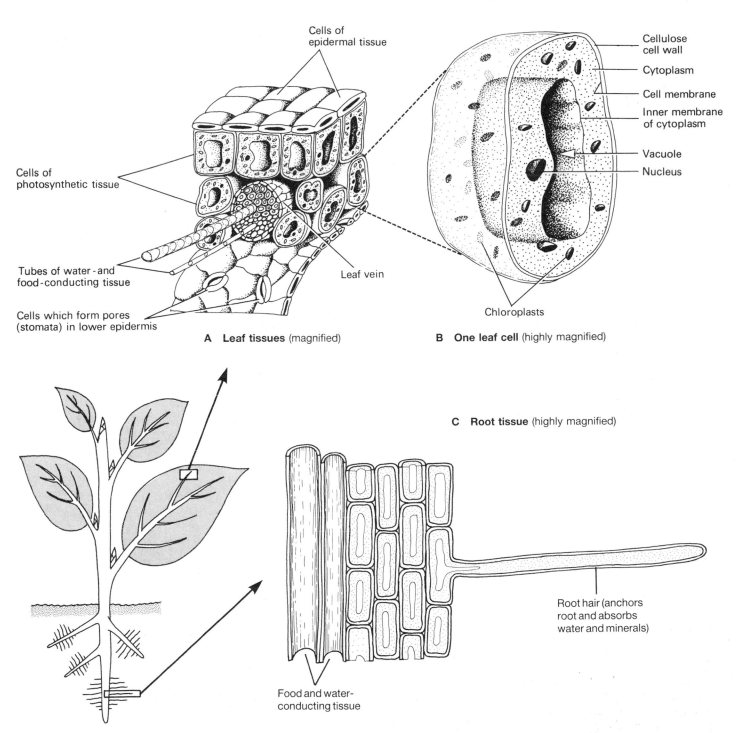

Cells of
epidermal tissue

Cells of
photosynthetic tissue

Tubes of water - and
food-conducting tissue

Cells which form pores
(stomata) in lower epidermis

Leaf vein

A   Leaf tissues (magnified)

Cellulose
cell wall

Cytoplasm

Cell membrane

Inner membrane
of cytoplasm

Vacuole

Nucleus

Chloroplasts

B   One leaf cell (highly magnified)

C   Root tissue (highly magnified)

Root hair (anchors
root and absorbs
water and minerals)

Food and water-
conducting tissue

Fig. 2.4   A selection of plant cells and tissues

zygote. The zygote grows into an organism composed of billions of cells specialized into tissues, organs, and organ systems. During this growth and development period the organism is called an **embryo**.

The cells making up an embryo in its early stages of growth are unspecialized, and when isolated from each other they look something like Figure 2.1. As growth proceeds, groups of cells adopt the shape and form of the different types of permanent tissue. This is called **cell differentiation** because the cells develop their *different* shapes and functions. When a cell is fully differentiated it usually loses the power of division. As growth proceeds cell division consequently becomes localized to restricted areas of the embryo.

17

Growth in animals

Growth occurs all over an animal's body until it reaches adult size (Fig. 2.7A). When an animal becomes an adult, growth does not stop altogether. Cells in certain areas remain unspecialized and have the power to divide and grow. These areas supply cells which can differentiate into any kind of tissue and bring about further growth, or repair worn and damaged tissue. In humans, for example, skin is replaced as fast as it wears away by a layer of rapidly dividing cells just below the skin surface.

Figure 2.5 illustrates **growth curves**; these result from plotting growth rates on a graph. The shape of a growth curve depends on factors such as; the type of growth measured, the species of animal investigated, the physical and hereditary features of its parents, together with health and diet.

Figure 2.5 also illustrates **continuous** and **discontinuous growth**. The human nervous system grows continuously until about 13 years of age. Growth of the body as a whole, however, is discontinuous: it grows rapidly for a while then stops, and continues again later.

Discontinuous growth is a feature of many insects. A locust or cockroach grows rapidly until its cuticle (exoskeleton) gets too small for its owner. This problem is overcome by moulting, or **ecdysis**, which is the periodic shedding of an old cuticle and its replacement by a new, larger one, after which growth can continue.

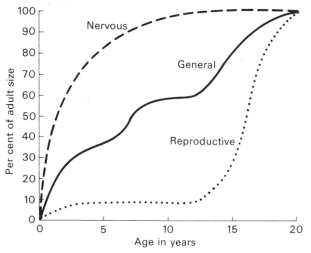

Fig. 2.5 Growth of human body parts. Different parts of the body grow at different rates. The body in general (muscles, skeleton, lungs, blood volume, and intestines) grows in spurts: in infancy (first year) childhood (5–7 years), and towards puberty (11–16 years). The nervous system (brain, spinal cord, eyes, and skull) grows very rapidly from birth to 6 years, when it reaches 90% of adult size. Reproductive organs remain small until puberty when they grow rapidly to adult size

Animal growth is investigated further in exercise D, and question 7.

Growth in plants

If plants have light, warmth, air, water, and minerals, they can grow continuously throughout life. But growth is restricted to the parts of a plant's body where there are unspecialized cells capable of dividing and differentiating. Areas of rapid cell division are called **meristems**.

Meristems occur at the tips of the root and shoot (Fig. 2.6B). New cells produced at a meristem grow to maximum size, and differentiate into plant tissues (Fig. 2.6C).

Regions of growth and differentiation in plants can be detected by marking a seedling as shown in Figure 2.7B, and observing how the marks move apart.

Fresh weight and dry weight

The bodies of organisms contain up to 90% water. Therefore if you measure weight changes in a living organism (**fresh weight** changes), you will not be able to see how solid material in the body (**dry weight**) is changing with growth. To study changes in solid matter you must dry the organism before weighing it. Figure 2.8 shows how fresh weight and dry weight change as seeds germinate and grow.

Fresh weight (curve A) increases continuously as a seed absorbs water, grows roots and leaves, and begins making food by photosynthesis. Dry weight (curve B) drops at first, as a seed uses up (respires) its stored food to make root and leaf tissue. There is no gain in dry weight until photosynthesis begins.

There are two types of meristem in a plant. **Apical meristems** occur at the tips of the shoot (Fig. 2.6). **Lateral meristem** occurs inside the stem of a plant. New cells produced at a meristem grow to maximum size, then differentiate into plant tissues (Fig. 2.6C).

Apical meristems are responsible for producing new, or **primary tissues**, which extend the length of the shoot and root. Apical meristems in the shoot are protected, during winter months, by buds. Root apical meristems are continually protected by a root cap (Fig. 7.1).

Lateral meristem, or **cambium**, is a region of active cell division responsible for producing **secondary tissues** which increase the circumference (girth) of a stem (Fig. 7.2). Cambium is particularly important in bushes and trees where it produces the wood which makes up the bulk of the plant stem.

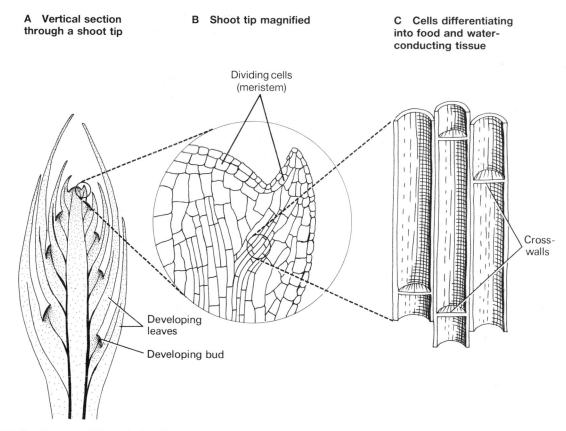

A Vertical section through a shoot tip

B Shoot tip magnified

C Cells differentiating into food and water-conducting tissue

Dividing cells (meristem)

Developing leaves

Developing bud

Cross-walls

Fig. 2.6 Cell division and differentiation in a shoot tip

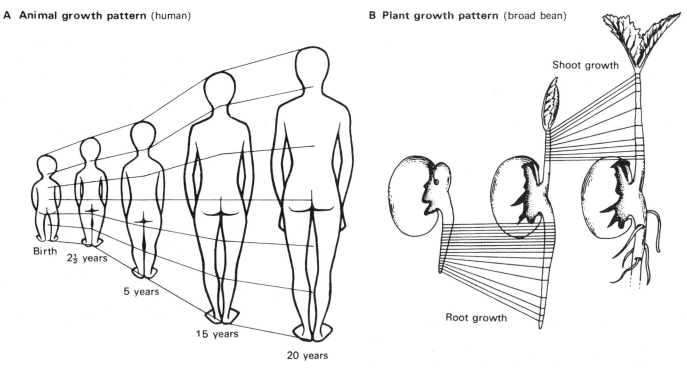

A Animal growth pattern (human)

B Plant growth pattern (broad bean)

Birth

2½ years

5 years

15 years

20 years

Shoot growth

Root growth

Fig. 2.7 Comparison between animal and plant growth patterns

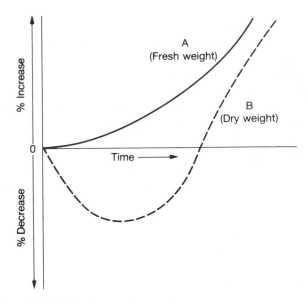

Fig. 2.8 Weight changes in germinating seedlings

2.6 Cell division by mitosis

Mitosis is the mechanism by which a cell passes an exact copy of each of its chromosomes to the daughter cells produced when it divides.

As explained in Section 2.1, chromosomes contain the hereditary instructions for making an organism. Mitosis ensures that every cell in an organism contains these instructions, so they can be used to build tissues, organs, and organ systems. Figure 2.9 illustrates mitosis in plant cells.

1. Before a cell divides, its chromosomes become shorter and thicker (Fig. 2.9A).

2. The chromosomes become attached to threads called **spindle fibres** (Fig. 2.9B).

3. The spindle fibres contract pulling each chromosome into two parts, which move to opposite ends of the cell (Fig. 2.9C).

4. The cell divides into two. Each has the same number of chromosomes as the original parent cell (Fig. 2.9D).

Fig. 2.9 Mitosis in a plant cell

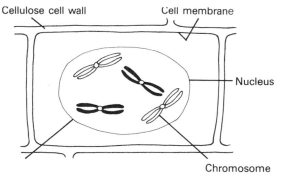

A Chromosomes become shorter and thicker

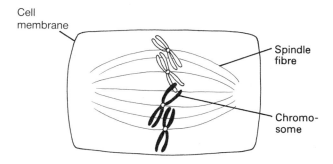

B Chromosomes become attached to spindle fibres

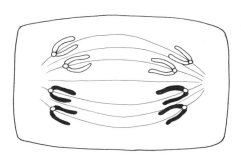

C Each chromosome splits into two parts which separate and move to opposite ends of cell

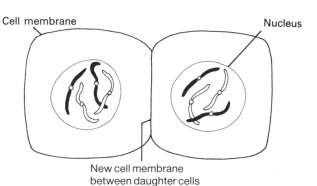

D Two daughter cells form, with same number of chromosomes as parent cell

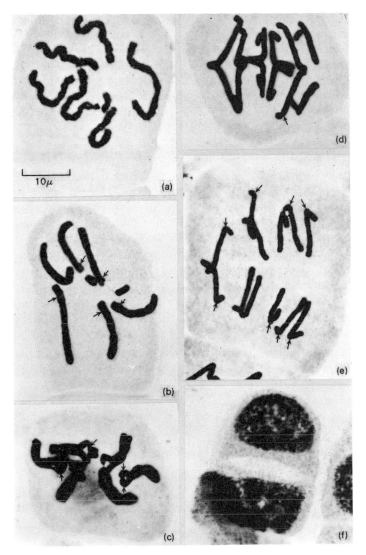

Mitosis in crocus root tip. (a-c) Chromosomes become shorter and thicker. (d) Chromosomes are pulled into two parts. (e) The two parts move away from each other. (f) A nucleus forms in each daughter cell

2.7 Osmosis and diffusion in cells

Water is essential for life, mainly because the chemical reactions of life (metabolism) take place between substances dissolved in water. Water moves into and out of cells and from one cell to another, by a process called **osmosis**. Osmosis is a special kind of diffusion, so it is first necessary to explain what diffusion is.

Diffusion is the movement of molecules from where they are concentrated to where they are *less* concentrated. In technical terms, molecules diffuse down **concentration gradients** and the 'steeper' the gradient the faster diffusion occurs.

If you let a drop of ink fall gently into a cup of water, ink molecules will diffuse outwards from the blob of ink where they are concentrated into regions where they are less concentrated. At the same time water molecules will diffuse from pure surrounding water where they are concentrated, into the ink blob where they are less concentrated. Diffusion will continue until the ink and water molecules are evenly dispersed throughout the cup.

Diffusion occurs for two reasons. First, there is a great deal of empty space between the molecules of all substances. This space is greatest in gases, much less in liquids, and least of all in solids. Second, all molecules are in a state of constant random movement so that they collide and intermingle all the time.

Think once more about a drop of ink in a cup of water. Both the ink and the water molecules are in constant motion which causes them to intermingle. The ink molecules move into the empty spaces between the water molecules and the water molecules move between the ink molecules.

In solids diffusion occurs extremely slowly, if at all, because the molecules are packed tightly together. Liquids and gases diffuse freely because their molecules are widely spaced.

Consider next the difference between *equal* volumes of pure water and and a strong sugar solution. The water will have more water molecules (i.e. a high concentration) than the sugar solution because, in the sugar solution, sugar molecules take up some of the space. Similarly, a weak sugar solution will have a higher concentration of water molecules than a strong sugar solution.

With this in mind, think what will happen in the situation illustrated in Figure 2.10A. Here, a strong sugar solution is separated from a weak sugar solution by a membrane through which both water and sugar molecules can pass. Sugar will diffuse from the strong solution to the weak solution until it is uniformly distributed on both sides of the membrane. Likewise, water will diffuse from the weak to the strong solution until it is uniformly distributed on both sides.

Next consider what must happen in Figure 2.10B. A strong sugar solution is here separated from a weak sugar solution by a membrane that will allow water but *not* sugar to pass through. Such a membrane is said to be **semi-permeable**. Under these circumstances the sugar molecules cannot diffuse from high to low concentration because they are 'imprisoned' behind the semi-permeable membrane. Only the water molecules can move: they will diffuse from the weak to the strong solution until they are uniformly distributed on both sides of the membrane.

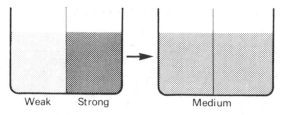

A   Diffusion   If a weak sugar solution is separated from
a strong sugar solution by a membrane permeable to
sugar and water, sugar will diffuse from the strong to
the weak solution, and water will diffuse from the
weak to the strong solution, until both solutions are of
equal strength.

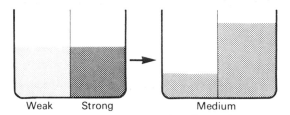

B   Osmosis   If a weak sugar solution is separated from
a strong sugar solution by a membrane permeable only
to water, water will diffuse from the weak to the strong
solution until both solutions are of equal strength.

Fig. 2.10   Diffusion and osmosis

Water movements of this type are called osmosis.
**Osmosis can be defined as: the diffusion of
water molecules through a semi-permeable
membrane from a weak to a strong solution.**
(Although generally used to describe diffusion of
water, osmosis can also be applied to the diffusion of
any solvent across a semi-permeable membrane in
response to a concentration gradient.)

Theoretical explanation of osmosis

Figure 2.11 shows how to demonstrate osmosis
with an **osmometer**, made with a commercially
available semi-permeable membrane called Visking
tubing (exercise E). This figure illustrates the generally accepted theory that such membranes have
microscopic pores through which small molecules
like water can pass but larger molecules like sugar
cannot. Thus, in an osmometer, Visking tubing acts
as a molecular net, holding back large molecules like
sugar but allowing small molecules like water to pass
through.

It is believed that osmosis occurs in living cells
because the cell membrane (described in section 2.1)

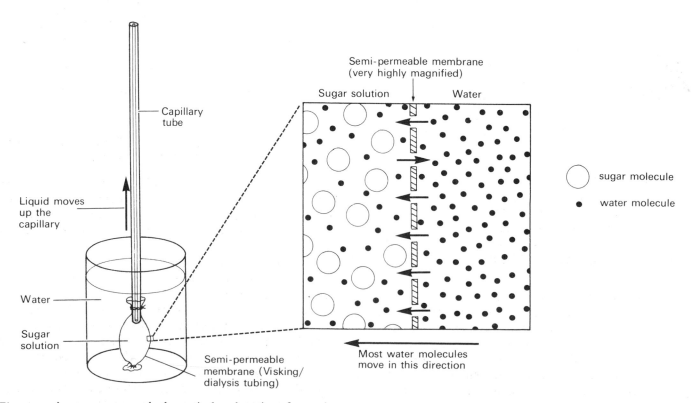

Fig. 2.11   An osmometer, and a theoretical explanation of osmosis

is semi-permeable, having microscopic pores like those in Visking tubing. In fully-formed plant cells there is a second important semi-permeable membrane; a membrane called the **tonoplast** which lines the cell vacuole (Fig. 2.12). Consequently, water entering plant cells by osmosis passes through the cell membrane, the cytoplasm, and finally the tonoplast before entering the vacuole. It is customary, however, to treat the cytoplasm and the two membranes on either side of it as if they were a single semi-permeable membrane. When water reaches the vacuole it mixes with a liquid called **cell sap**.

Osmotic potential
The osmotic potential of a solution can be defined as: a measure of the pressure with which water molecules could diffuse out of the solution if it were separated from another solution by a semi-permeable membrane.

Osmotic potential depends upon the amount of dissolved material (solute) in a solution. The greater the concentration of solute, the *lower* the osmotic potential of the solution. Pure water has the highest possible osmotic potential, and so it follows that addition of solutes lowers this potential.

In osmosis, therefore, water molecules will diffuse across a semi-permeable membrane from regions of high osmotic potential to regions of lower osmotic potential.

Osmosis in plants
Plant cells behave like osmometers: they absorb or lose water by osmosis depending upon the concentration of solutes in their sap. Cells with a low osmotic potential (high concentration of solutes) will absorb water from cells with a higher osmotic potential, and also from liquid which bathes all the cells of a plant, provided this liquid has a higher osmotic potential.

As cells take in water by osmosis they undergo several changes. The stream of water entering a cell causes it to swell, and this causes the cellulose cell wall to stretch slightly. When this wall can stretch no further it becomes taut and firm.

Wall pressure and turgor pressure
A fully stretched cell wall is exerting a restraining inward force called **wall pressure** on cell contents. This wall pressure is balanced by an equal but oppositely-directed force called **turgor pressure** in the cell contents, and when the cell wall can stretch no further the cell is said to be **fully turgid** (Fig. 2.12A).

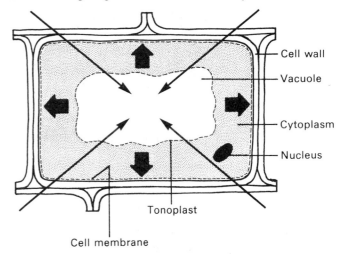

A Cell becoming turgid as it takes in water by osmosis

Cell wall
Vacuole
Cytoplasm
Nucleus
Tonoplast
Cell membrane

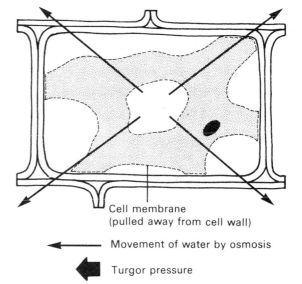

B Cell becoming plasmolysed as it loses water by osmosis

Cell membrane
(pulled away from cell wall)

← Movement of water by osmosis

◄ Turgor pressure

Fig. 2.12 Turgidity and plasmolysis. (Pores shown in the semi-permeable membrane are not drawn to scale.)

As turgor pressure builds up it squeezes water molecules out of the cell and, when the cell is fully turgid, water leaves it under the influence of turgor pressure at the same rate as water enters it by osmosis.

Turgidity and support
Most of the time plant cells take in water not from each other but from the liquid which fills the spaces between the cells and penetrates the tiny cavities between the cellulose fibres which make up cell walls. Since most cells are bathed in this liquid, and since it almost always has a higher osmotic potential than cell sap, it follows that the bulk of a plant will consist of fully turgid cells. Turgid cells make a plant firm, maintain its shape, and allow it to function

efficiently. All young seedlings, herbaceous (non-woody) plants, and plant structures like leaves and flowers, depend entirely upon the turgidity of their cells for support.

Plasmolysis

If a cell is surrounded by a liquid with a lower osmotic potential than its own sap, the cell loses water by osmosis. Turgor pressure drops to zero as water is drawn from the cell. The vacuole and cytoplasm contract, which usually pulls the cell membrane away from the cell wall (Fig. 2.12B).

The effects of water loss are called **plasmolysis**, and cells in this state are said to be **plasmolysed**. Plasmolysed cells are deflated and soft owing to lack of turgor pressure. The technical name for this condition is **flaccid**. Flaccid cells give no support to a plant and wilting occurs; that is, the leaves, flowers, and other non-woody tissues droop and become limp.

Plasmolysis occurs very rarely in nature. It can happen if land is flooded with sea water because this leads to a sudden steep decline in the osmotic potential of water surrounding the roots. This, in turn, leads to loss of water by osmosis from the plants. Seashore plants do not suffer in this way because their cell sap has a lower osmotic potential than sea water.

Osmosis in animals

Animal cells take in and lose water by osmosis in the same way as plant cells. But animal cells are not surrounded by a cellulose cell wall like plant cells. Animal cells can therefore expand much further than plant cells as they absorb water by osmosis. In fact, free-floating cells like red blood cells will expand until they burst if placed in a liquid which has a much higher osmotic potential than their cell contents.

However, cells rarely burst inside an animal's body, for two reasons. First, animals can control the osmotic potential of fluids surrounding their cells so that it does not become high enough to cause damage. In vertebrate animals control of osmotic potential, called **osmoregulation**, is carried out mainly by the kidneys and liver. Second, most animal cells are packed together in groups (tissues) so that as they expand they press against each other with ever-increasing force until further expansion is prevented. At this stage the tissue is fully turgid.

Jellyfish, earthworms, caterpillars, and other animals without hard skeletons depend to a large extent on the turgidity of their tissues for support.

Support in plants and animals is discussed more fully in chapters 7 and 12.

Investigations

A *Looking at cells*

The following techniques can be used to prepare slides for examination under a microscope.

Onion epidermis Cut up an onion bulb to obtain a piece of one of the white, fleshy 'leaves' from inside the bulb. A piece of epidermis may be stripped from this by breaking and tearing it into smaller pieces. Mount the epidermis, without folding it, on a slide and add a few drops of iodine. Place a cover slip over it and observe the cell shapes and nuclei.

Section cutting Cut thin slices with a razor-blade from bottle cork, potato, and other soft plant stems and roots. Select the thinnest slices and mount them on a slide in 25% glycerine solution (which reduces air bubbles in the tissue). Observe the variety of cell shapes.

Animal cell Scrape the inside of the cheek with a sterilized spatula and mix the collected tissue with a drop of methylene blue stain on a slide using mounted needles. Observe individual epidermal cells noting their shapes (where they are not distorted by folding), and the presence of nuclei.

B *Observation of chromosomes during mitosis, and subsequent cell growth*

1. Cut off about 15–20 mm from a root of an onion bulb which has been suspended with its base just touching water for several days. (This will have encouraged root growth).

2. Place the piece of root on millimetre graph paper and slice off the terminal 5 mm with a razor-blade, and then a further 5 mm length.

3. Place these two pieces in separate watch-glasses containing 10 parts of acetic orcein stain to one part normal hydrochloric acid.

4. *Warm* (do not boil) the preparation for five minutes and then place the root pieces on separate labelled slides with a few drops of the same solution.

5. Break up the tissues of each preparation with the tip of a mounted needle without altering the relative arrangement of cells.

6. Place a cover slip on the preparation and cover with a paper towel. Press down with the thumb on the cover slip through the towel taking care not to move the cover slip sideways.

7. Observe cells from the root-tip portion for the various stages of mitosis.

8. Are there any signs of mitosis in cells from the *upper* 5 mm root segment? Compare the size and shape of cells in both preparations to confirm that cells do in fact grow after the stage of rapid cell division at the tip of the root.

C   *An investigation into growth patterns of leaves*

1. Grow a bean seedling to the last stage shown in Figure 2.7B. Make an ink marker by stretching cotton thread across the gap between the ends of a piece of wire bent into a U-shape. Use this, dipped in finger-print ink (or indian ink), to print a 2 mm grid on a young leaf attached to the plant. Copy the leaf outline and its grid on tracing paper. On the next and subsequent days for at least a week, compare the leaf with its tracing, making new tracings if a difference is observed.

2. Does the leaf grow evenly all over? If not, discover where the major growth regions are by studying the sequence of tracings.

3. Repeat with different plants.

D   *Investigating fresh and dry weight changes in plants*

1. Soak 40 pea or bean seeds in water, and wrap them in wet paper towelling. To do this fold a paper towel in half and wet it. Place five soaked seeds in a row near the top edge, roll up the towel, and secure it with a rubber band. Repeat until all 40 seeds are wrapped, then place them in water (Fig. 2.13), in a warm dark place.

2. At regular intervals (3–5 days depending on growth rate), remove five seedlings and find their average weight.

3. Continue until all the seeds have been weighed, then plot weight changes on a graph. What conclusions can be drawn from these results?

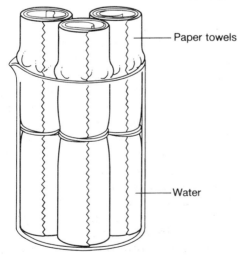

Fig. 2.13   How to prepare seeds for exercise D

E   *To verify that osmosis takes place through a semi-permeable membrane*

1. Prepare the osmometer apparatus shown in Figure 2.11 using a length of Visking/dialysis tubing as the semi-permeable membrane.

   a) Cut a length of tubing – about 10 cm – and tie a knot in it at one end.

   b) Fill the tube with strong sugar solution and tie it to a length of capillary tubing as illustrated.

   c) Insert the tube in a beaker of water and observe the movement of sugar solution up the capillary.

F   *To observe plasmolysis in living cells*

Strip a small piece of coloured epidermis from a rhubarb petiole (leaf stalk) and mount it in strong sucrose solution on a miscroscope slide. The plasmolysis of cell contents may be clearly observed under low magnification owing to the movement of coloured cytoplasm in the cells. Use Figure 2.12 as a guide. Restoration of turgor may be seen by washing the plasmolysed cells in water.

See also question 4.

Questions

1. Growth is an increase in both the number and size of cells. This process is distributed unevenly throughout the bodies of animals and plants, giving each a different growth pattern. The diverging lines on each sequence of drawings in Figure 2.7A and B show how growth occurs in animals and plants. The youngest organism in each sequence is marked off into regions of equal length, but these regions soon show different rates of development.

   a) Find the human body region in Figure 2.7A which grows very little compared with the rest, and suggest reasons why this is so.

   b) The diverging lines in Figure 2.7B indicate the major growth regions in plants. Where are they located?

   c) From the answers to the above questions try to explain the basic difference between plant and animal growth patterns. Define growth.

2. *a)* What is a cell?

   b) Name three specialized animal cells and three specialized plant cells.

   c) What is the name for a group of specialized cells which work together?

   d) What is an organ? Name a plant and an animal organ.

3. *a)* Describe five structures common to *both* plant and animal cells.

b) Describe three ways in which plant cells are different from animal cells.

4. *a*) What is osmosis?

b) What do the words turgid, plasmolysed, and flaccid mean when used to describe cells?

c) What would a plant look like if its cells became flaccid?

d) Why is cell turgidity important to earthworms and caterpillars?

e) What is osmoregulation in animals and which organs are concerned with it?

5. Carry out the osmosis experiment illustrated below.

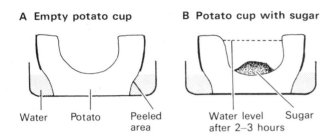

A Empty potato cup   B Potato cup with sugar

Water   Potato   Peeled area   Water level after 2–3 hours   Sugar

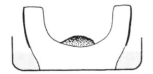

C Boiled potato cup with sugar

a) Explain in detail why water gathers in the hollowed portion of potato **B**.

b) Why is potato **A** necessary in this experiment?

c) Explain why water does *not* gather in the hollowed-out portions of potatoes **A** and **C**.

6. Study the apparatus illustrated in Figure 2.11 on page 22.

a) What does the sugar solution in this experiment correspond to in a cell?

b) What does the Visking tubing correspond to in a cell?

7. Below is a record of weight changes in a turkey from hatching to almost full size.

Age in months	0*	1	2	3	4	5	6	7	8	9	10
Weight in kilograms	0.05	0.5	2.1	3.9	6.2	8.6	10.6	12.7	14.6	14.8	14.9

* Just hatched

Produce a graph of these results in *two* different ways:

a) with age along the horizontal axis and weight along the vertical axis;

b) with age along the horizontal axis, and on the vertical axis *the amount of growth since the last measurement* (e.g. the amount of growth in the first month was 0.5 kg – 0.05 kg = 0.45 kg). During which period does the turkey show the most rapid weight increase? During which period does its weight increase at a *decreasing* rate? Which graph shows these two growth periods most clearly? Why doesn't the other graph show them so clearly? What does this exercise tell you about the importance of looking at results in more than one way?

Summary and factual recall test

The outer boundary of a cell is called the (1) and it is (2) permeable. All the living substance in a cell except the nucleus is called (3). This substance contains important substances such as (4–name four). Cytoplasm also contains living structures called (5). Examples are (6–name two). Nuclei contain rod-shaped (7) made of a complex chemical called (8), which controls (9).

Plant cells differ from animal cells in the following ways (10–name three).

Multicellular organisms consist of groups of cells called (11), which are specialized in performing a particular function and achieve greater efficiency through the division of (12). Examples of such cell groups are (13–name two). Where several of these groups work together they form an (14), an example of which is (15).

Cell size is limited by the ratio of (16) to (17), because (18).

Growth results from an increase in both the (19) and (20) of cells. The main difference between animal and plant growth is (21).

During mitosis a cell passes a copy of its (22) to daughter cells. This ensures that every cell in an organism contains (23) instructions.

Osmosis is the (24) of (25) molecules through a (26) membrane from a (27) to a (28) solution. A cell taking in water by osmosis inflates and presses outwards against its (29) cell wall with a force called (30) pressure. This force is opposed by (31) pressure, and when the two forces are equal the cell is fully (32). Osmosis stops at this stage because (33). The following are supported by the turgidity of their cells (34–list four examples).

A cell surrounded by liquid with a lower osmotic potential than its sap will (35) water by osmosis. The cell membrane will pull away from the cell wall, a condition called (36).

3

Food and diet

Every cell of every tissue and organ in the body has been made from, and is maintained in a healthy state by the food which is eaten.

If a person eats too little food her body becomes weak and thin because the tissues are used up as a source of energy; growth and repair processes slow down so that wounds fail to heal properly; and the body loses the ability to fight off infections. On the other hand, if a person eats too much her body becomes inflated with stored food. This condition, known as **obesity**, may result in damage to the heart and circulatory system. Finally, if a person eats only a moderate amount of food but of the wrong type to suit her body's requirements, she is likely to suffer from diseases such as scurvy and beri-beri.

To remain healthy, fairly precise amounts of the right kinds of food must be consumed. This is known as living on a **balanced diet**. Failure to do so results in **malnutrition**, a term which describes the effects on the body of eating too little, too much, or the wrong kinds of food.

This chapter describes various types of food and their different functions in the body. It then shows how these foods can be put together to make meals which form a balanced diet. But first it is necessary to describe a function common to all types of food, which is to supply energy.

Effects of malnutrition. Compare the well-fed child on the left who is 18 months old and weighs 11 kg, with the under-fed child on the right who is 3 years old and weighs 6.5 kg

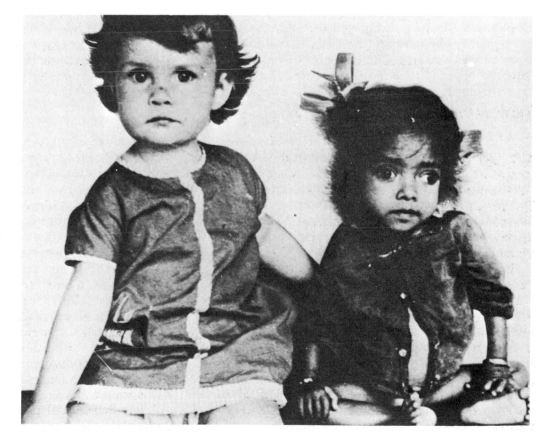

3.1 Energy value of food

Most types of food contain stored energy. But this energy cannot be used by the body until it is released by the chemical reactions of respiration, described in chapter 8.

There is a simple method of measuring the amount of energy in each type of food so that its energy value may be stated in so many units. When a measured quantity of food is burned in air it gives off exactly the same amount of energy, in the form of heat, as it does when the same quantity is respired inside an organism. The food is burned inside a device called a **calorimeter**. This ensures that all the heat released during burning is transmitted to a known quantity of water, thereby raising its temperature. All forms of energy are now measured in units called **joules**. It takes 4.2 joules of heat energy to raise the temperature of 1 g of water by 1°C. After a food has finished burning, its energy value is calculated by the formula: temperature rise × 4.2 × mass of the water in the calorimeter (see exercise B).

The energy value of food is usually measured in **kilojoules** (abbreviated to kJ). 1 kilojoule = 1000 joules. It must be remembered, however, that these figures indicate a food's *potential* rather than its *actual* energy value. For example, a slice of bread has an energy value of about 1000 kJ. But the fibre (bran) in each slice passes undigested through the gut, so never yields its energy to the body. In addition, the body cannot use all of the energy released when the digested bread is respired since some is lost as heat. In humans, for instance, up to 85% of respiratory energy is lost in this way.

3.2 Types of food and their composition

There are five main types of food: carbohydrates, fats and oils, proteins, minerals, and vitamins. They each have different chemical compositions, different properties, and different functions in living organisms.

Carbohydrates

The most familiar carbohydrates are sugars and starches. Sweet fruits, honey, jam, and treacle are examples of sugary foods. Examples of starchy foods are bread, potatoes, rice, and spaghetti. Cellulose is a less familiar carbohydrate. It is a major constituent of plants, where it forms a wall around each cell. But cellulose can only be digested by herbivores, and consequently the enormous supplies of cellulose in the world are only available to carnivores and omnivores indirectly, i.e. when they eat herbivores.

Functions of carbohydrates Carbohydrates are the chief source of energy for living things, which is why they are sometimes called the 'fuel' of life. On average, 1 g of carbohydrate can yield 17 kJ of energy when respired. The principal carbohydrate used in respiration is glucose. In fact, most carbohydrate foods are converted by the body into glucose before they are consumed in respiration.

Carbohydrates are also important as food reserves which are stored within organisms. Many plants store large quantities of **starch**: in wheat and maize seeds, and in potato tubers. In animals, the main carbohydrate food reserve is **glycogen**, which is very similar to starch in its chemical composition. If more carbohydrates are eaten than are necessary to satisfy the body's immediate energy requirements the excess food is converted into glycogen and stored in the liver and muscles. The body can store only limited amounts of glycogen, however, and when this limit has been reached, any excess carbohydrate in the diet is converted into fat or oil and stored in special tissues described below. Carbohydrates are made of carbon, hydrogen, and oxygen atoms, which are joined together so that the hydrogen and oxygen atoms are always present in the ratio of 2:1. Glucose for example, which is one of the simplest carbohydrates, has the chemical formula $C_6H_{12}O_6$.

Fats and oils

Examples of fatty and oily foods are butter, lard, suet, dripping, olive oil, and cod-liver oil. The main difference between fats and oils is that oils are liquid at 20°C, while fats are solid at that temperature.

Functions of fats and oils On average, 1 g of fat or oil can yield up to 38 kJ of energy when respired, which makes them a very important source of energy. But they are less easily digested than most other foods, which offsets their energy value a little.

Fats and oils are also important as food reserves in the body. In mammals, for instance, they are stored in special cells known collectively as **adipose tissue**. This tissue occurs under the skin, and around muscles, the heart, the kidneys, and several other body organs. These fat deposits have two main functions. First, they form a food reserve with an extremely high energy yield per gram and are therefore less bulky than carbohydrates with the same energy value. Second, the layer of fat under the skin insulates the body against loss of heat and is especially well developed in arctic anmals such as seals and polar bears.

The scientific name for fats and oils is **lipids**. Like carbohydrates, lipids consist of carbon, hydrogen,

These are examples of starchy, sugary, oily, and fatty foods. They are a rich source of energy

These are examples of protein foods. They are needed for growth and repair of tissues

and oxygen, but their molecules differ from carbohydrates in that they contain relatively little oxygen. Lipids are made of fatty acids, which are linked in various combinations with glycerol (glycerine).

Proteins

Meat, liver, kidney, eggs, fish, and certain beans contain much protein. Proteins are used for growth, and for replacing and repairing worn-out and damaged tissues. Protein foods are needed particularly by growing children, pregnant women, and people recovering from injuries or sickness. Proteins are not normally used to provide energy but can yield up to 17 kJ of energy per gram.

During digestion proteins are broken down into chemicals called **amino acids**. There are about twenty-six amino acids but only ten of them are essential to human health. All animal proteins contain all the essential amino acids, and so they are called **first-class proteins**. Plants contain what are called **second-class proteins**, because no single plant protein has all the essential amino acids. However, all of them can be obtained by including a wide variety of plant foods in the diet. Unfortunately in certain countries plant foods are unobtainable in sufficient variety and the people who live there have to rely almost exclusively on one type.

For example, rice is the staple food in the Far East, and maize in South America. In such areas **protein deficiency disorders** such as **kwashiorkor**, which involves degeneration of the liver and pancreas, are common (see section 3.6).

Vitamins

Vitamins are chemicals which animals (including humans) need, in minute amounts, to grow and to remain healthy. Some vitamins are needed in only a few thousandths, or even millionths of a gram each day.

Functions of vitamins There are 13 major vitamins: A, C, D, E, K, and eight different B vitamins. They have no energy value. Their role is to take part in the chemical reactions of metabolism, for the most part in conjunction with enzymes. Lack of a vitamin causes the reaction in which it takes part to slow down. Since most metabolic reactions are part of a long sequence of events, an alteration in the rate of any one of them can have widespread effects on the body. These effects are known as **vitamin deficiency diseases**. Some of these are described below.

Vitamin A keeps the skin and bones healthy, helps prevent infection of the nose and throat, and is necessary for vision in dim light. Lack of vitamin A causes poor night vision, and increases the chances of infection of the nose and throat. Vitamin A is found in carrots, milk, fish-liver oils, and green vegetables.

Vitamin B_1 helps the body obtain energy from food. Lack of it reduces growth, and causes **beri-beri**, a disease in which the limbs are paralysed. Vitamin B_1 is found in yeast, wholemeal bread, nuts, peas, and beans.

Vitamin B_2 enables the body to obtain energy from food. Lack of it causes stunted growth, cracks in the skin around the mouth, an inflamed tongue, and damage to the cornea of the eye. Vitamin B_2 is found in liver, milk, eggs, yeast, cheese, and green vegetables.

Vitamin B_{12} enables the body to form protein and fat, and to store carbohydrate. Lack of it causes **pernicious anaemia** (failure to produce haemoglobin for red blood cells). Vitamin B_{12} is found in liver, meat, eggs, milk, and fish.

Vitamin C is destroyed by cooking, grating, or mincing food. It disappears from food if it is stored for long periods. Vitamin C helps wounds to heal, and is needed for healthy gums and teeth. Lack of it causes **scurvy**, a disease in which the gums become soft, the teeth grow loose, and wounds fail to heal properly. Vitamin C is found in oranges, lemons, black currants and green vegetables.

Vitamin D enables the body to absorb calcium and phosphorus from food. These chemicals are needed to make bones and teeth. Lack of vitamin D causes **rickets** (soft weak bones, which bend under pressure). Vitamin D is found in liver, butter, cheese, eggs, and fish.

Vitamin pills Millions of pounds are spent each year on vitamin pills. Are they necessary for good health? The answer for most people is NO! The best way to satisfy vitamin needs is from food; food contains things which pills cannot provide, like dietary fibre, minerals etc. This does not mean that you must eat much more expensive, or fattening foods.

Some precautions The following precautions will reduce loss of vitamins when preparing food.

1. Juices left over from cooking contain vitamins that have drained out of food, so use them to make sauces, gravy, or soups.

2. Store vegetables in a cool dark place, and eat them as fresh as possible. Wash *before* cutting them up, and don't leave them soaking in water for long periods. Cook them as lightly as possible.

These foods contain more than enough vitamins to supply your needs for one day

Special vitamin needs There are some circumstances in which extra vitamins are needed.
Vitamin B_{12} A strict vegetarian diet, like that eaten by Vegans, contains very little vitamin B_{12}, since this is found mostly in meat, fish, and dairy produce.
Vitamin C Convalescents need extra vitamin C, especially after surgery. Some people believe that extra vitamin C helps prevent colds, although there is no conclusive evidence that this is so.
Vitamin D Most of the body's vitamin D is made in skin exposed to sunlight. Therefore in prolonged dull weather extra supplies are needed.

Minerals

Humans require about fifteen different mineral elements in their diet. Most of these are supplied by meat, eggs, milk, green vegetables, and fruits. Minerals have no energy value but they do have important functions in the body, which are listed in Table 1.

Sodium chloride, or common salt, is constantly and rapidly lost from the body in perspiration, urine, and faecal matter. In temperate climates an intake of at least 4 g daily is required to make up for this loss. Some of this salt is already in food when it is eaten; the rest must be added to the diet if health is to be maintained. More salt is needed on a hot day because perspiration rate is higher. Strenuous exercise under these conditions can result in salt loss at such a high rate that muscle cramp and heat exhaustion may occur unless extra salt is taken into the body. Too much salt in the diet can contribute to high blood pressure problems.

Table 1 Mineral requirements of humans
($1 \text{ mg} = 0.001 \text{ g}$)

Mineral	Daily requirement mg	Function in the body
Sodium chloride (common salt)	5–10	Blood plasma is almost 1% salt. Needed for digestion and passage of impulses along the nerve fibres
Potassium	2 ⎫	Necessary for muscle contraction
Magnesium	0.3 ⎭	
Calcium	0.8	A major constituent of bones and teeth
Iron	0.01	A major constituent of haemoglobin (red substance of blood needed to transport oxygen to body tissues)
Iodine	0.00003	Needed by the thyroid gland. Deficiency causes goitre.

The child on the right has rickets, a vitamin-deficiency disease caused by an inadequate supply of vitamin D in her diet

3.3 Why we need water

At least two-thirds of the human body consists of water. Water has no food value, but it is still one of the most essential components of living matter. One of the most important functions of water is to act as the medium in which all the chemical reactions of metabolism take place, and for this reason life cannot continue in the absence of water. Water is also the medium in which soluble foods and excretory wastes are transported throughout the body. In higher animals the function of transporting such substances is carried out by the blood, which is mostly water.

In temperate climates at least two litres of water are lost daily by the human body as urine and perspiration. This must be replaced by drinking, and by eating foods which contain water. Try to drink six or eight glasses of liquid a day.

3.4 Dietary fibre (roughage)

Dietary fibre is the part of cereal grains, fruits, and vegetables which is not digested. It passes almost unchanged through the gut. Fibre in grains is called **bran** (Fig. 3.1). Fibre in fruit and vegetables is the **cellulose walls** of plant cells.

Benefits of a high-fibre diet
Studies have shown that people with low-fibre diets are more likely to suffer from constipation, heart disease, diabetes, and cancer of the bowel, than those with a high-fibre diet. It is not clear how fibre helps prevent all these diseases. The facts discovered so far can be summarized as follows:

1. Low-fibre food passes slowly through the gut; it tends to form hard dry lumps which get stuck – causing constipation. These lumps can decay, and

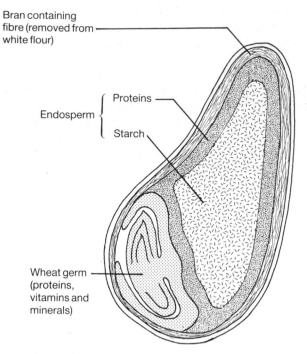

Bran containing
fibre (removed from
white flour)

Endosperm {
  Proteins
  Starch

Wheat germ
(proteins,
vitamins and
minerals)

Fig. 3.1 Parts of a wheat grain. (White bread is made from the endosperm after bran and wheat germ have been removed)

form swellings which must be removed surgically.

2. Fibre adds a firm, bulky mass to food. This allows gut muscles to grip and move the food easily along the intestine.

3. Fibre absorbs poisonous wastes from the gut. This may be linked with the fact that high-fibre diets reduce the risk of bowel cancer.

4. Fibre helps avoid obesity by adding indigestible bulk to food. This satisfies the appetite, but doesn't make you fat. Fibre may also reduce the rate at which other foods are absorbed from the gut, so they pass unused out of the body.

High-fibre diets

Wholemeal bread is preferable to white because it contains 8.5% bran, compared with white bread's 3%. Wholemeal spaghetti and brown rice also contain more bran than their white equivalents. Bran breakfast cereals have up to 25% bran in them.

Other high-fibre foods are; potatoes cooked in their skins, berries, nuts, bananas, apples, oranges, and peaches. High fibre vegetables include sweetcorn, spinach, and broccoli.

How much fibre?

For most people 30 g of fibre a day is sufficient. This, for example, can be obtained from four slices of wholemeal bread, two apples, and a bowl of bran cereal.

3.5 A balanced diet

Good health depends, to a large extent, on eating the correct *amount* of food, and the correct *proportions* of each type of food. When these requirements are met, a diet is said to be **balanced**. But it must be balanced in two different ways.

1. The amount of food eaten each day should provide no more and no less than the amount of energy used during that day. This means that food eaten must 'balance' energy used.

2. A correct balance must be achieved between the proportions of energy-rich foods (carbohydrates and fats), body-building foods (proteins), and protective foods (vitamins and minerals) in a diet.

Daily energy requirements

The amount of energy a person uses each day varies according to age, sex, bodysize, occupation, and special conditions such as pregnancy (Fig. 3.2).

During deep sleep, an adult male expends energy at a rate of 4.2 kJ per minute. When awake and resting in a chair, he expends about 10 kJ per minute; he expends 21 kJ walking; and running very fast, he expends 42 kJ per minute.

If, over a long period, insufficient food is eaten to supply daily energy requirements, the body uses its stored fat and then protein, and soon becomes thin and unhealthy. If excess food is eaten it is converted into fat, and the body becomes over-weight (obese). The effects of eating too little and too much are described in section 3.6.

Balanced proportions of food

It is possible to meet total daily energy requirements by eating nothing but pure white sugar. But anyone who did this would not remain alive and healthy for long; her mineral, vitamin, and protein reserves would quickly be exhausted.

A balanced meal consists of about one part protein, one part fat, and up to five parts carbohydrate. It should also include foods rich in vitamins and minerals.

Carbohydrates

Adults should eat between 300 g and 500 g of carbohydrate each day, depending on their daily energy expenditure (Fig. 3.2). Avoid eating too much sugar (see section 3.6). Healthy sources of carbohydrate include wholemeal bread, potatoes, brown rice and pasta, bananas, peaches, apples, and other fruits.

Fats

Adults should eat between 80 g and 100 g of fats daily. A small amount of fat is required to supply the fat-soluble vitamins A, D, and E, and essential fatty acids. Fats are not essential beyond these needs but, when used for frying and baking, they make food more tasty (Fig. 3.3).

Fat also reduces the bulk of food needed to supply energy requirements, since 1 g of fat contains twice the energy of 1 g of carbohydrate.

Eating large amounts of fat is dangerous to health (see section 3.6), so foods such as butter, cream, cheese, and the fat around meats should be eaten in moderation. Remember also that there are large amounts of 'hidden' fat in foods like burgers, sausages, crisps, peanuts, and chocolate.

Fig. 3.2 The amount of energy used each day

Age, sex, occupation	Energy used in 1 day	Diet (daily)
15 years	**Males** 12 600 kj **Females** 9600 kj	
Adult (light work)	11 500 kj 9450 kj	80–100g of protein a day; 300 g of carbohydrate (except for those doing heavy work who should eat far more), and 50–100 g of fat . Large amounts of sugar should be avoided as sugar increases tooth decay.
Adult (moderate work)	12 100 kj 10 500 kj	
Adult (heavy work) 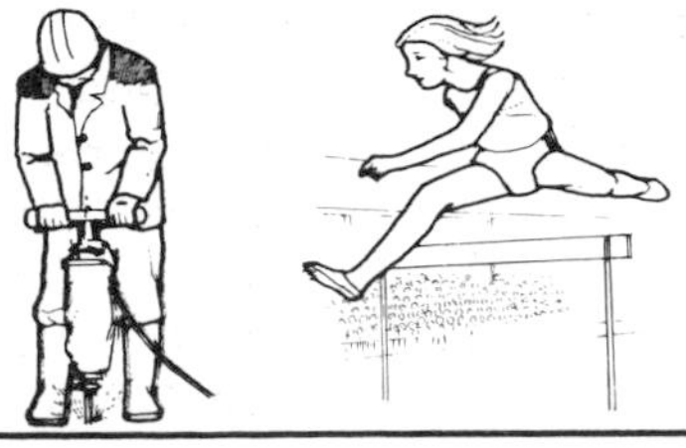	15 000 kj to 20 000 kj 12 600 kj	

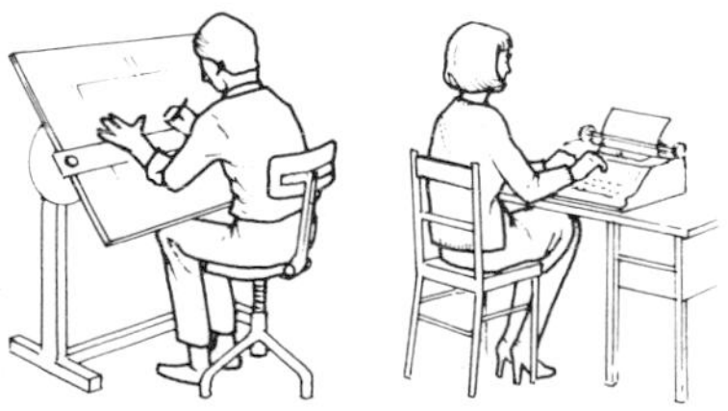

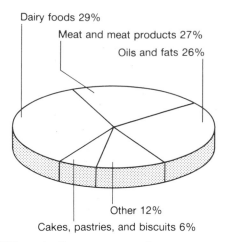

Dairy foods 29%
Meat and meat products 27%
Oils and fats 26%
Other 12%
Cakes, pastries, and biscuits 6%

Fig. 3.3 Where the fat we eat comes from

The pie chart above shows how easy it is for us to eat too much fat without realizing it. Remember this when you are next tempted by an in-between-meal snack! Remember too that snacks count when you are trying to achieve a balanced diet.

Proteins

Adults should eat 80 g to 100 g of protein food daily. 60% of this should be **first-class protein** (meat, eggs, fish). If these are not available, a wide variety of plant proteins should be eaten, such as beans and peas, wholemeal bread, brown rice and pasta, and nuts.

3.6 **Diet problems**

Eating too little

Protein energy malnutrition (PEM) is a term used to describe diets which lack proteins, and energy-giving carbohydrates. PEM causes a group of diseases common among children in poorer parts of the world.

In rural areas of East Africa, PEM results from a staple diet of cassava, maize, and bananas. Meat and other proteins are very expensive, so farmers prefer to sell their animal products rather than eat them. According to UNICEF, in East Africa up to 20% of children under five suffer from the results of PEM.

Kwashiorkor and **marasmus** are diseases which result from PEM. Symptoms of kwashiorkor include swelling of the body, especially the belly, flaking skin, and red hair. Symptoms of marasmus include thinness and poor muscle development, so that bones show through the skin. Both these disorders result in poor mental development and reduced resistance to disease.

Anorexia nervosa is an obsessive desire to become slim. It is increasingly common in teenage girls, and young women in more affluent countries. Sufferers go to all lengths to avoid eating, and so endanger their health. Some have died from this condition. It is sensible to eat less if you are overweight, but it is also very important not to become obsessed by slimming.

Kwashiokor

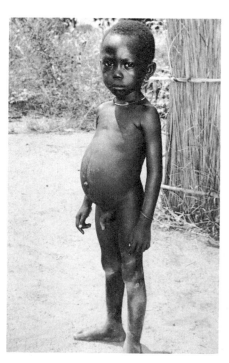

Marasmus

Obesity

Eating too much

Eating too much can lead to weight increase. Figure 3.4 gives ideal weights of adults according to height, and shows when a person is overweight.

About half of Britain's population is overweight, and this is dangerous (Fig. 3.5). There is evidence that obesity contributes to heart disease, high blood pressure, diabetes, gall bladder disease, cancer of the bowel in both sexes, and cancer of the breasts and womb lining in women. Excess fat and sugar in the diet are the main causes of obesity.

Excess fat Many experts think that animal fats are particularly dangerous because they lead to high levels of **cholesterol** in the blood. Cholesterol contributes to the blockage of arteries, which causes heart disease (section 6.7, and Fig. 3.6). Also, it is estimated that diets high in animal fats and low in fibre cause one third of all cancers.

Excess sugar Section 4.2 describes how sugar causes tooth decay. It is also linked with the onset of diabetes (section 13.7). Sugar can cause obesity because, like fat, you can eat a lot without feeling full. The most fattening foods of all contain both sugar *and* fat. Ice cream and chocolate are examples. Sugar is also present in large amounts in cakes, fruits tinned in syrup, and sugary soft drinks.

Fig. 3.4 Half the population of Britain is overweight

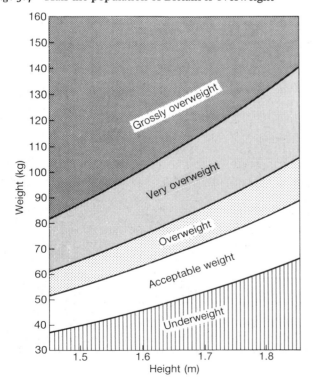

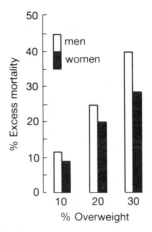

Fig. 3.5 Obesity is dangerous

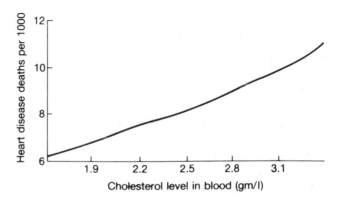

Fig. 3.6 High levels of cholesterol are linked with heart disease

Excess salt There is evidence which links high salt intake with high blood pressure, a major factor in heart disease and strokes, and with stomach cancer. On average the British consume 2.5 teaspoons of salt daily (12 g). The recommended level is 5 g. If you wish to cut down salt intake avoid; foods 'in brine' or 'smoked', packet soups and sauces, preserved meat like bacon, and salted crisps.

3.7 **Food additives**

Additives are chemicals which food manufacturers put into their products, but which have no nutritional value. Additives are found in **processed foods**, that is, foods which have been through a manufacturing process. Table 1 lists some processed foods, and some which are free of additives.

35

Why additives are used

The food industry uses additives for many reasons (Table 3).

Antioxidants These stop fats in food going rancid.

Colourings These are used to make food look attractive and cheaper to manufacture. They restore and enhance colour lost in processing. Since they may be potentially dangerous their use is difficult to justify. Tinned peas and strawberries owe their vivid colours to chemicals, and the colour of blackcurrant drink owes nothing to blackcurrants. Foodstuffs would, however, be more expensive if natural colours and flavours were always used.

Emulsifiers and stabilizers Emulsifiers allow water and oil to be mixed together as an emulsion. Stabilizers keep them from separating.

Flavourings These restore flavours lost in processing. They can also make a food appear to have an ingredient which it doesn't. For example 'strawberry flavour' means a food contains a strawberry flavoured chemical, and not strawberries.

Preservatives These slow down the rate at which foods go bad. They allow foods to be transported long distances, and stored in shops for some time before they are unfit to eat.

Why additives cause concern

1. Evidence suggests that certain additives may cause allergies such as asthma, headaches, behaviour problems in children, damage to organs such as the kidneys and liver, and certain cancers (Table 3).
2. Children are more at risk than adults. Children have a low body weight and so receive additives in relatively large doses. Also the body systems which remove unwanted substances are not well developed in children.
3. Additives are tested on mice and rats, which have short life spans. Only a few additives have been tested over long periods on humans.

Under European law, every food additive has to be shown on a packet under its own 'E' number. The name, and effects, of these additives can be checked in an official guide.

Most additives are probably harmless, and those such as preservatives help keep food prices down and increase the shelf-life of foods.

Table 2 Some foods with, and without additives

Processed (with additives)	Additive-free foods
Packet soups	Fresh vegetables
Packet cake mixes	Fresh fruit (except those
White bread	with added skin
Burgers	colourings)
Fish fingers	Free-range eggs
Fruit-flavoured drinks	Fresh meat
Flavoured crisps	Real fruit juices
Tinned meats	Real-fruit yoghurt
	Muesli

Table 3 Uses and possible ill-effects of food additives

Additive	Uses	Possible ill-effects
Amaranth E123	Red colouring in jams, jellies, ketchup, 'fruit' drinks	Tumour production and allergic reactions
Tartrazine E102	Yellow colouring in 'fruit' drinks	Behaviour problems in children
Monosodium glutamate 621	Flavouring in cup-a-soups and Chinese foods	Allergic reactions, psychological effects (depression)
Sulphites	Preservative in beers, wines, fruit juices, dried fruits	Interferes with vitamin A and B metabolism

Most additives are probably harmless.

3.8 Feeding the world's billions

During the time it takes you to read this sentence, four people, most of them children, will have died of starvation. How can this be happening when there is plenty of food in the world? We have more than enough to feed everyone, including the 80 million additional mouths which have to be fed each year.

Unequal distribution of food

Part of the answer to this question is that food supplies are not shared equally throughout the world. People in rich countries eat far more than they need, and build up huge 'grain mountains' and 'milk lakes' (compare Ghana with Denmark in Fig. 3.7). Americans, for example, represent only 6% of the world's population, yet they consume 35% of the world's resources. But unequal distribution of food is not the only problem.

Destructive farming methods

Many people believe that famines are caused by drought. They assume that when the rains begin again all will be well. But this is not always true. Drought is obviously an important factor, but an underlying cause of famine lies with the way land is used and abused by those who depend on it.

Cash crops instead of food In many under-developed countries an increasing proportion of the best farm land is devoted to growing tobacco, coffee, tea, and other crops which earn money in the export-market. Over the years, however, the cost of imported tractors and fertilizers needed to tend these crops increases. If the world price for a cash crop falls, or if there is a poor harvest, then the crisis begins: food production has fallen while population has increased, but there are now fewer export earnings to buy imported food. If these events coincide with a drought, people starve.

Deforestation When the best land has been cultivated, forests are cleared to make way for more crops. This has two serious consequences in tropical countries. First, exposed top soil is washed away by the next rains, and the soil is then baked hard by the sun. Second, fewer forests mean less rainfall, because forests hold on to moisture and release moisture through transpiration. These factors help maintain the climatic conditions which cause rain.

Animal versus plant proteins Farmers in poor countries try to keep as many cattle as possible because the protein they yield is worth more than plant proteins. But a cattle farmer needs 30 times as much land to produce the same amount of protein as a plant crop. There is a temptation, therefore, to put more cattle on to land than it can sustain. Soon all vegetation disappears and deserts develop.

Some possible solutions

1. When famine strikes, rich countries respond with food and money. But the starving can come to depend on food aid unless it is soon replaced by long-term help. All to often much of the money goes into cash crops, industries, and city development, instead of food production.

2. The most successful aid is used to improve health, crop and livestock care, and irrigation. But this can go wrong unless it is preceded by detailed research, and administered with care.

3. The cost of agriculture can be reduced by using draft animals for ploughing. These are cheaper and less likely to 'break down'.

4. Soil erosion can be reduced by terracing on hillsides, and by planting belts of trees as wind breaks.

5. New sources of food can be found. Many wild plants and animals are being bred on farms because of their ability to live on poor soils in arid regions. The oryx, for example, is one of the few meat producing animals that can live in deserts. New preparations like **textured vegetable protein** can be used to convert cheap plant foods, like soya flour, into imitation meat. This means that the land presently used to grow wheat to feed cattle, can be put to more efficient use.

6. Above all else, poor countries should concentrate on food production. China, for instance, suffered famines for centuries. But it has concentrated on agriculture for food production rather than profit. Now Chinese farmers feed 22% of the world's population on 7% of the world's land.

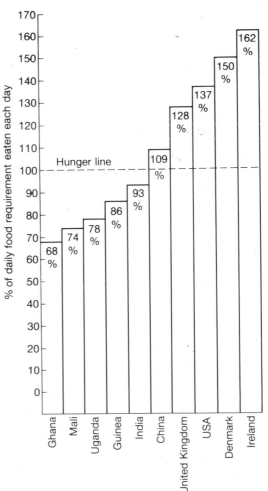

Fig. 3.7 The gulf between well-fed and hungry people. These are average figures so they do not show the extremes of under-nourishment and over-consumption.

3.9 Biotechnology and food production

Biotechnology is the harnessing of biological processes to produce useful substances. There is nothing new in this idea. Yeasts have been used for centuries in wine and bread making, as have the bacteria used to make cheese.

What is new is the enormous effort now being made to use living organisms in the large scale manufacture of drugs, fuels, and basic foodstuffs.

Microbes and food production

Most biotechnology is based on the use of microbes, such as bacteria, algae, yeasts, and fungi. Microbes are living miniature factories. They are capable of producing almost any organic substance at a fraction of what it costs to make the same chemicals in an industrial process.

Food from oil

One of the first successes in modern biotechnology was the discovery of microbes which live on low grade hydrocarbons found in crude oil. These microbes reproduce at a phenomenal rate in a liquid 'broth' derived from oil. When extracted, dried, and compressed, they yield a high grade protein animal food.

Food from algae

A number of microscopic algae, capable of thriving in a liquid little richer than tap water, can yield considerable amounts of proteins, vitamins, and many other useful substances.

Starch to sugar

Microbes can also be used to reproduce enzymes which change low value maize starch into high grade fructose syrup. This is a very sweet substance which can replace cane and beet sugar.

Increasing crop yields

All plants need nitrogen to make proteins for growth. They obtain nitrogen from nitrates in the soil, therefore plants grow best on soils rich in these chemicals.

Some algae, however, can use nitrogen gas in the *air* to make protein. It may soon be possible to transfer this ability from algae to crop plants such as cereals. The plants could then yield far more food without the addition of expensive nitrogen fertilizers to the soil.

At present, food production using biotechnology is in its infancy. The main problem is developing large scale industrial processes which duplicate small scale laboratory experiments.

Investigations

A Food tests

Range of food substances required

Solutions of glucose, sucrose, and starch; ground-up suspensions in water of pea, bean, carrot, and grape; milk; castor-oil beans, peanuts, and cooking oil; pieces of potato, bread, and boiled white of egg.

1. *Tests for carbohydrates*

a) *Test for glucose* Place equal quantities of Benedict's solution and glucose solution (about $2\,cm^3$ of each) in a test-tube, lower the tube into a beaker of boiling water and note the brick-red precipitate of cuprous oxide which appears. Repeat the test with carrot and grape juice.

b) *Test for starch* Add a few drops of iodine solution to a starch solution and note the blue-black colour which appears. Repeat with bread, and add iodine to the cut surface of bean, maize, and potato. Scrape stained tissue from the last three substances on to separate microscope slides and observe the shape of starch grains under high magnification.

2. *Tests for fat and oil*

a) *Emulsion test* Add a few cm^3 of ethanol to a small amount of fatty food, such as ground-up castor-oil bean, peanut, suet, or cooking oil, in a test-tube and shake the mixture. After allowing it to settle pour off the ethanol into an equal quantity of water noting that a white emulsion is formed.

b) *Translucent paper mark* Press pieces of peanut and castor-oil bean on to paper and note that the translucent mark which appears will wash out with acetone but not with water.

3. *Biuret test for proteins*

Add $2\,cm^3$ of 2% sodium hyroxide solution to milk and mix. Add a few drops of 1% copper sulphate solution and note the violet colour which appears. It is not necessary to boil the mixture. This is a test for soluble protein. Repeat it with pea and bean suspensions.

4. *Test for vitamin C*

Place $2\,cm^3$ of the dye Phenol-indo-2:6-dichlorophenol in a test tube. Add 1% vitamin C solution a drop at a time from a graduated pipette. Note the quantity of vitamin needed to decolorize the dye. This result can be used as a standard for estimating the vitamin C content of fruit juices (e.g. lemon), and of extracts from fresh and old vegetables.

B *Investigating the energy value of foods*

1. Half-fill a number of boiling tubes with water (put the same amount in each).

2. Obtain foods which will burn – e.g. bacon fat, cake, bread, cheese, etc. Weigh out the same quantity of each (only a few grams are needed).

3. Burn a weighed sample of a food under a boiling tube of water. This can be done with matches while the food is stuck on a mounted needle, or held on a small spoon or wire gauze. Measure the temperature rise in the water.

4. As a control, burn the same number of matches, but no food, under another tube containing the same amount of water.

5. Compare the temperature rise in each tube. What do your results tell you about the amount of energy in each food? How could this apparatus be improved to capture more of the heat released as a food burns?

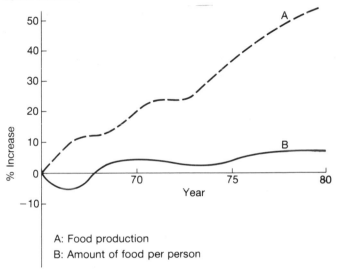

A: Food production
B: Amount of food per person

Fig. 3.8 (See question 2)

Questions

1. *a)* What is meant by a balanced diet?

b) Why is it important that one day's food should supply no more, or no less, than one day's energy?

2. Study Figure 3.8. Why does the amount of food per person remain about the same despite increased food production? How could this problem be solved?

3. The table below shows the composition of four different foods.

a) Which foods are likely to cause scurvy if used as a staple diet?

b) Which food would best prevent anaemia? Give a reason for your answer.

c) Which foods would you recommend for a young child? Give three reasons for your answer.

d) If you eat one of these foods, and no others, which would be most likely to result in kwashiorkor? Give a reason for your answer.

e) Why could foods B to D be termed 'unbalanced'?

Food	A	B	C	D
Energy value – kilojoules	280	500	100	150
Carbohydrates g	4.8	0	5.0	9.0
Fats g	3.5	13.5	0.2	0.1
Proteins g	3.2	25.0	0.9	0.5
Iron mg	0.01	1.5	0.25	0.05
Calcium mg	12.5	0.8	3.0	0.5
Vitamin A mg	4.5	600	0	0
Vitamin B_1 mg	3.8	35.0	25.0	10.0
Vitamin C mg	130	2000	0	0
Vitamin D mg	0.005	0.1	0	0

Summary and factual recall test

Examples of carbohydrate foods are (1–name six). These foods are the body's chief source of (2). They yield (3) kJ of energy per gram, and if eaten in excess are converted into (4), which is similar to starch, and is stored in the (5) and (6), or is converted into (7). Examples of fatty foods are (8–name five). These yield (9) kJ of energy per gram. Fat is stored in special cells which make up (10) tissue. This occurs in the following· parts of the body (11–name three). Fat under the (12) insulates the body against (13) loss. Fat is a better food reserve than carbohydrate because (14–give two reasons). Examples of protein foods are (15–name six). The body uses proteins for (16–describe two functions). Animal proteins are called first-class because (17). Vitamins have no (18) value, but lack of them causes diseases such as (19–name four). Humans need about (20) different minerals, which are found in (21–name four foods). High-fibre diets help prevent diseases such as (22–name four). A healthy diet is 'balanced' in two ways: (23), and (24). The amount of food a person needs depends on (25–name five factors). Obesity is dangerous because (26–give five reasons). Food additives cause concern because (27–give two reasons). People are starving because of (28) distribution of food and destructive farming methods such as (29–name two examples). Biotechnology means (30). Microbes are used to make food from (31–name three substances).

4

Feeding and the digestive system

Animals are the killers and thieves of this world. But they do not kill and steal out of malice; they do it to obtain food because, unlike plants, they cannot sit in the sun and make food by photosynthesis. Herbivores, carnivores, and omnivores all obtain food at the expense of other organisms.

Most of the food which animals eat is no use whatsoever to their bodies in the form in which it is eaten. In the first place, most foods are insoluble and so cannot pass through cell membranes into cells. In the second place, most foods are chemically different from the substances that make up body tissues. They must therefore be processed before the body can use them.

4.1 Digestion, absorption, and assimilation

An animal is able to make full use of the food it eats after the following events have taken place.

1. In large animals, including humans, food is first chewed up into pieces small enough to swallow. This is called **mastication** and is described in more detail later.

2. Food enters the gut, or **alimentary canal**. This is a tube running from the mouth to the anus, in which substances called **digestive enzymes** break food down into simple water-soluble chemicals. This process is called **digestion**. It is also known as **extracellular digestion** because it takes place outside the cells of the body.

Fig. 4.1 Diagram of extracellular digestion

Digestion takes place in an intestine, shown here as a straight tube through the animal from mouth to anus. Enzymes are released into this tube, and food is absorbed into the body through the walls of this tube

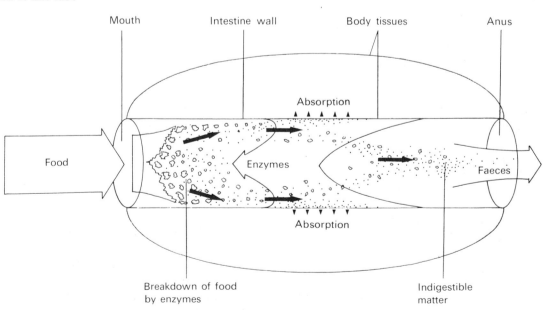

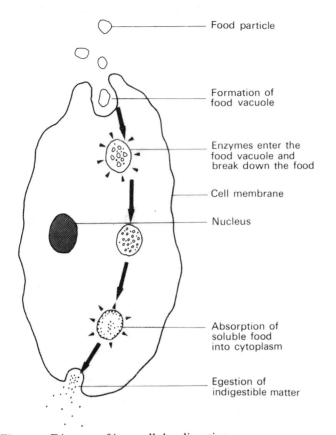

Food particle

Formation of food vacuole

Enzymes enter the food vacuole and break down the food

Cell membrane

Nucleus

Absorption of soluble food into cytoplasm

Egestion of indigestible matter

Fig. 4.2 Diagram of intracellular digestion

Digestion takes place within food vacuoles, from which soluble food is absorbed directly into the cytoplasm of the cell

3. The soluble food passes through the walls of the gut into the blood stream. This is called **absorption**.

4. Blood transports digested, soluble food to all parts of the body. The food enters cells and is transformed into substances which take part in metabolism. This is called **assimilation**. Figure 4.3 illustrates how this sequence of events changes protein in human *food* into proteins which make up human *body tissues*.

Any solid substances in food which cannot be digested, such as plant fibres, are expelled from the body as **faecal matter**, or **faeces**. This is called **egestion** (Fig. 4.1).

In simple organisms such as *Amoeba* and other protozoa, digestion takes place *inside* a cell; it is known as **intracellular digestion** (Fig. 4.2). The sequence of events is similar to that already described, except that digestion takes place in a bubble of liquid called a **food vacuole** inside the cell.

4.2 Digestive enzymes

All enzymes are **catalysts**; that is, they speed up chemical reactions which would otherwise pro-

ceed very slowly. Digestive enzymes are only one example of many types of enzyme which exist in living things. The reactions which these enzymes speed up involve splitting complicated molecules into simpler ones. Figure 4.4 illustrates one theory of how this may happen.

It is thought that the enzyme combines briefly with molecules of food and while in this state the food undergoes a rapid chemical change in which its molecules are split apart into chemically simpler substances. These substances separate from the enzyme leaving it immediately available for another identical reaction. In other words, enzymes are not used up in the reactions which they control but are used countless times in rapid succession.

Amino acids sequence in protein food

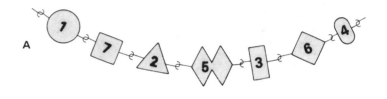

A

Amino acids split apart by digestive enzymes

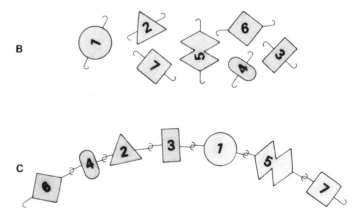

B

C

Amino acids rearranged to form human protein

Fig. 4.3 Digestion, absorption, and assimilation in humans. **A** shows the sequence of amino acids in protein before it is digested. **B** shows the amino acids separated by digestive enzymes. The amino acids are absorbed into blood and transported to body cells. **C** Inside a cell the amino acids are connected together in a different sequence to make a protein found in the human body.

The following list describes the more important features of digestive enzymes, most of which they share with enzymes in general.

1. Small amounts of enzyme can bring about a chemical change in relatively large amounts of another substance.

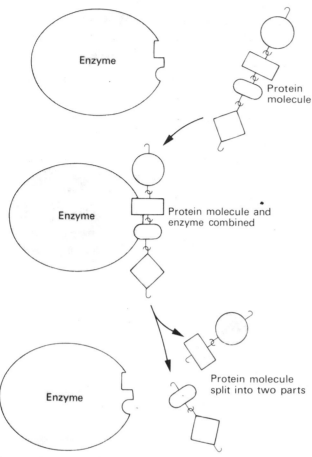

Fig. 4.4 Diagram of how digestive enzymes work

In this example a protein molecule, made up of linked amino acids, combines with an enzyme molecule and undergoes a chemical change which breaks the link between two of the amino acids. This splits the protein into two smaller molecules. Protein and enzyme molecules are not actually shaped like those in the diagram. The shapes indicate that each enzyme combines with only one specific part of the food molecule.

2. Enzymes are **specific**. This means that each enzyme is limited to reactions with only one, or very few, types of food substance. Because of this, digestive enzymes can be divided into groups according to the foods which they digest. **Amylases**, sometimes called **diastases**, are a group of enzymes which break down starchy foods into sugars such as glucose. **Lipases** are a group of enzymes which break down fats and oils into their component fatty acids and glycerol. **Proteases** are enzymes which break down proteins into their component amino acids.

3. The speed of reactions involving enzymes is influenced by temperature. A rise in temperature causes an increase in reaction speed, up to what is called the enzyme's **optimum efficiency**. At temperatures above this point the reaction goes faster for a time but eventually stops because excessive heat destroys enzymes.

4. Some enzymes work best in acid conditions, others in neutral conditions, and others in conditions which are alkaline. In technical terms, each enzyme requires a specific pH level for optimum efficiency.

5. Certain enzymes require the presence of vitamins of the B complex before they can function. These vitamins are known as **co-enzymes**.

A well-known commercial use of digestive enzymes is in 'biological' washing powders. These make use of an enzyme's ability to break down foods, and other substances which commonly stain clothes, into soluble chemicals. Biological washing powders are useful in removing stains caused by blood, gravy, egg yolk, chocolate etc.

4.3 Teeth

Teeth of mammals are set within sockets in the jawbone (Fig. 4.5). Each tooth consists of: a **crown**, the part above gum level; a **neck**, the part surrounded by gum; and a **root**, the part embedded in bone. The crown is the biting surface. Its outer layer consists of **enamel**, which is the hardest substance found in animals. The root and inner portion of the crown consist of a bone-like substance called **dentine**, except for a central **pulp cavity**. This contains nerves, and blood vessels that supply the growing tooth with food and oxygen.

Fig. 4.5 Structure of a tooth

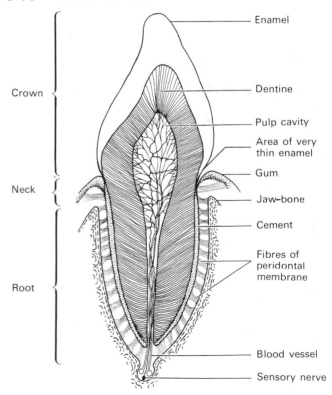

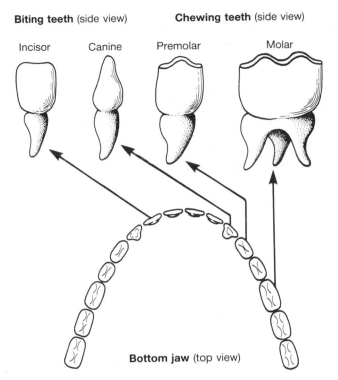

Biting teeth (side view) **Chewing teeth** (side view)

Incisor Canine Premolar Molar

Bottom jaw (top view)

Fig. 4.6 There are four types of tooth. Incisors and canines are used for biting. Premolars are used for chewing, which crushes and pulverizes food making it ready for digestion.

Tooth decay and gum disease

Tooth decay is caused by **plaque** – a sticky film of food, saliva, and bacteria which forms on teeth after meals. If you eat sweet foods between meals plaque on your teeth absorbs the sugar like a sponge. Bacteria in the plaque then transform this sugar into acid, which dissolves away tooth enamel, eventually making a hole (Fig 4.7).

In addition plaque builds up where the teeth and gums meet and can cause a space to form between the gum and a tooth (Fig. 4.7). Bacteria in this space damage fibres which hold the teeth in place. This causes the teeth to become loose and fall out. This is the main cause of tooth loss in middle age.

Make your teeth last a lifetime

Two things will help you keep your teeth: avoid sugar between meals, and brush your teeth regularly using a fluoride toothpaste.

Sugar If no sugar is present in plaque very little acid is produced, and most of the acid is neutralized by saliva so hardly any damage is done to the teeth. In addition, the surface of a tooth can 'heal' by forming new enamel. Healing occurs much faster if fluoride is present in the mouth; so always use a fluoride toothpaste.

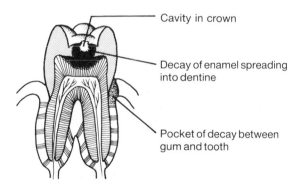

Cavity in crown

Decay of enamel spreading into dentine

Pocket of decay between gum and tooth

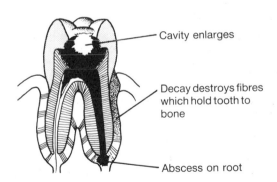

Cavity enlarges

Decay destroys fibres which hold tooth to bone

Abscess on root

Fig. 4.7 Tooth decay and gum disease. Decay is caused by bacteria which digest food stuck to teeth, changing it into acid which dissolves enamel. Bacteria may also get in between the gum and the root. This causes gum disease; which loosens teeth so they fall out

Downward strokes clean the upper teeth

Upward strokes clean the lower teeth

A back-and-forth action cleans the molars

An up-and-down action cleans behind the teeth

Fig. 4.8 How to brush your teeth

Brushing teeth Brush your teeth after breakfast and before going to bed, as explained in Figure 4.8. It is especially important to remove plaque from the edges of the gums in order to prevent gum disease.

Fluoride in drinking water

The addition of fluoride to drinking water is called **fluoridation**. Many Governments have decided to introduce fluoridation for the following reasons.

1. Studies in the United States of America and the United Kingdom have shown that there is little tooth decay in areas where fluoride occurs naturally in the water supplies, provided there is at least one part of fluoride to every million parts of water (1 ppm).

2. Experiments have shown that if fluoride is added to drinking water to bring it up to 1 ppm there is a big reduction in the amount of tooth decay, especially among children.

3. Fluoridation costs little, and does not require people to change their habits.

4.4 Digestion in the mouth

In mammals, food is first broken down by the mechanical action of teeth. Chewing is a physical process of reducing food to small particles. Thorough chewing of food is important for at least three reasons. First, small pieces of food are more easily swallowed than large lumps. Second, the surface area of food is vastly increased when it is broken into fragments, and since digestive juices work from the surface of food inwards, thorough chewing speeds up the rate of digestion. Third, chewing mixes food with saliva.

Saliva is a neutral or slightly alkaline fluid produced by three pairs of salivary glands (Fig. 4.9). In humans these glands produce about 1.5 litres of saliva daily. Saliva consists of water, the digestive enzyme **salivary amylase** (sometimes called ptyalin), **mucin**, and several other substances including sodium, potassium, bicarbonate, and chloride.

Functions of saliva

1. The water and mucin in saliva moisten, soften, and lubricate dry food so that it is more easily chewed and swallowed.

2. The enzyme salivary amylase breaks down starch into the soluble sugar maltose. This reaction is the first stage of carbohydrate digestion. Food remains in the mouth for only a few seconds and little digestion occurs here, but after swallowing the action of saliva continues for some time in the stomach. It is finally stopped when the acid in stomach digestive juice penetrates the food and destroys the amylase.

3. Amylase, and another enzyme in saliva called **lysozyme**, help to remove carbohydrate food and bacteria from between the teeth and thus help to prevent tooth decay.

4. Saliva moistens the mouth, tongue, and lips, which facilitates talking. Saliva production stops during nervousness or severe illness, making speech difficult.

5. By dissolving food, saliva makes it possible for the chemicals within it to reach the taste buds in the tongue. This is important because taste buds are not stimulated by dry food.

6. Bicarbonate in saliva acts as a **buffer**, which means that it keeps saliva at a more or less constant level of weak alkalinity. This helps prevent tooth decay by reducing the strength of mouth acids that dissolve tooth enamel. During sleep saliva production slows down considerably.

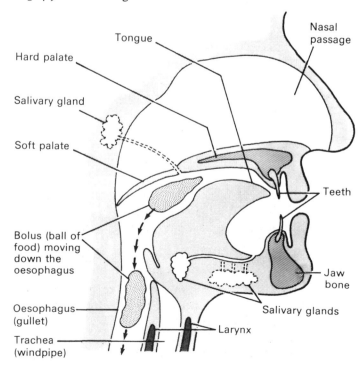

Fig. 4.9 Swallowing

Nasal passage

Tongue

Hard palate

Salivary gland

Soft palate

Teeth

Bolus (ball of food) moving down the oesophagus

Jaw bone

Salivary glands

Oesophagus (gullet)

Trachea (windpipe)

Larynx

4.5 Swallowing and peristalsis

Immediately before swallowing, food is rolled into a ball, or **bolus**, by the tongue and is then thrust to the back of the mouth. When this is done clumsily, or while breathing in through the mouth, food can be drawn into the windpipe, or **trachea**, causing a fit of coughing which usually clears any blockage. Normally, the following sequence of events prevents the risk of a blocked trachea.

First, the soft palate is pushed upwards to shut off entry to the nasal cavity. Second, muscles pull the top of the trachea upwards and under the back of the tongue. This action also pulls the larynx upwards causing the familiar 'bobbing' action of the 'Adam's apple', another name for the larynx. Third, the entrance to the trachea is reduced in size by the contraction of a ring of muscle around its circumference. Fourth, the **epiglottis**, a flat piece of cartilage and skin, drops over the tracheal entrance, forming a bridge over which food passes as it goes into the **oesophagus**, a tube leading to the stomach (Fig. 4.9).

Movement of food along the intestine

Food does not simply fall through the intestine under the influence of gravity. In fact a person can drink and eat while standing on his head, because food is moved along the intestine by wave-like muscular contractions known as **peristalsis** (Fig. 4.10).

These contractions start as soon as food enters the oesophagus. Circular muscles in the oesophagus wall contract immediately behind the bolus pushing it towards the stomach. These waves of contraction continue throughout the whole intestine at speeds of up to 20 cm per second. Movement of food is assisted first by the lubricating action of salivary mucin and then by **mucus**, a similar substance produced by **goblet cells** in the intestine wall (Fig. 4.12).

4.6 Digestion in the stomach

Perhaps the first person to investigate the stomach was an Italian called Spallanzani who in 1770 was curious enough, and brave enough, to swallow pieces of sponge tied to lengths of string. The sponges were pulled back up again, squeezed out, and the liquid from them examined. (This procedure could be dangerous and is not recommended to readers.)

In 1822 an army doctor called Beaumont treated a gunshot wound in one of his patients. The man's injury healed, but left a small hole right through the body wall into his stomach. Over an eleven year

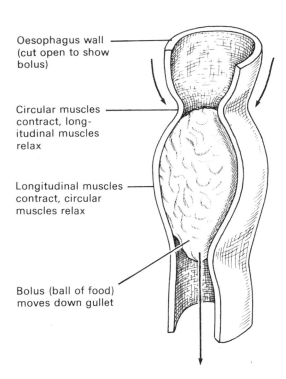

Oesophagus wall (cut open to show bolus)

Circular muscles contract, longitudinal muscles relax

Longitudinal muscles contract, circular muscles relax

Bolus (ball of food) moves down gullet

Fig. 4.10 Peristalsis (muscular contractions which move food along the gut)

period the doctor took samples of stomach contents through this hole and from them discovered a great deal about what happens to food in the stomach. Eventually, and perhaps not surprisingly, the 'patient' grew tired of the experiments and refused to let them continue!

The experiments of Spallanzani and Beaumont were among the first to provide evidence that the stomach produces chemicals, now called digestive juices, which dissolve food. Beaumont went on to investigate the conditions in which these juices work and the types of food that they digest. His results, and those of more advanced studies carried out by other workers, can be summarized as follows.

The stomach is an enlarged, bag-like region of the gut. In humans it is J-shaped when standing, and U-shaped when lying down. There are rings of muscle called **sphincters** around the entrance and exit to the stomach. These muscles act like valves: when they contract movement of food is prevented and when they relax food movements are resumed. As swallowed food approaches the stomach its uppermost sphincter relaxes and its lower sphincter contracts. The food enters the stomach, and the stomach walls expand until it reaches its normal capacity. In humans this is approximately 1 litre. More than this amount of food can be eaten at one meal but it produces an over-full sensation.

Once food is in the stomach, the circular muscles in its walls begin wave-like contractions similar to peristalsis which pass down the stomach from the oesophagus end about three times a minute. These contractions have the important effect of churning up food and mixing it thoroughly with a substance called **gastric juice** which flows out of thousands of tiny hollow pits, the **gastric glands**, in the stomach lining. The semi-liquid result of this churning process is called **chyme**.

When food reaches the liquid consistency of chyme the lower sphincter of the stomach relaxes periodically releasing small quantities of food at a time into the next region of the gut.

Gastric juice
Adult humans produce about three litres of gastric juice daily. It consists of the following substances.

Hydrochloric acid produced to create the acid conditions necessary for the action of gastric enzymes.

Protease produced to begin the digestion of proteins in food; it requires the acid conditions created by hydrochloric acid to do this.

Rennin an enzyme produced only in the stomachs of young mammals; it solidifies the proteins in milk and in this state they are retained in the stomach long enough for the protease enzyme to begin digesting them.

In humans, the digestive system below the stomach consists of two regions: the **small intestine** and the **large intestine**. The small intestine has a smaller diameter than the large intestine. The small intestine consists of two regions: the **duodenum** which is about 30 cm long, and the **ileum** which is about 8 metres long.

4.7 Digestion and absorption in the small intestine

Digestion in the duodenum
The duodenum receives fluid called **bile** through the **bile duct** from the **gall-bladder** in the liver. It also receives **pancreatic juice** through the **pancreatic duct** from the **pancreas**.

Bile Bile is a greenish-yellow liquid which is manufactured in the liver partly from substances resulting from the breakdown of old red blood cells. Bile is stored in the gall bladder, and is released whenever food enters the duodenum (Fig. 4.13).

Bile contains dissolved substances including: yellow-green bile pigments (2.5%) which have no digestive function; organic bile salts (6%) which have no digestive function; organic bile salts (6%); and inorganic substances of which sodium bicarbonate (0.8%) is the most important.

Organic bile salts have several functions. They react with fat-soluble vitamins (i.e. A, D, E, and K) and cholesterol, an important constituent of cell membranes. This makes them water-soluble and therefore easier for the intestine to absorb. Bile salts assist in the digestion of fats and oils in two ways. First, they reduce the surface tension of fats and oils so they disintegrate into a mass of tiny oil droplets, called an **emulsion**. This increases the surface area of fat exposed to digestive enzymes. Second, bile salts activate the fat-digesting enzyme **lipase** produced by the pancreas and the small intestine wall. Up to 95% of bile salts are absorbed by the small intestine and returned to the liver for re-use.

The **sodium bicarbonate** in bile is extremely important because it neutralizes stomach acid and creates an alkaline medium which is necessary for efficient functioning of all enzymes produced in the small intestine.

Pancreatic juice The pancreas is a long, narrow organ situated between the duodenum and the stomach. The pancreatic juice which it produces is colourless and contains sodium bicarbonate, which makes it alkaline. It also contains many powerful enzymes, in fact they are so powerful that in combination with bile and other enzymes from the ileum wall they are capable of completing the whole digestive process without the aid of mouth and stomach enzymes.

Pancreatic juice contains a protease enzyme which continues the digestion of proteins. It also contains an amylase enzyme similar in function to the amylase in saliva, together with a lipase enzyme which breaks down fats and oils to fatty acids and glycerol.

All of these enzymes continue their digestive activities within food as it passes into the ileum.

Digestion in the ileum
Within the ileum all digestive processes are completed and the soluble products are absorbed into the blood-stream. The ileum is therefore both a digestive and an absorptive organ.

Glands in the duodenum and ileum walls secrete many important enzymes which together form a fluid called intestinal juice (**succus entericus**). (Fig. 4.11 and 4.12.)

Intestinal juice contains several more protease enzymes which complete the breakdown of proteins to amino acids. In addition, there are three amylase enzymes which complete the breakdown of carbohydrate foods to glucose, and lipase enzymes which continue the breakdown of fats and oils to fatty acids and glycerol.

Absorption in the duodenum and ileum

The end product of enzyme activity upon food is a liquid called **chyle**. This liquid is a mixture of many soluble substances including amino acids, fatty acids, glycerol, monosaccharide sugars, vitamins, and minerals together with insoluble but finely emulsified fats and oils. All of these are absorbed into the blood-stream through the duodenum and ileum walls.

The whole of the ileum's internal surface is covered by finger-like projections called **villi** (singular villus) which are about 1 mm long (Figs. 4.11 and 4.12). Each square millimetre of ileum has up to forty villi and there are about five million in the ileum as a whole. The presence of villi gives the ileum a far greater internal surface area available for absorption than if it had a smooth lining. Under very high magnification ($\times 40\,000$) the surface of each individual cell lining the ileum is seen to be folded into **micro-villi**, giving the ileum an estimated total internal surface area of thirty square metres.

Inside each villus there is a dense network of blood capillaries, and a single **lacteal**, or lymph vessel, which is closed at its upper end (Fig. 4.12). It is into these two kinds of vessels that digested food passes after it has been absorbed into the cells covering the outer surface of each villus. In general the soluble foods, amino acids, sugars, vitamins, minerals, and some fatty acids and glycerol, pass into the blood capillaries of the villi. These capillaries join together to form a larger blood vessel called the **hepatic portal vein** which carries the food to the liver. The bulk of fatty acids and glycerol passes through the villi walls into the lacteals. The remaining emulsion of fat and oil droplets is absorbed into the lacteals of the villi in the following manner. Fat and oil droplets pass between the micro-villi of the cells covering each villus, and are absorbed whole into these cells by a process like amoeboid feeding. The droplets then pass through the cytoplasm of the cells and out the other side into the lacteals. They then pass into the main lymphatic system which eventually discharges them into the blood.

Movement of absorbed food out of the villi into the blood-stream and lymphatic system is assisted by periodic contractions of villi during which they suddenly become shorter and fatter, then relax slowly. This occurs about six times per minute in each villus.

4.8 Functions of the large intestine

The large intestine consists of the **colon**, **rectum**, and **anus**. The colon receives indigestible material from the ileum. This material consists of plant fibres, cellulose, bacteria, dead cells dislodged by friction from the intestine walls, mucus, and considerable quantities of water. It remains in the colon for about thirty-six hours, during which time most of the water and salt are absorbed from it. The semi-solid residue, called **faecal matter** or **faeces**, is lubricated by mucus and periodically ejected from the body via the rectum and through the anus.

Fig. 4.11 Structure of the small intestine wall

A Region of small intestine cut open to show villi

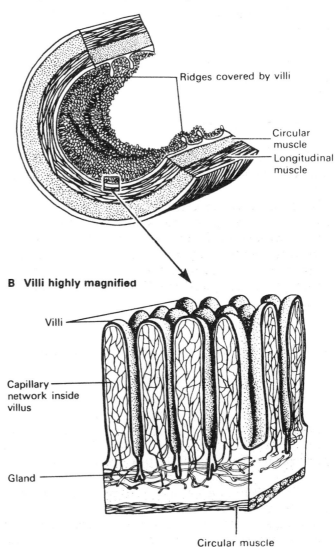

B Villi highly magnified

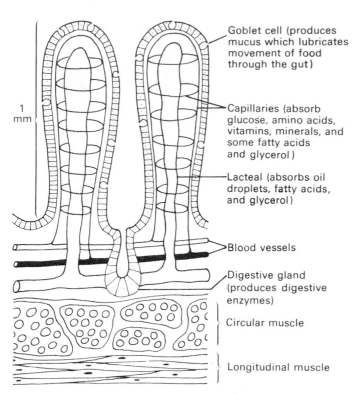

Goblet cell (produces mucus which lubricates movement of food through the gut)

Capillaries (absorb glucose, amino acids, vitamins, minerals, and some fatty acids and glycerol)

Lacteal (absorbs oil droplets, fatty acids, and glycerol)

Blood vessels

Digestive gland (produces digestive enzymes)

Circular muscle

Longitudinal muscle

1 mm

Fig. 4.12 Diagram of the ileum wall and villi

A variety of shapes of villi from the ileum. **F** is finger-like; **L** is leaf-like; and **C** is long and curved, a form known as convoluted

Caecum and appendix

At the point where the ileum joins the colon there is a tube called the **caecum**, from which arises a sac called the **appendix**. In ruminants and other herbivores such as rabbits and horses these organs are large and contain the bacteria which help digest cellulose. In humans these organs have no function and are referred to as **vestigial structures**. This term is used to describe organs which in the course of evolution have become reduced in size and have lost their original functions.

The presence of a vestigial caecum and appendix may indicate that long-extinct ancestors of the human species were capable of digesting cellulose material from a vegetable diet.

4.9 Summary of digestion and assimilation

The purpose of digestion is to make food soluble so that it can be transported by the blood to the cells of the body. Here it is **assimilated**, i.e. used by the cells for energy, growth, and repair. Digestion and assimilation of foods can be summarized as follows:

Carbohydrate foods are digested by amylase enzymes in saliva, in the pancreas, and from glands in the small intestine wall. These enzymes break down carbohydrates into glucose, which is assimilated and used as a source of energy. Any excess glucose is stored as glycogen in the muscles and liver, and as fat around internal organs and under the skin.

Fatty and oily foods (lipids) are first emulsified by bile. They are then digested by lipase enzymes, from the pancreas and the small intestine wall, into fatty acids and glycerol. Some of these are assimilated into cells to help build cell walls and other components. Most are converted into fats and oils, and stored in the body as food reserves.

Protein foods are digested by protease enzymes in the stomach, in the pancreas, and from the small intestine wall, into amino acids. These are assimilated into cells and used as building materials for growth and repair, and for making enzymes.

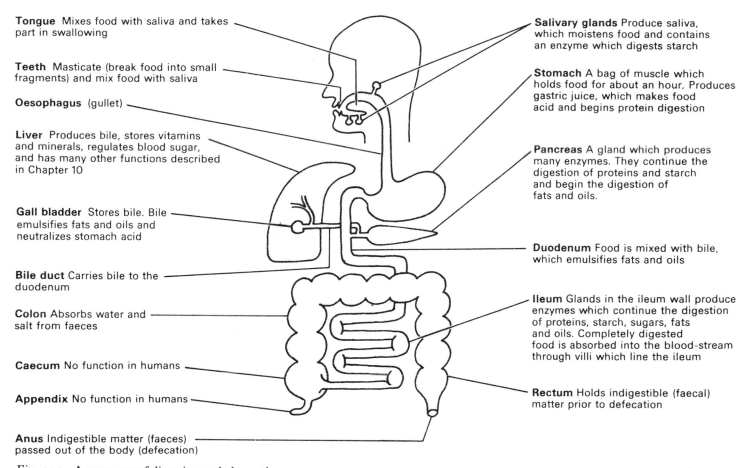

Tongue Mixes food with saliva and takes part in swallowing

Teeth Masticate (break food into small fragments) and mix food with saliva

Oesophagus (gullet)

Liver Produces bile, stores vitamins and minerals, regulates blood sugar, and has many other functions described in Chapter 10

Gall bladder Stores bile. Bile emulsifies fats and oils and neutralizes stomach acid

Bile duct Carries bile to the duodenum

Colon Absorbs water and salt from faeces

Caecum No function in humans

Appendix No function in humans

Anus Indigestible matter (faeces) passed out of the body (defecation)

Salivary glands Produce saliva, which moistens food and contains an enzyme which digests starch

Stomach A bag of muscle which holds food for about an hour. Produces gastric juice, which makes food acid and begins protein digestion

Pancreas A gland which produces many enzymes. They continue the digestion of proteins and starch and begin the digestion of fats and oils.

Duodenum Food is mixed with bile, which emulsifies fats and oils

Ileum Glands in the ileum wall produce enzymes which continue the digestion of proteins, starch, sugars, fats and oils. Completely digested food is absorbed into the blood-stream through villi which line the ileum

Rectum Holds indigestible (faecal) matter prior to defecation

Fig. 4.13 A summary of digestion and absorption

Investigations

A *An investigation of enzymes*

1. Materials: saliva, obtained by rinsing the mouth with water and collecting it in a test-tube; 10% pepsin solution; 10% pancreatin solution; 5% starch solution; coagulated egg white, obtained by mixing the white of one egg with 500 cm³ of water and heating it, while stirring, until a white suspension is obtained; Benedict's reagent; dilute hydrochloric acid; dilute sodium hydroxide solution. Refer to the food tests on page 38 whenever necessary.

2. Verify that saliva digests starch into maltose as follows. Mix 2 cm³ each of starch solution and saliva in a test-tube and place the tube in a beaker of water warmed to about 20°C. After five minutes, during which the tube must be shaken several times, test the contents for starch, and use the glucose test to detect maltose.

3. Verify that pepsin digests protein as follows. Mix 2 cm³ each of pepsin solution and coagulated egg white in a test-tube with one drop of dilute hydrochloric acid. Observe that the white suspension of egg becomes a clear liquid when the tube is left in a beaker of warm water for a few minutes and shaken from time to time.

4. Devise and carry out controlled experiments to prove that it was the pepsin and not the dilute hydrochloric acid which digested the egg white within a few minutes in 3 above.

B *Investigating differential permeability of Visking tubing*

1. Warm a beaker full of water to about 37°C.

2. Tie a tight knot at one end of a length of Visking tubing. Use a syringe to fill the Visking tubing with a mixture of starch and glucose solution. Close the Visking tubing with a paper clip.

3. Rinse the outside of the Visking tubing with tap water to remove all traces of starch and glucose. Place the filled tubing in a large test tube of warm water. Withdraw some of the water in contact with the Visking tubing and test it for starch and glucose, as explained at the end of Chapter 3.

4. After 20 minutes withdraw some more water in contact with the Visking tubing and test it for starch and glucose. If these tests are negative repeat them after 10 minutes.

5. The Visking tubing represents the gut wall. The starch represents undigested food and the glucose represents digested food. Like the gut wall, Visking tubing will allow small molecules, like glucose and water, to pass through but not large molecules such as starch. This is called **differential permeability**. If starch remains unchanged in the gut it will not pass through into the body. Starch (and other foods) must first be broken down by digestive enzymes into molecules small enough to pass through the gut wall into the blood stream.

C *How to demonstrate peristalsis*

1. Obtain a length of bicycle inner tube (approximately 1 m) and a round stone roughly the same diameter as the tube.

2. Push the stone into the tube. Hold the end of the tube with one hand. With the thumb and forefinger of the other encircle the tube just behind the stone. Use these fingers to push the stone through the tube.

3. a) What does the tube represent?
b) What does the stone represent?
c) Which set of muscles do the fingers encircling the tube represent?

Questions

1. a) What are digestion, absorption, assimilation, and egestion?
b) Where does each of these functions take place in the body?
c) Why must food be digested before the body can use it?

2. Below is a diagram of the human digestive system. Name all the parts indicated with a letter. Which letter indicates the part of the digestive system where:
a) bile is temporarily stored
b) acid is added to food
c) faeces leave the body
d) most absorption takes place
e) vitamins and minerals are stored
f) water is absorbed?

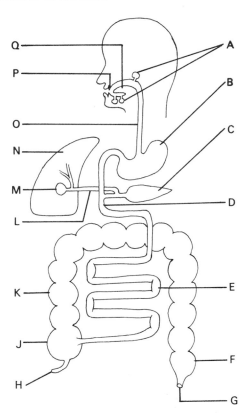

Summary and factual recall test

Digestion is the process by which food is broken down into chemically (1) molecules which are (2) in water. Digestion is necessary because most foods are (3) when eaten and so cannot be (4) into cells.

Amylases are a group of (5) which break down (6) foods into (7) such as (8); lipases break down (9) and (10) into (11) and (12); and proteases break down (13) into (14).

Intracellular digestion takes place (15) cells and is typical of protozoa such as (16). Extracellular digestion takes place (17) cells and typically occurs within a tube called the (18), which runs from the (19) to the (20).

Thorough chewing of food is important because (21–three reasons). The main action of saliva is to break down (22) into (23). Food is moved along the intestine by (24)-like (25) contractions called (26). Movement of food in and out of the stomach is controlled by muscles called (27). The stomach wall produces (28) juice, which contains (29) acid, and a (30) enzyme.

The duodenum receives (31) from the liver. Its main function is to break down fats and oils into (32) which are known as an (33). The pancreas produces (34) enzymes which are released into the (35).

The ileum wall is covered with millions of (36). These (37) the surface area available for (38) of soluble food. The main function of the large intestine is to remove (39) and (40) from indigestible substances, called (41).

5

Plant nutrition

How are plants able to make food for their own use, as well as for humans and other animals, and yet never eat anything for themselves?

At first, it was assumed that plants obtained all that they needed to grow from the soil. Then it was discovered that: if a plant seedling is placed in a container of soil and given nothing but water, the plant gains weight rapidly, but the soil loses practically nothing. Could it be only water that plants take from soil?

In the mid-nineteenth century, it was found that plants develop very poorly when grown in distilled water, compared with others whose roots are in a solution of various minerals. So, it is minerals, and not just water, which plants take from soil.

Meanwhile, it had been discovered that plants need light, and the green pigment chlorophyll to grow. And, more important still, when a plant has light, it takes in carbon dioxide from the air and releases oxygen.

The parts of the puzzle can now be brought together. To make food plants need light, chlorophyll, carbon dioxide, water, and minerals. Since light is necessary, this food-making process is called **photosynthesis**.

5.1 The mechanism of photosynthesis

Photosynthesis is the process by which green plants make sugar from carbon dioxide and water, using sunlight energy absorbed by chlorophyll. Oxygen gas is a bi-product. Plants use the sugar, together with minerals from the soil, to make all the substances they need.

During photosynthesis, light energy is absorbed by chlorophyll in the green parts of a plant and used to combine carbon dioxide and water. Oxygen is released into the air. This process can be written out in words:

Fig. 5.1 Summary of photosynthesis

$$\text{Carbon dioxide} + \text{Water} \xrightarrow[\text{Light}]{\text{Chlorophyll}} \text{Glucose} + \text{Oxygen}$$

or as a chemical equation:

$$6CO_2 + 6H_2O \xrightarrow[\text{Light}]{\text{Chlorophyll}} C_6H_{12}O_2 + 6O_2$$

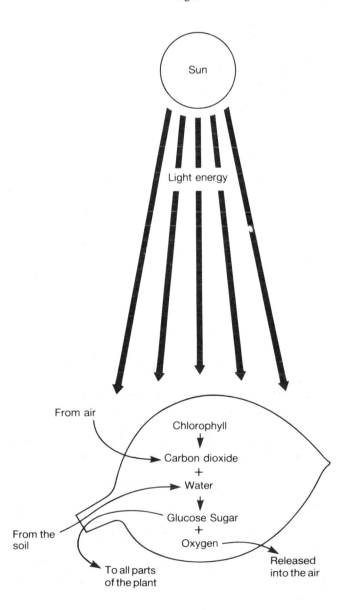

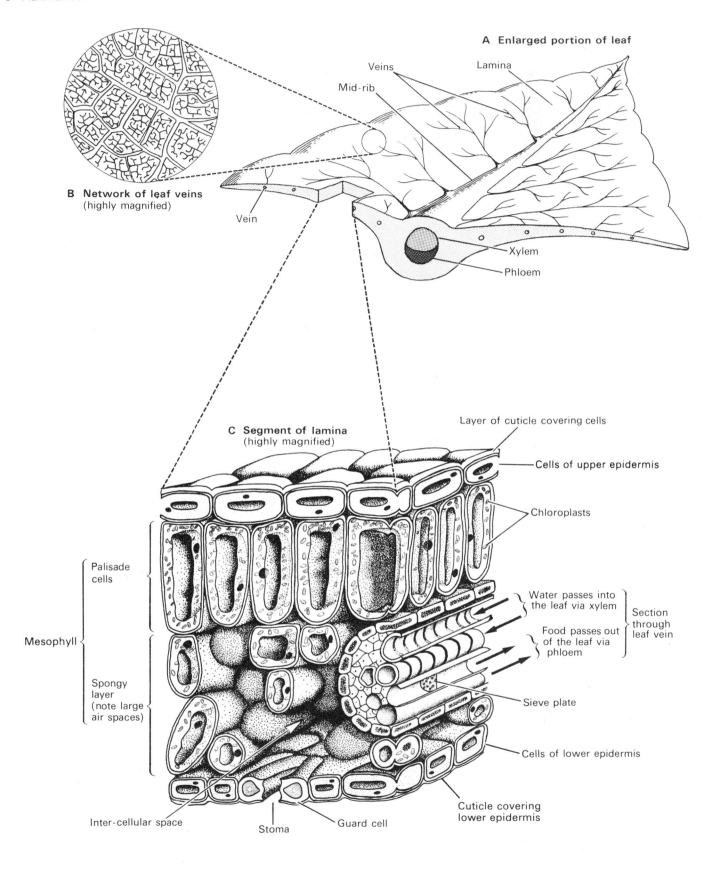

A Enlarged portion of leaf

Veins

Lamina

Mid-rib

Xylem

Phloem

B Network of leaf veins
(highly magnified)

Vein

C Segment of lamina
(highly magnified)

Layer of cuticle covering cells

Cells of upper epidermis

Chloroplasts

Palisade
cells

Mesophyll

Water passes into
the leaf via xylem

Section
through
leaf vein

Food passes out
of the leaf via
phloem

Sieve plate

Spongy
layer
(note large
air spaces)

Cells of lower epidermis

Cuticle covering
lower epidermis

Inter-cellular space

Stoma

Guard cell

Fig. 5.2 Structure of a leaf

A Cross-section of a leaf

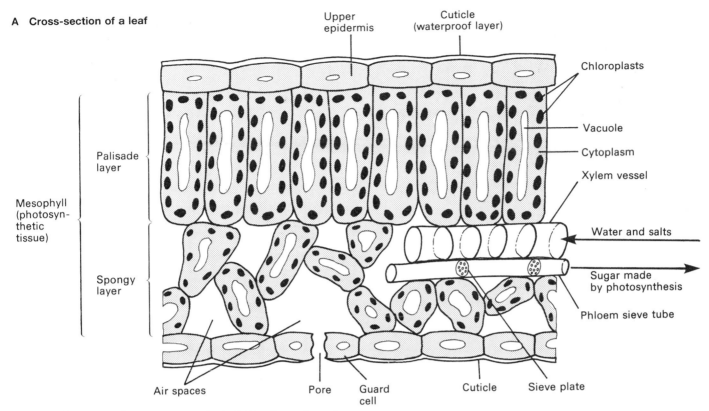

Fig. 5.3 Diagram of the structure of a leaf (simple version of Fig. 5.2C)

B Leaf pore (stoma)

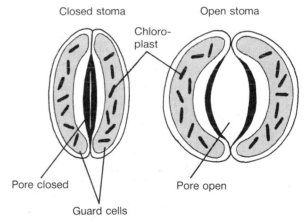

The equation for photosynthesis is misleading because it does not indicate the immense complexity of the mechanism. Photosynthesis actually consists of at least fifty separate reactions which follow in sequence, each catalysed by a different enzyme system, and connected with many side reactions.

5.2 Leaf structure in relation to photosynthesis

All the green parts of a plant carry out photosynthesis in daylight hours, but leaves are the principal photosynthetic organs. The following description refers mainly to the structure of leaves found on the group of flowering plants known as dicotyledons (section 1.4).

External features of a leaf
A leaf usually consists of the following parts:

Petiole The petiole is the narrow stalk of a leaf by which it is attached to the stem. In some leaves it is almost non-existent and the lamina extends to the stem.

Lamina The lamina is the photosynthetic portion of a leaf. It is usually thin and flat and its shape can be a fair guide to the species of plant. Its flat, thin shape allows a large area of chlorophyll to be presented to the light; in fact most leaves are so thin that some light passes right through them. This means that chlorophyll deep inside a leaf is not completely shaded from the light by the outer layers.

In certain plants the petiole can bend so that the lamina is approximately at right angles to the sun's rays throughout the day. Observe a geranium on a window-sill at intervals during the day. Movements of this kind are called **tropisms**.

Mid-rib and veins   Leaf veins consist of many tiny tubes which convey water into the leaf and carry food from it. These tubes enter through the petiole and pass in a thick bundle, called the **mid-rib**, down the leaf centre. Smaller bundles, the **veins**, spread out from the mid-rib to all parts of the leaf giving the net-veined appearance typical of dicotyledons.

Internal structure of a leaf

Figures 5.2 and 5.3 illustrate in detail the internal structure of a typical dicotyledonous leaf.

Epidermis   The epidermis is the outermost layer of cells of a plant. This layer is one cell thick, and consists of regularly shaped, often flattened cells which cover the plant like a skin. In leaves there is usually a distinct difference between the upper and lower epidermis. The upper epidermis is often covered by a continuous layer of waxy **cuticle** which protects the plant against disease organisms such as parasitic fungi. The cuticle is also waterproof and to some extent limits the loss of water from the leaf by evaporation.

Plants which live in hot, dry, and windy conditions typically have a very thick cuticle, which gives their leaves a hard and stiff texture, e.g. *Ficus elastica*, the ornamental rubber plant. The lower epidermis is characterized by the presence of pores called **stomata** (singular stoma) at regular intervals over its surface. Each stoma is made up of a pair of crescent-shaped **guard cells** which can change shape, thereby opening and closing the pore (Fig. 5.3B).

Stomata are the openings of an extensive system of air spaces between the cells of the leaf. These spaces allow gases to diffuse in and out of the leaf.

In between the upper and lower epidermis of a leaf is the photosynthetic tissue, called the **mesophyll**. Mesophyll cells are the only ones in the leaf which contain chlorophyll, apart from the guard cells of certain species. There are two types of mesophyll:

Palisade mesophyll   The palisade layer of a leaf consists of elongated, cylindrical cells which are situated immediately below the upper epidermis. The long axis of each cell is at right angles to the leaf surface. This feature permits light to penetrate deep into the photosynthetic tissue without passing through very many cell walls, so avoiding loss of light by absorption and reflection. Palisade cells contain more chlorophyll than any other type of cell in the leaf. The chlorophyll is contained within organelles called **chloroplasts** which can move within the cells to areas in which illumination is strongest.

Spongy mesophyll   Spongy mesophyll consists of irregularly shaped cells with large air spaces between them, and it has the appearance of a sponge when viewed under the microscope. These cells have fewer chloroplasts than palisade, and they receive light at a lower illumination.

5.3   The leaf in action

To understand the activities within a leaf it is necessary to remember four important points. First, plants respire at *all* times. Second, during daylight hours plants are involved in both respiratory and photosynthetic activities. This is obvious because photosynthesis must by definition be interrupted by darkness. It cannot and need not be continuous because more than sufficient food is made in daylight to last through the night. Respiration, on the other hand, must occur at all times since life depends upon it. Third, photosynthesis is the *reverse* of respiration. Fourth, photosynthesis uses the products of respiration, and respiration uses the products of photosynthesis. The last two facts are illustrated below.

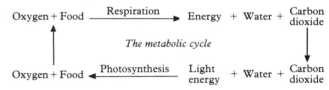

In darkness leaves take in oxygen and release carbon dioxide owing to the continuous process of respiration. As dawn breaks photosynthesis begins in the mesophyll cells and, since this is the reverse of respiration, the following sequence of events takes place. The release of carbon dioxide from the leaf slows down and eventually stops because this gas is absorbed in increasing amounts by the mesophyll as a raw material of photosynthesis. Simultaneously, the absorption of oxygen from the air slows down and stops because this gas is produced in increasing amounts by leaf photosynthesis. For a brief moment the movement of gases in and out of the leaves stops altogether. This moment is called the **compensation point**. At this time photosynthetic and respiratory processes are equal and compensate each other. As dawn turns into day a further increase in sunlight intensity results in food production requiring more carbon dioxide than is supplied by respiration alone. Likewise, more oxygen is produced than is used by respiration. Leaves therefore absorb carbon dioxide and produce oxygen for the remainder of the day, until another compensation point is reached as the sun sinks.

5.4 The rate of photosynthesis

The rate at which photosynthesis occurs depends on the amount of light and carbon dioxide it receives, and on the temperature of its surroundings. The rate of photosynthesis can be determined by measuring the amount of oxygen a plant gives off.

Carbon dioxide and light

In the dark plants respire, and photosynthesis is impossible. If light intensity is slowly increased, the rate of photosynthesis increases up to a point at which it can go no further. This is called the **light saturation point** (Fig. 5.4).

If the plant is now given more carbon dioxide, photosynthesis increases again; which shows that the rate of photosynthesis was limited by carbon dioxide supply.

Alternatively, if a plant is placed in a light of fixed intensity and its carbon dioxide supply increased, the rate of photosynthesis increases up to the **carbon dioxide saturation point** (Fig. 5.5). Photosynthetic rate is now limited by light supply.

If both carbon dioxide and light supply are increased together, the rate of photosynthesis increases to a certain point and then levels out. At this point its rate is limited by neither carbon dioxide supply nor light. It is limited by the capacity of leaves to absorb carbon dioxide. It becomes physically impossible for carbon dioxide to diffuse any faster into the leaves.

Temperature

If a plant is kept at a low temperature and the light intensity is increased, the rate of photosynthesis does increase but the light saturation point is quickly reached (Fig. 5.6B). If this procedure is repeated at a higher temperature the rate of photosynthesis increases further, reaching the light saturation point more slowly (Fig. 5.6A). This means that temperature affects the rate of photosynthesis, which is higher at higher temperatures (up to an optimum point).

Commercial uses of these facts

It is possible to increase the yields of certain greenhouse plants grown at optimum temperature and light intensity, if their carbon dioxide supply is increased. This is called **carbon dioxide enrichment**. The technique is expensive, however, and so has limited usefulness.

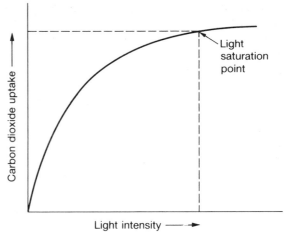

Fig. 5.4 Effects of increasing light intensity on the rate of photosynthesis

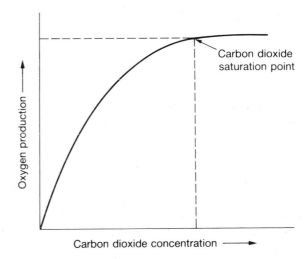

Fig. 5.5 The effects of increasing carbon dioxide concentration on the rate of photosynthesis

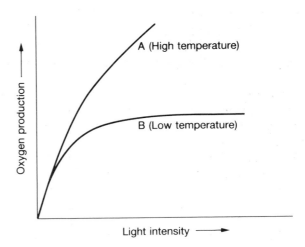

Fig. 5.6 The effects of temperature on the rate of photosynthesis

5.5 **The importance of photosynthesis**

This section examines some of the reasons why photosynthesis is of vital importance to the continuation of life on earth.

Production of oxygen

It has been explained that photosynthesis is the reverse of respiration. This explains why the supply of atmospheric oxygen is not exhausted by processes such as respiration and combustion.

At present the sum total of photosynthetic organisms on earth is sufficient to maintain atmospheric oxygen at a level which can support life and combustion. Should it ever fall below this level animals and many other organisms would suffocate.

Deforestation It is estimated that the forests of the world contribute, through photosynthesis, more than 60% of the world's oxygen supplies. They also help absorb the massive doses of carbon dioxide exhaled into the atmosphere by modern industrial societies.

If we continue to destroy forests at the present rate, decreased oxygen levels, and increased carbon dioxide levels could result in a catastrophe of world-wide proportions.

Food production

Plants manufacture all their body-building and energy-producing substances from simple raw materials, using the energy of sunlight. Animals, however, are not independent in this way; they live by eating plants and/or each other.

This fundamental difference between animals and plants (or, to be more exact, between heterotrophic and autotrophic organisms) has extremely important consequences. Consider what would happen if the sun were to go out (ignoring the fact that the earth would quickly freeze).

1. Green plants, deprived of sunlight energy, would be the first organisms to die.

2. Herbivores, deprived of plant food, would be the next organisms to die.

3. Carnivores, deprived of herbivores to eat, and omnivores, deprived of herbivores and plants to eat, would be the next to die.

4. Parasites, deprived of host organisms, would be the next to die.

5. Finally, saprotrophs, the organisms which live on the decaying remains of other organisms, would die when all these remains had been used up.

This means that the ability of plants to use the energy of sunlight to manufacture food leads indirectly to the feeding of the whole living world on the products of light, air, water, and minerals from the soil. Figure 5.7 summarizes this principle, and in addition shows that plants are responsible for the presence of fossil fuels in the earth: these are the remains of long-dead organisms. Cars, aeroplanes, home-heating systems, and industrial complexes all make use of sunlight energy which reached the earth and was absorbed by plants millions of years ago.

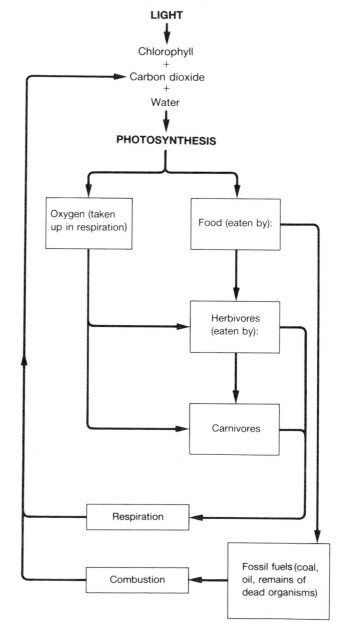

Fig. 5.7 Diagram showing the conversion of light energy by plants into substances which feed other living things and give rise to fossil fuels such as coal and oil

5.6 A plant's mineral needs

Plants cannot thrive on carbon dioxide and water alone. They need at least twelve minerals for healthy growth.

A plant's mineral needs can be discovered by growing seedlings with their roots in solutions, each of which lacks just one of the minerals necessary for healthy growth. These plants are compared with controls whose roots are in solutions containing *all* the necessary minerals. This is known as **water culture**.

This method shows exactly what happens when a plant lacks a particular mineral. The defects which develop are called **mineral deficiency symptoms**. They are of great value in agriculture since they indicate the type of treatment (e.g. addition of fer-tilizer) needed to improve mineral-deficient soils (Fig. 5.8). Minerals are used by plants mainly to manufacture proteins.

Certain of these minerals are called the **major elements**, because they are required in relatively large quantities. In the water culture method, for example, major elements must be present in several hundred parts per million for healthy plant growth. These are major elements: nitrogen, phosphorus, sulphur, potassium, calcium, and magnesium.

Other minerals are called **trace elements** because they are required in very small amounts. In water culture they need only be present in quantities as low as one part per million. Some of the trace elements are: manganese, copper, zinc, iron, boron, and molybdenum.

Fig. 5.8 Some mineral deficiency symptoms

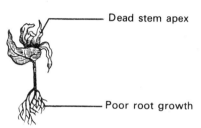

Dead stem apex

Poor root growth

No calcium
(growing points die, growth stops)

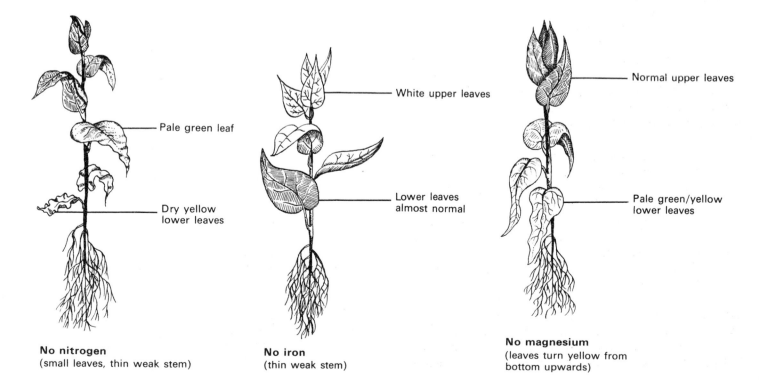

Pale green leaf

Dry yellow lower leaves

No nitrogen
(small leaves, thin weak stem)

White upper leaves

Lower leaves almost normal

No iron
(thin weak stem)

Normal upper leaves

Pale green/yellow lower leaves

No magnesium
(leaves turn yellow from bottom upwards)

Investigations

A   *Testing a leaf for starch*

The presence of starch in a leaf is an indication that photosynthesis is or has recently been taking place. Starch can be detected by the method illustrated in Figure 5.9.

1 Take a leaf from the experimental plant and boil it for 30 seconds to kill it and make its cells more permeable to iodine.

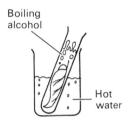

Boiling water

Leaf

2 Boil the leaf in alcohol (using a hot water bath to avoid risk of fire) to remove chlorophyll and make the starch/iodine reaction easier to see.

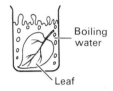

Boiling alcohol

Hot water

3 Soften the leaf by dipping it in boiling water again.

4 Spread the leaf on a white tile and drop iodine on it. A blue-black colour indicates that starch is present.

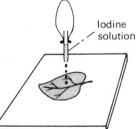

Iodine solution

Fig. 5.9   How to test a leaf for starch

B   *To verify that plants require light, carbon dioxide, and chlorophyll for photosynthesis*

Carry out the following experiments either on one plant with variegated leaves, e.g. *Tradescantia, Coleus,* or *Pelargonium,* or on separate plants. The plant must first be 'destarched' by keeping it in the dark for forty-eight hours, and then a leaf tested to verify that it is starch-free (test A above).

1. *Light*   Partly cover one or more leaves with light-proof paper (Fig 5.10A), while still attached to the plant. After the plant has been in the light for one day, detach the covered leaves and test them for starch. Only those parts exposed to light will assume a blue-black colour (Fig. 5.10A1).

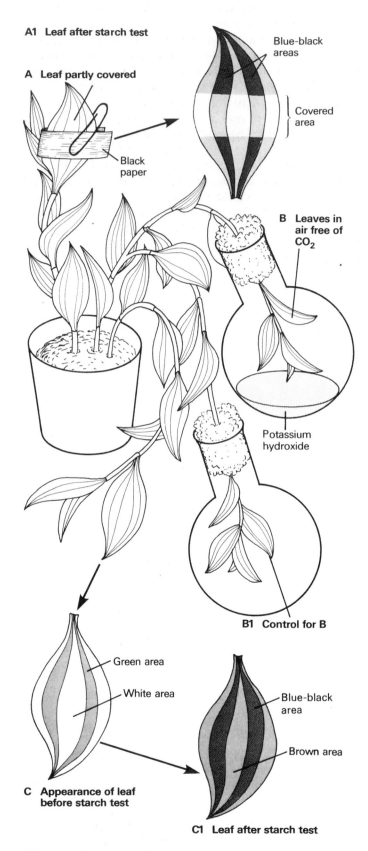

A1 Leaf after starch test

A Leaf partly covered

Blue-black areas

Covered area

Black paper

B Leaves in air free of CO_2

Potassium hydroxide

B1 Control for B

Green area

White area

C Appearance of leaf before starch test

Blue-black area

Brown area

C1 Leaf after starch test

Fig. 5.10   See exercise B

2. *Carbon dioxide* The plant must first be 'destarched' by keeping it in the dark for forty-eight hours, and then a leaf tested to verify that it is starch-free (test A above). Insert a leafy stem into a flask as shown in Figure 5.10B. The potassium hydroxide in the flask will absorb all carbon dioxide which diffuses through the cotton wool plug. Set up a control as shown in Figure 5.10B1. After the plant has been illuminated for one day test the experimental and control leaves for starch. Only control leaves will contain starch. Why is cotton wool used to plug the flasks instead of an airtight rubber bung? Why is the control experiment necessary?

3. *Chlorophyll Tradescantia* leaves usually have green areas and chlorophyll-free areas in alternate stripes. Draw the pattern of green areas of one leaf while it is still attached to the plant (Fig 5.10C). After the plant has been illuminated for one day detach the leaf and test it for starch. Only the green areas should contain starch (Fig. 5.10C1).

C *To verify the production of oxygen during photosynthesis*

Using the apparatus illustrated in Figure 5.11 demonstrate that a pond weed such as *Elodea* produces a colourless gas when illuminated. After the gas has accumulated for a day or so test it with a glowing wood splint. The splint will re-light, proving that the gas contains a higher proportion of oxygen than normal atmospheric air.

D *An investigation into the influence on the rate of photosynthesis of variations in light intensity, temperature, and carbon dioxide concentration*

These experiments involve bubble counting. That is, the number of bubbles given off by a sprig of *Elodea* over a given period is used as a rough indication of photosynthetic rate.

1. *Influence of temperature*

a) Set up at least four sets of apparatus as illustrated in Figure 5.12, ensuring that the *Elodea* in each is about the same size and of equal quality.

b) Each set of apparatus should be brought to a different temperature: 0°C (using ice and salt if necessary); 10°C; 20°C; and 30°C. Check that these temperatures are maintained.

c) The plants should be allowed to adjust to each temperature for at least five minutes in front of a bench lamp which is at a distance, tested prior to the experiment, and known to produce a steady flow of bubbles from a plant at 20°C.

d) Record the number of bubbles produced per minute over a five minute period at each temperature. Graph the results.

e) Do variations in temperature affect photosynthetic rate? If so, suggest why. What are the temperature limits within which the rate is highest?

2. *Influence of light intensity*

a) The same sprig of *Elodea* can be used throughout this exercise. Set it up in the apparatus illustrated in Figure 5.12 and keep the water bath at a constant 20°C. Cover half of the water bath with black paper so that the *Elodea* receives light only from a bench lamp.

Fig. 5.11 Verification of oxygen production in photosynthesis

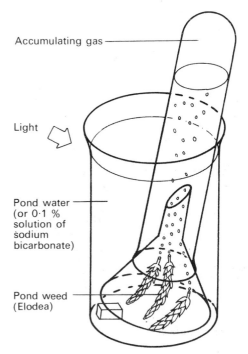

Accumulating gas

Light

Pond water (or 0·1 % solution of sodium bicarbonate)

Pond weed (Elodea)

Fig. 5.12
Apparatus
for exercise D

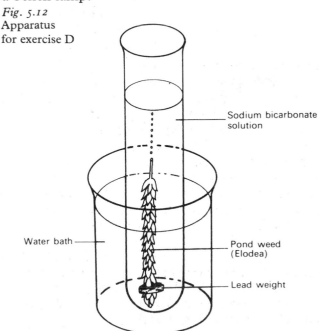

Sodium bicarbonate solution

Water bath

Pond weed (Elodea)

Lead weight

b) Begin the experiment with the apparatus 80 cm from the lamp. Allow it to adjust to these conditions for five minutes before a five minute bubble count is made. Move the apparatus to 40 cm from the lamp, wait one minute and then count bubbles for five minutes. Repeat this procedure with the apparatus at 20 cm and then 10 cm from the lamp.

c) Using this method, light intensity is increased four times each time the apparatus is moved closer to the light. Thus, if 80 cm is given a relative light intensity of one, the relative intensity of the other distances will be as follows:

Distance from light	80 cm	40 cm	20 cm	10 cm
Relative light intensity	1	4	16	64

d) Plot a graph of your results with relative light intensity on one axis and rate of bubbling on the other.

e) Do variations in light intensity influence photosynthetic rate? If so, is there a point at which further increase in light intensity produces no further increase in photosynthesis? Under these conditions what factor could be preventing further increase?

3. *Influence of carbon dioxide concentration*

a) Set up four sets of apparatus as illustrated in Figure 5.12 each with a fixed temperature of 20°C and at a distance from a bench lamp which produces a steady flow of bubbles.

b) Each plant should be in a tube with a different concentration of sodium bicarbonate solution: e.g. 0.1%, 0.3%, 0.5%, plus a control in distilled water.

c) After five minutes of adjustment, make a five minute bubble count and graph the results.

d) Do variations in carbon dioxide concentration affect the rate of photosynthesis? Do these results and results from experiments 1 and 2, suggest methods by which horticultural yields may be increased?

E *Designing controlled experiments*

These exercises provide an opportunity to solve problems by designing controlled experiments.

1. *Materials*

Bromothymol blue indicator; large test-tubes; beakers; sprigs of *Elodea*; carbonated water (e.g. from a soda syphon). All of these materials should be used in conjunction with the apparatus shown in Figure 5.12, after reading the notes below.

2. *The problem*

The problem is to devise experiments which give conclusive evidence that plants: (a) use up carbon dioxide, but (b) only when exposed to light, and (c) that they produce carbon dioxide in the dark.

3. *Points to note*

a) Bromothymol blue is an indicator; that is, its colour depends upon (i.e. indicates) the degree of acidity/alkalinity to which it is exposed. Discover and note its various colours by dropping some into weak acid, distilled water, and weak sodium hydroxide solution.

b) Carbonated water is a solution of carbon dioxide (i.e. a weak solution of carbonic acid). When placed in carbonated water *Elodea* can carry out photosynthesis very rapidly. Verify this by dropping some indicator into carbonated water, and by comparing the production of bubbles from *Elodea* in carbonated and distilled water.

c) Using the materials listed, devise experiments to solve 2 *(a)*, *(b)*, and *(c)* above, noting that bromothymol blue is not poisonous to plants, and that somewhere in the experiment it is necessary to show that the indicator is not affected by exposure to light.

F *Investigating plant mineral requirements*

1. Prepare a number of *Tradescantia* cuttings of about the same length with their lower leaves removed.

2. Using apparatus illustrated in Figure 5.13, place the cuttings with their stems immersed in solutions which lack just one of the minerals essential for healthy growth.

3. Growth of these cuttings can be compared with others grown in solutions containing all the minerals that plants require.

Tradescantia cuttings can be used because they root quickly and, since they have no stored food, depend entirely upon photosynthesis and minerals in the solutions right from the start.

Complete stock solution:

Calcium nitrate	2.0 g
Potassium nitrate	0.5 g
Magnesium sulphate	0.5 g
Potassium phosphate	0.5 g
Ferric chloride (*aq*)	a few drops
Distilled water	1 litre

(Dilute stock solution 1 : 5 with distilled water.) Make deficient solutions by omitting ferric chloride, and magnesium sulphate. Substitute potassium nitrate for calcium nitrate, and chloride salts for nitrate salts.

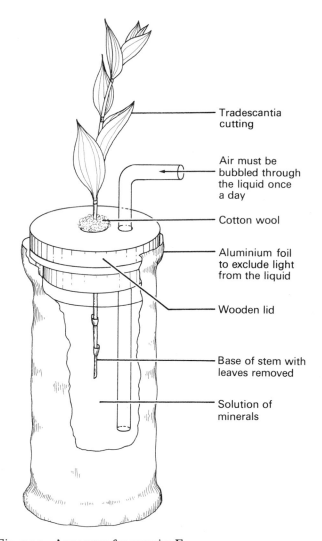

Tradescantia cutting

Air must be bubbled through the liquid once a day

Cotton wool

Aluminium foil to exclude light from the liquid

Wooden lid

Base of stem with leaves removed

Solution of minerals

Fig. 5.13   Apparatus for exercise F

Summary and factual recall test

During photosynthesis plants take in (1) gas which enters through pores called (2) mainly in the (3) epidermis of leaves, and they take in (4) which enters through the (5). The green pigment (6) absorbs (7) energy which drives this process, the equation for which is (use either words or chemical symbols):

$$(8) + (9) + (10) \xrightarrow{\;(11)\;} (12) + (13)$$

The rate of photosynthesis in plants depends on the amount of (14) and (15) they receive and the (16) of their surroundings.

Questions

1. What substances does a plant use to make food by photosynthesis, and what food does this process make?

2. What part do light and chlorophyll play in photosynthesis?

3. *a)* What are stomata, and where are most of them found on a plant?
   b) What gas enters stomata during photosynthesis and what gas passes out of stomata during photosynthesis?

4. Without photosynthesis the air would quickly become unbreathable. Why is this so?

5. Five test-tubes were filled with pond water, prepared as shown below, then placed in bright light.

Tube A	Tube B	Tube C	Tube D
Pond weed 25°C	Pond weed Water snail 25°C	Pond weed 10°C	Pond weed Tube enclosed in aluminium foil 25°C

   a) Which tube would produce the most oxygen?
   b) Explain why each of the other tubes would produce *less* oxygen.
   c) What does this investigation tell you about the effects of temperature and carbon dioxide concentration on photosynthetic rate?

6. The graphs below show the changing concentrations of nitrates and green algae found in the upper regions of temperate seas throughout the year.
   a) Suggest two reasons why the numbers of algae increase at A.
   b) Suggest one reason why algae decrease at B.
   c) Why do algae numbers decrease in winter.
   d) Suggest why nitrate concentration increases in winter.

Chlorophyll is contained within organelles called (17) which are found mostly in the (18) cells of the (19) layer of leaves. (20) vessels transport (21) to these cells from the roots, and (22) tubes transport (23) from the leaves to the plant's (24) and (25)-storage areas.

All life on earth ultimately depends on photosynthesis for two main reasons, which are (26) and (27). Deforestation could endanger our planet because it could decrease the amount of (28) in the atmosphere and increase the amount of (29).

6

Transport in mammals

6.1 The need for a transport system

Living things constantly absorb useful substances like oxygen and food, which must then be distributed throughout their bodies. In addition, they produce a continuous stream of waste materials, such as carbon dioxide, which must be removed from their bodies before they accumulate to harmful levels.

The distribution of food and oxygen throughout the body, and the removal of body wastes, is performed by a transport system.

Transport in unicellular organisms

In microscopic organisms such as *Amoeba,* the volume of the body is so small that useful substances can be distributed, and waste materials removed, by a process called **diffusion.**

Diffusion is the movement of substances from where they are plentiful to where they are scarce.

Oxygen, for example, is usually plentiful in the water surrounding an *Amoeba,* but less plentiful inside the cell where it is continually used up during respiration. Because of the cell's smallness, oxygen can diffuse from the water into the *Amoeba* as fast as it is needed for respiration.

Carbon dioxide is more plentiful inside an *Amoeba* than outside because it is continually produced by respiration. The carbon dioxide diffuses out from the *Amoeba* into the water with sufficient speed to prevent it accumulating to harmful levels within the cell (Fig. 6.1).

Transport in multicellular organisms

In large multicellular organisms, however, body volume is so great that diffusion alone is far too slow a process for the adequate distribution of oxygen and food, and the removal of waste. The cells in a multicellular animal relying on diffusion alone would be like people in a tightly packed crowd: those in the middle would not get enough oxygen. But most large organisms do not rely on diffusion, they have a **transport system** of some kind to distribute useful sub-

stances around their bodies, and to carry away wastes.

In multicellular organisms diffusion still plays an important part, but in combination with a transport system.

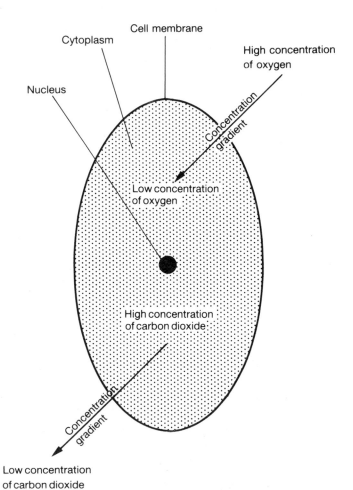

Fig. 6.1 Diffusion in a unicellular organism. Substances diffuse down concentration gradients from where they are concentrated to where they are less concentrated

6.2 **Composition of blood**

The average person has about 5.5 litres of blood. Although blood is a liquid, about 45% of it is made up of solid particles held in suspension. The remaining 55% is a straw-coloured fluid called **plasma**. The solid matter in blood consists of **red cells**, sometimes called red blood corpuscles or **erythrocytes**; **white cells** or **leucocytes**, which are actually colourless; and tiny particles called **platelets**.

Red cells

Human red cells are tiny bi-concave discs, i.e. concave on both sides (Fig 6.2A). They measure 0.008 mm in diameter and are 0.002 mm thick, which means that 125 of them side by side would make a row 1 mm long. They are called 'cells' but they have no nucleus. This is one reason why they live for only about four months, after which they are broken down in the spleen and the liver. Some of their component chemicals, notably iron, are re-used to make new red cells. The new red cells are made in the bone marrow, particularly at the ends of the long arm and leg bones, in the ribs, and in the vertebrae. More than two million red cells are destroyed and replaced every second in the human body.

When seen individually under a microscope red cells are not red at all; they are golden yellow. It is only when they are seen massed together in a drop of blood that the red colour becomes apparent. This colour is due to the presence in each cell of a substance called **haemoglobin**.

Haemoglobin is the substance which enables blood to transport large quantities of oxygen from the lungs to the body tissues. As blood flows through vessels in the lungs it meets oxygen which has entered the body from the atmosphere. Haemoglobin in the red cells reacts with this oxygen to form an unstable substance called **oxyhaemoglobin**. When blood leaves the lungs and flows through vessels among the body cells in which there is little oxygen, the oxyhaemoglobin breaks down and releases the oxygen, which diffuses into the cells and is used in respiration.

These reactions, which are summarized below, form the body's oxygen transport system. This is described in more detail in section 6.5.

There are about 5 million red cells in one cubic millimeter of human blood. This means that there are some 30 million million red cells in an average person's blood.

Fig. 6.2 Blood cells

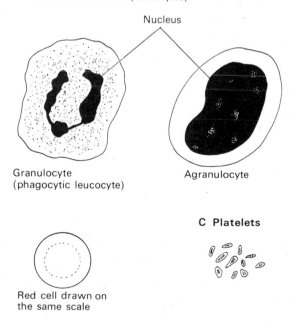

A Red blood cells

Red cells as they appear in a blood clot

Cross-section of red cell showing bi-concave shape

B White blood cells (leucocytes)

Nucleus

Granulocyte (phagocytic leucocyte)

Agranulocyte

C Platelets

Red cell drawn on the same scale

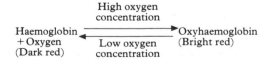

High oxygen concentration

Haemoglobin + Oxygen (Dark red) → Oxyhaemoglobin (Bright red)

Low oxygen concentration

White cells or leucocytes

Leucocoytes are colourless and have a nucleus (Fig. 6.2B). There are only about 8000 of them per cubic millimeter of human blood, which means there are about 75 million in the whole body. They are much larger than red cells, measuring up to 0.02 mm in diameter.

About 75% of leucocytes have large lobed nuclei, and are irregularly shaped, rather like an *Amoeba*. They have tiny granules in their cytoplasm and because of this they are called **granulocytes**. This type of leucocyte is made in the bone marrow but by different cells from those which make red blood cells.

Granulocytes are **phagocytic**, a word which means 'cell eater'. They 'eat' or engulf other cells, particularly bacteria. Granulocytes form the body's chief defence mechanism against disease-causing bacteria. They gather in wounds and destroy bacteria before they can enter the body. They are also capable of leaving the blood vessels altogether by squeezing between the cells which make up the vessel walls. In this way granulocytes can reach infected areas anywhere in the body.

The remaining 25% of leucocytes or white cells are rounded in shape, have very large round nuclei, and a thin layer of cytoplasm without granules. These cells are **agranulocytes** (Fig 6.2B). This type of leucocyte does not originate in the bone marrow, but in the lymphatic system, and for this reason they are sometimes called **lymphocytes**. Agranulocytes can move like *Amoebas* but they are usually not phago-cytic. The function of agranulocytes is to produce chemicals called **antibodies** which help prevent disease. This is described in chapter 21.

Platelets

Platelets are irregularly shaped objects about 0.003 mm in diameter (Fig. 6.2C). There are about 250 000 platelets per cubic millimetre of blood. They originate in the bone marrow, probably as fragments which become detached from cells. They have no nucleus.

Platelets play an important part in the system which causes blood to clot in wounds (section 6.6). They also release a chemical called seratonin, which causes blood vessels to constrict, so reducing bleeding.

Plasma

Plasma is the liquid portion of blood. It consists of 92% water and contains many important dissolved substances, including the products of digestion, such as glucose, fatty acids, glycerol, amino acids, vitamins, and minerals; plasma proteins such as albumin, fibrinogen, and antibodies; substances called hormones; and waste materials, such as urea and carbon dioxide.

Owing to the presence of the protein **fibrinogen** plasma can clot: that is, it can turn into a semi-solid jelly. This characteristic is extremely important because it is responsible for sealing off damaged blood vessels in wounds, thereby preventing excessive loss of blood.

Plasma can be forced through blood vessel walls under high pressure, carrying with it food and oxygen from the blood stream. Once it is outside the blood vessel walls plasma forms a liquid called **tissue fluid** which bathes every cell in the body. A large quantity of plasma in the form of tissue fluid is constantly circulating among body cells supplying food and oxygen, and removing waste products.

It is important to realize that plasma and the tissue fluid derived from it form the environment which keeps body cells alive. In a sense, these fluids are equivalent to the pond and sea water in which unicellular organisms live, which supplies their food and oxygen, and into which they excrete waste. The difference is that the chemical composition and temperature of tissue fluid is controlled within very precise limits by the liver, kidneys, and other organs. As a result, body cells live and grow in a watery environment which in a healthy person is always perfectly suited to their needs.

Red and white cells of human blood. p is a typical phagocyte; l is a typical lymphocyte; and m is a lymphocyte belonging to a type called monocytes.

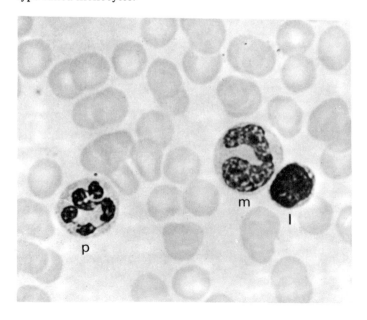

6.3 Circulation of blood in mammals

In the early seventeenth century it was believed that blood flowed from the heart to the body organs and back again via the same blood vessels on the return journey. This is sometimes called the 'tidal theory' because it suggests that blood ebbs and flows like the tide.

In 1628 an English doctor called William Harvey produced experimental evidence suggesting that blood flows away from the heart in one set of vessels which are now called **arteries** and towards the heart in a separate set of vessels now called **veins**. One of Harvey's most famous experiments is illustrated in Figure 6.3. This shows that blood will flow in only one direction in a vein: always towards the heart. Furthermore, it shows that it will only flow one way because valves are present at intervals along the veins. The existence and structure of these valves was already known in Harvey's time (Fig. 6.4), but he was the first to demonstrate their function.

From this and other evidence, Harvey proposed the theory that the heart and blood vessels form a **circulatory system**, rather than a tidal system. He believed that the heart pumps blood along a circular route: outwards from the heart along the arteries, through the body organs, and back to the heart along the veins. Unfortunately, Harvey could not prove that blood circulates in this way because he was unable to find any connection between arteries and veins at their furthest point from the heart.

The connections between arteries and veins were discovered some seventy years later by the Italian biologist Marcello Malpighi. While he was examining the lungs of a frog under a microscope, Malpighi discovered a network of extremely narrow blood vessels which completed the circuit between the artery leading into the lung and the vein leading out of it. These tiny blood vessels are now called **capillaries**. Harvey did not use a microscope for his work and could never have found capillaries with the unaided eye because they are only 0.005 mm to 0.02 mm in diameter.

Fig. 6.3 Simplified version of Harvey's experiment to show that blood flows towards the heart in a vein

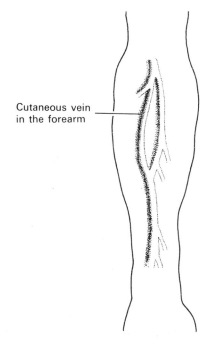
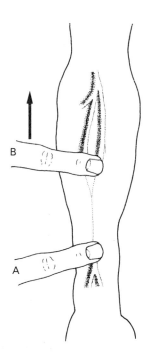
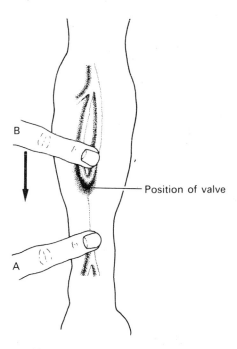

1 Hold the arm downwards for a few seconds. The veins will become prominent.

2 Place one finger (A) firmly on the wrist. Sweep blood upwards with another finger (B). Lift finger B from the arm. Blood will not flow back into the empty vein, because it never flows *away* from the heart.

3 Move finger B down the vein. A bulge will appear showing where a valve prevents blood flowing away from the heart. Remove finger A and watch the empty vein fill again.

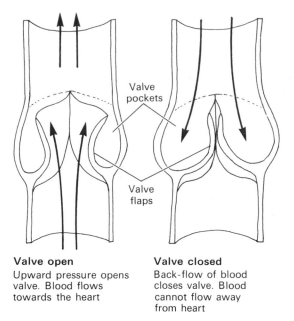

Valve open
Upward pressure opens valve. Blood flows towards the heart

Valve closed
Back-flow of blood closes valve. Blood cannot flow away from heart

Fig. 6.4 Diagram of how vein valves prevent blood flowing away from the heart

To summarize: **the circulatory system of mammals consists of the heart, which pumps blood into vessels called arteries, which divide inside body tissues and organs into extremely fine vessels called capillaries, which in turn empty into veins that carry blood back to the heart** (Fig 6.5).

Double circulation in mammals

Figures 6.5 and 6.6 show the double circulatory system in mammals. Their blood circulates through two systems: the **pulmonary circulation** which conveys blood to and from the lungs, and the **systemic circulation** which conveys blood to and from all other parts of the body. These two systems are connected at the heart, but before the connections can be traced it is necessary to understand the structure of the heart (Fig. 6.7).

In mammals the heart consists of four chambers. There are two chambers on each side of the heart and these are completely separated by a central wall. The uppermost chambers on each side of the heart are called **atria** (singular atrium), and they have relatively thin muscular walls. (Some biologists prefer to call these chambers auricles.) Below each atrium is a thick-walled chamber called a **ventricle**. A system

Fig 6.5 Diagram of double circulation in a mammal

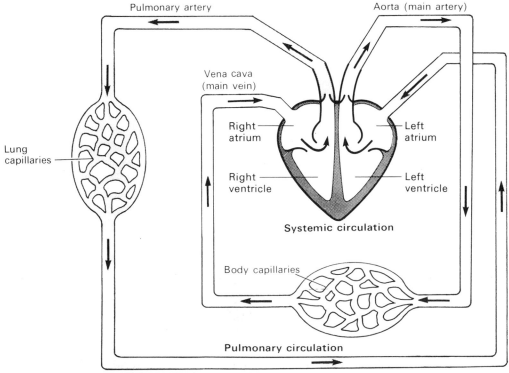

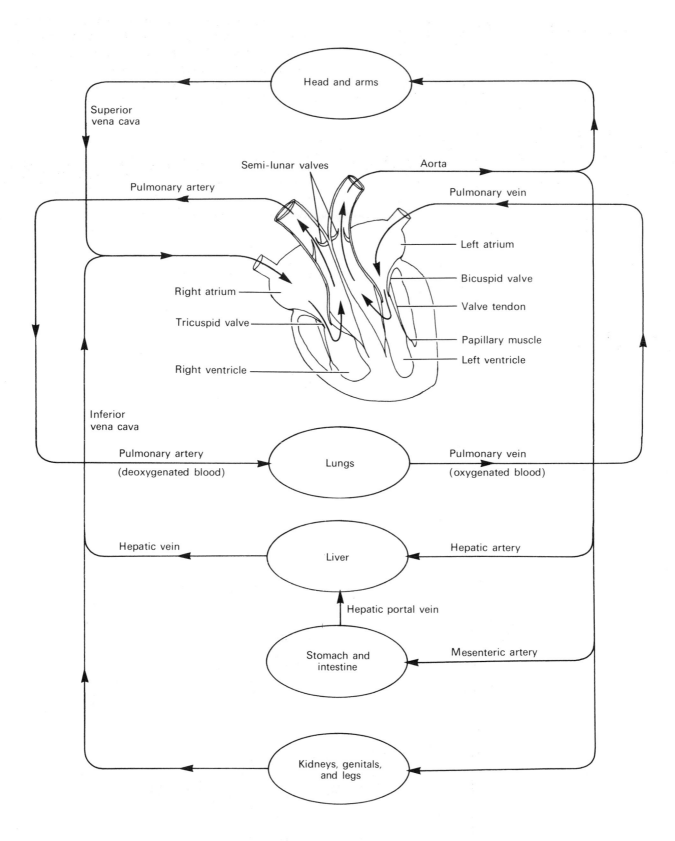

Fig. 6.6 Diagram of the main arteries and veins in a mammal, and the structure of a mammalian heart

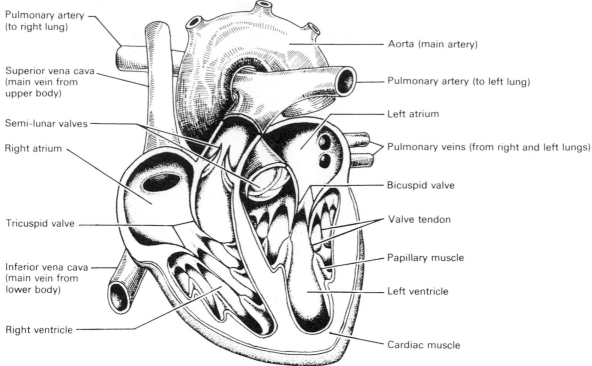

Pulmonary artery (to right lung)

Superior vena cava (main vein from upper body)

Semi-lunar valves

Right atrium

Tricuspid valve

Inferior vena cava (main vein from lower body)

Right ventricle

Aorta (main artery)

Pulmonary artery (to left lung)

Left atrium

Pulmonary veins (from right and left lungs)

Bicuspid valve

Valve tendon

Papillary muscle

Left ventricle

Cardiac muscle

Fig. 6.7A   Structure of the mammal heart

Fig. 6.7B   Diagram of the heart showing direction of blood flow

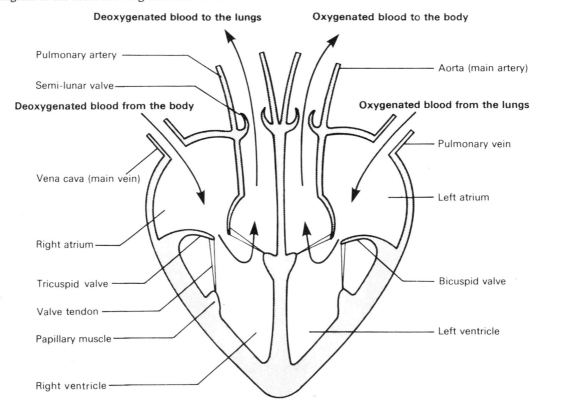

Deoxygenated blood to the lungs

Oxygenated blood to the body

Pulmonary artery

Semi-lunar valve

Deoxygenated blood from the body

Oxygenated blood from the lungs

Aorta (main artery)

Pulmonary vein

Vena cava (main vein)

Left atrium

Right atrium

Tricuspid valve

Valve tendon

Papillary muscle

Bicuspid valve

Left ventricle

Right ventricle

of valves on each side of the heart permits blood to flow from the atria into the ventricles, but not in the reverse direction. Traditionally, the two sides of the heart are always described from the animal's point of view. (The right atrium is the atrium on the animal's right-hand side and *not* the viewer's right-hand side.)

Using Figures 6.6 and 6.8 it is possible to trace the double circulation of blood starting from where it enters the right atrium. This blood has come from the systemic circulation: that is, from all the organs except the lungs. On its journey it has lost its supplies of oxygen and so it is called **deoxygenated blood**. The deoxygenated blood flows into the right atrium until it is full. Then the muscles in the atrium walls contract (in time with the left atrium) forcing blood down into the right ventricle, which is relaxed at this stage. A fraction of a second later when it is full the right ventricle contracts (in time with the left ventricle) forcing blood into the pulmonary artery. Blood is prevented from flowing back into the right atrium by the **tricuspid valve** (so called because it consists of three valve flaps). Blood is prevented from flowing back into the right ventricle by pocket-like **semi-lunar valves** located at the point where the pulmonary artery leaves the heart. Blood flows

through the pulmonary artery to capillaries in the lungs where it absorbs more oxygen, and after this it is called **oxygenated blood**. This oxygenated blood returns from the lungs through the pulmonary vein, which empties into the left atrium. When full the left atrium contracts forcing blood into the left ventricle, and then the left ventricle contracts forcing blood into the main artery of the body, which is called the **aorta**. Blood is prevented from flowing back into the **bicuspid valve**, which has two valve flaps, and it is prevented from flowing back from the aorta into the ventricle by another set of semi-lunar valves. The left ventricle has thicker muscular walls than the right ventricle. This gives the left ventricle the extra muscular power necessary to pump oxygenated blood all around the systemic circulation, which is far more extensive than the pulmonary circulation.

When deoxygenated blood returns via the main veins (the **superior** and **inferior vena cavae**) to the right atrium, the double circuit has been completed.

The heart

The heart's pumping action is driven by **cardiac muscle** in the walls. Cardiac muscle differs from other types of muscle in at least three important

Fig. 6.8 Diagram of the heart's pumping action. (Only the left side of the heart is shown)

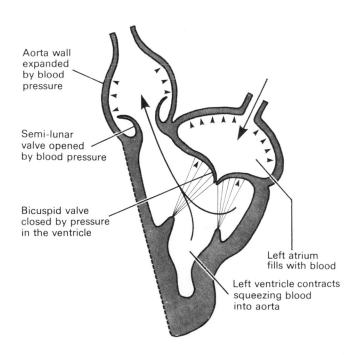

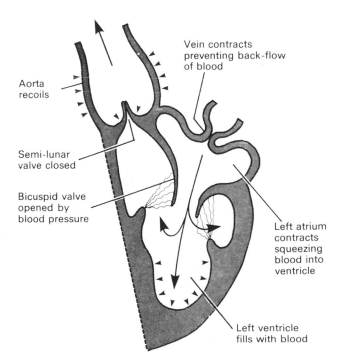

A Ventricle contracts (systole)

B Ventricle relaxes (diastole)

ways. First, it is made up of branching muscle fibres connected to each other in the form of a network, and not unbranched parallel fibres as in the muscles which move the arms and legs. This characteristic allows contractions to begin at one point in the heart and spread outwards in all directions. Second, cardiac muscle contracts and relaxes rhythmically in what are called 'beats'. This natural rhythm is generated within the muscle itself and not by impulses from the nervous system, William Harvey saw this in 1623 when he cut strips from a living heart and watched the pieces beat for some time all on their own. Third, cardiac muscle does not fatigue despite continuous rapid contractions over many years. The heart beats about 60 to 70 times a minute in a resting adult human, increasing to 150 or more during strenuous exercise. This adds up to an average of over 100 000 beats a day, in the course of which the heart pumps about 14 000 litres of blood. Compare this performance with the type of muscle attached to the skeleton by opening and closing one hand 70 times in a minute, and noting how tired it becomes.

Blood is driven around the body mainly by the pumping actions of the heart. But the flow of blood is greatly assisted by arteries and veins as well.

Arteries
Artery walls consist of an inner membrane one cell thick surrounded by a heavy layer of interwoven muscle fibres and elastic fibrous material. On the outside is a further layer of fibrous material (Fig. 6.9A).

When ventricles contract blood is forced out of the heart under high pressure, and this pressure stretches the aorta and artery walls outwards (Fig. 6.8A). Later, when the ventricles relax and the semi-lunar valves close, the elastic fibres in the artery walls recoil (like someone letting go of a stretched elastic band). Therefore a powerful force presses inwards on the blood, pushing it away from the heart. It must move in this direction because the semi-lunar valves have closed behind it, preventing it flowing back into the heart (Fig. 6.8B). This sequence of events begins at the heart and moves outwards along the arteries. Every time a ventricle contracts a volume of blood moves through the arteries preceded by a wave of artery wall expansion, and followed by a wave of elastic recoil and muscular contraction which forces the blood onwards. These waves are the 'pulse' which can be felt with the fingers in arteries at the wrist and neck.

Arteries divide to form smaller vessels called **arterioles** which have a thinner muscular/elastic layer. Arterioles divide many times to form a dense network of capillaries which have walls only one cell thick (Fig. 6.9B).

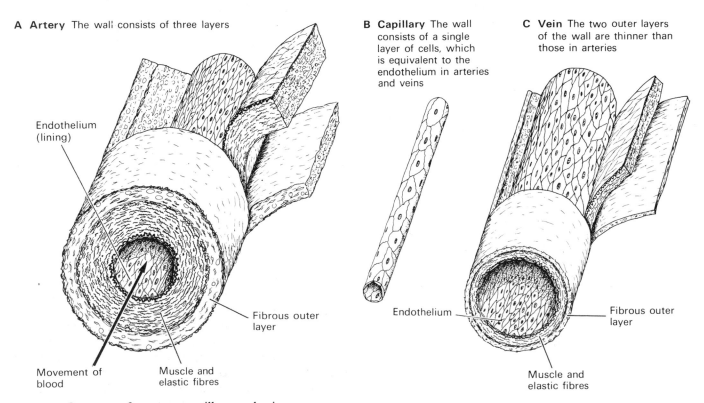

A Artery The wall consists of three layers

Endothelium (lining)

Movement of blood

Muscle and elastic fibres

Fibrous outer layer

B Capillary The wall consists of a single layer of cells, which is equivalent to the endothelium in arteries and veins

C Vein The two outer layers of the wall are thinner than those in arteries

Endothelium

Fibrous outer layer

Muscle and elastic fibres

Fig. 6.9 Structure of an artery, capillary, and vein

Capillaries

The capillary network of the body is so extensive that if the rest of the body were dissolved away an outline of it and all its organs would still be visible owing to the mass of capillaries left behind. In fact, the capillaries are so densely packed together that none of the body's 25 million million cells is more than a fraction of a millimetre from one of these tiny vessels.

Capillary walls are extremely thin, and are permeable to (i.e. let through) water and all dissolved substances except the larger protein molecules. Water and dissolved substances are forced through the capillary walls by the pressure of blood and the resulting tissue fluid passes between the cells of the body. All the blood's functions depend on the permeability of capillary walls. The other parts of the circulatory system simply deliver high pressure oxygenated blood loaded with food to the capillaries, and remove low pressure deoxygenated blood from them.

Capillaries unite to form wider vessels called **venules**, and these eventually unite to form veins.

Veins

Compared with arteries, veins have a thinner muscular elastic layer in their walls, and a wider passage or **lumen** for the movement of blood (Fig. 6.9C).

Blood in the veins is at low pressure, and its flow towards the heart is assisted in two ways. First, veins contain valves which permit blood to flow only towards the heart (Fig. 6.4). Second, many veins are situated between the large muscles of legs, thighs, arms, etc. Veins are squeezed flat when these muscles contract, which forces blood towards the heart. Muscles are never completely still for long even when a person is asleep, so blood flows along the veins at all times.

6.4 Tissue fluid and the lymphatic system

It has been explained that each tissue and body organ contains a dense network of capillaries. These are usually called **capillary beds** (Fig. 6.10). It has also been explained that a liquid called tissue fluid is forced under pressure through the capillary walls. This process tends to occur at the artery end of a capillary bed, since blood pressure is greatest at this point.

Tissue fluid

When tissue fluid is being forced out of the

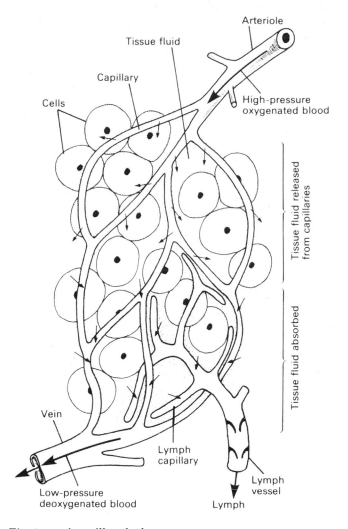

Fig. 6.10 A capillary bed

capillaries the capillary walls act as a filter holding back red blood cells, most of the white cells, and large protein molecules. The following substances which pass through the capillary walls make up the tissue fluid: water with dissolved oxygen, glucose, fatty acids, glycerol, amino acids, vitamins, minerals, and hormones.

Tissue fluid flows away from the capillaries and passes among the body cells, which extract oxygen and food from it and at the same time release carbon dioxide and other waste materials into it. Meanwhile, blood in the capillaries drains to the other end of the capillary bed and loses most of its pressure. At this stage the blood is a highly concentrated solution of protein molecules because it has lost water and soluble materials at the artery end of the bed. At the same time tissue fluid has become a weak solution, having given up most of its soluble contents to the body cells as it passed among them. Under certain circumstances concentrated solutions absorb water from weaker solutions by a process called **osmosis** (explained in chapter 2), and this is what happens at the vein end of a capillary bed. The blood absorbs some of the tissue fluid by osmosis through the capillary walls. These processes are illustrated in Figure 6.10, and summarized diagrammatically in Figure 6.11. Any tissue fluid not absorbed in this way passes into the lymphatic system.

Lymphatic system

Lymph vessels, usually called **lymphatics**, begin as tiny blind-ended tubes within the capillary beds and are about as numerous as the blood capillaries (Fig. 6.10). Tissue fluid which is not absorbed

into the blood stream drains into these lymphatics and is then called **lymph**. The small lymphatics drain into larger ones, which are similar to veins in some ways. They have valves which ensure that lymph flows in only one direction, and they are situated mainly among muscles which squeeze the lymph along as they contract.

At intervals along the lymphatics there are structures called **lymph nodes**. These contain a system of narrow channels through which the lymph drains. Large phagocytes are attached to the walls of these channels and their function is to engulf bacteria and dead cells from the lymph. There are large lymph nodes in the groin, under the arms, and in the neck (Fig. 6.12).

The large lymphatics unite into two main lymphatic ducts which empty their contents into the blood stream at the subclavian veins near the heart (Fig 6.12). All tissue fluid eventually drains back into the blood.

Fig. 6.11 Diagram showing the formation and absorption of tissue fluid
A is tissue fluid containing dissolved food and oxygen which has been forced out of the artery end of a capillary by high blood pressure. **B** is tissue fluid moving, partly by osmosis, into the vein end of a capillary. **C** is tissue fluid moving into the lymphatic system

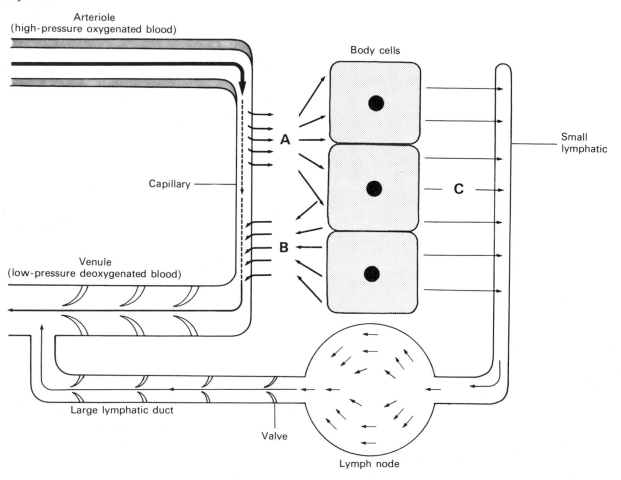

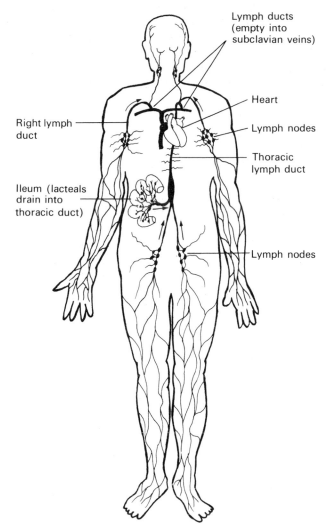

Lymph ducts
(empty into
subclavian veins)

Right lymph
duct

Heart

Lymph nodes

Thoracic
lymph duct

Ileum (lacteals
drain into
thoracic duct)

Lymph nodes

Fig. 6.12 Diagram of the human lymphatic system (greatly simplified)

6.5 The main functions of blood

Transport of oxygen from the lungs

Oxygen is transported from the lungs to the body cells by the red cells of the blood in the form of oxyhaemoglobin. The ability of haemoglobin to carry oxygen in this form depends on the presence in every haemoglobin molecule of four atoms of iron. Each of these four atoms attracts and combines loosely with one molecule of oxygen. There is enough iron in the haemoglobin of one red cell to carry about 1 000 000 000 oxygen molecules.

When red cells reach the lungs they are deoxygenated. They contain no oxygen because they have left it all behind on their travels around the body tissues. But when red cells leave the lungs they are oxygenated, because they have *picked up* oxygen on their travels. On average it takes only 45 seconds for each red cell to collect a full load of oxygen from the lungs deliver it to the body cells, and return for more.

Transport of carbon dioxide to the lungs

Carbon dioxide is produced constantly by cell respiration. It diffuses out of the cells, into tissue fluid, and then into blood passing through capillary beds in the tissues and organs. Once it is in the blood, carbon dioxide is transported to the lungs in one of three ways.

1. About 85% of the carbon dioxide is transported in the form of **sodium bicarbonate** dissolved in the plasma. This substance is produced by a sequence of chemical reactions, some of which take place inside red cells. During its formation, however, the reactions within the red cells release a by-product which triggers off the breakdown of oxyhaemoglobin, so that oxygen is released from the blood. Consequently, whenever blood enters a region of the body with a high carbon dioxide content, oxygen is automatically released from red cells.

2. About 5 to 10% of carbon dioxide transported by the blood is in the form of a substance called **carbamino-haemoglobin**. This is formed from carbon dioxide which diffuses into red cells where it reacts with a certain amino acid contained in haemoglobin.

3. About 5% of carbon dioxide transported by the blood is in the form of carbonic acid dissolved in the plasma.

When blood loaded with carbon dioxide reaches the lungs, it takes in oxygen and oxyhaemoglobin forms in its red cells. This reaction triggers off the simultaneous breakdown of carbamino-haemoglobin in the red cells and sodium bicarbonate in the plasma, so that carbon dioxide is released from both these substances. Consequently, the absorption of oxygen by red cells automatically releases most of the blood's carbon dioxide. The carbonic acid in the plasma changes into carbon dioxide gas.

Transport of urea to the kidneys

A number of chemical reactions in the liver concerned with the metabolism of protein release poisonous substances as waste products. These poisons are quickly converted into **urea**, which is relatively harmless. Urea travels in the blood stream in solution until it reaches the kidneys, where it is removed and excreted from the body in urine.

Transport of digested food

The soluble products of digestion pass into the blood stream through villi which line the ileum (section 4.7). From here, the soluble food is transported to the liver in the hepatic portal vein (Fig. 6.6). The liver releases food into the blood as the body requires it.

Distribution of heat, and temperature control

Whatever the weather conditions, mammals and birds can maintain a high and relatively constant body temperature. This is possible partly because the blood transfers heat from places where it is produced, such as the muscles, and distributes it fairly evenly throughout the body. In addition, there is a mechanism which helps to control the rate at which heat is lost through the body surface by controlling the amount of blood which flows through capillaries close to the surface of the skin.

Protection against diseases

Section 6.2 and chapter 21 describe how phagocytes and antibodies help destroy germs which enter the body, and establish immunity to certain diseases.

6.6 Blood groups and blood clotting

Blood from one person can only be transfused into certain other people. This is because there are several different types of blood and when some types are mixed the red cells stick together in large clumps. This is called **agglutination.** Later the red cells split open spilling their contents into the plasma, and if this happened inside a patient's body during a blood transfusion it could seriously damage the kidneys.

Blood groups

The four types of blood are called **blood groups,** and are known by the letters A, B, AB, and O. Before a blood transfusion can take place it is necessary to make sure that the donor's and recipient's blood will mix together without agglutination. Blood groups which mix without agglutination are said to be **compatible.**

Blood compatibility depends upon chemicals called **antigens** on the surface of the red cells, and chemicals called **antibodies** in the plasma. There are two types of antigen: A and B; and two types of antibody: anti-A and anti-B.

Blood group A   has A antigen on its red cells and anti-B antibody in its plasma.

Blood group B   has B antigen on its red cells and anti-A antibody in its plasma.

Blood group AB   has A and B antigens on its red cells and no antibodies in its plasma.

Blood group O   has no antigens on its red cells and both anti-A and anti-B antibodies in its plasma.

Blood transfusions

Anti-A plasma agglutinates A red cells, and anti-B plasma agglutinates B red cells. So these combinations of plasma and red cell are incompatible as far as blood transfusion is concerned.

These facts have given us a rule for blood transfusions: **the donor's red cells must be compatible with the recipient's plasma.** Thus, a recipient with anti-A plasma can only receive blood with either B red cells, or red cells without antigens; and a recipient with anti-B plasma can only receive blood with either A red cells, or red cells without antigens.

Blood can be safely transfused as follows:

Table 1   Transfusion of blood groups

Blood group	Can be transfused into	Can receive blood from
A	A and AB	A and O
B	B and AB	B and O
AB	AB only	All groups
O	All groups	O only

People with Group O blood are called **universal donors.** Their red cells have no antigens and so cannot be agglutinated by blood of any other group. People with Group AB are called **universal recipients.** Their plasma has no antibodies therefore it does not agglutinate blood from the other groups.

Blood clotting

Cuts bleed for a while and then the blood changes into a thick jelly called a **blood clot.** The clot closes the injured blood vessels and prevents further loss of blood. Later the clot dries out and forms a scab which prevents germs and dirt entering the wound. White blood cells below the blood clot destroy any germs which enter the cut.

A complicated series of chemical reactions occur in blood before it clots. The main substance involved in these reactions is a chemical called **fibrinogen.** Fibrinogen is a soluble blood protein which is made in the liver.

When a blood vessel is damaged and blood is exposed to the air, platelets release a chemical which changes fibrinogen into a mass of tangled threads made of an *insoluble* protein called **fibrin**. Fibrin threads radiate from the platelets like the spokes of a wheel. The threads form a thick mesh in which red blood cells are trapped. This forms a blood clot which blocks the wound.

Haemophilia Haemophilia is an inherited blood disorder. Male children can inherit it from their mothers. The blood of a haemophiliac clots very slowly or not at all. There are two reasons why this can happen. Either the haemophiliac has too few platelets, or his platelets cannot produce the chemical which causes fibrinogen to change into fibrin.

6.7 Diseases of the circulatory system

Diseases of the circulatory system kill more people than any other illness. But these diseases are decreasing in number because of modern discoveries about their causes, treatment, and prevention.

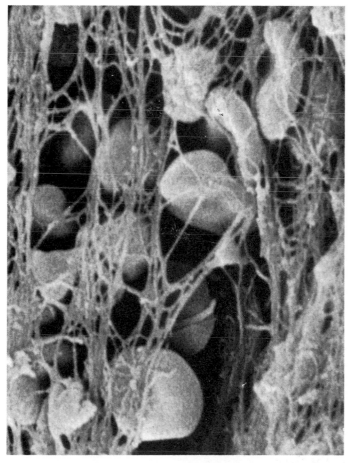

A blood clot. Note red blood cells trapped in a tangled web of fibrinogen fibres

Types of circulatory disease
Thrombosis Thrombosis is the formation of a small solid lump of blood inside a blood vessel. The solid lump of blood is called a blood clot, or thrombus. A blood clot, or any other solid particle floating in the blood stream is known as an **embolism.**

Hardening of the arteries As people get older the elastic and muscle layers of their arteries are gradually replaced by inelastic fibrous tissue. The artery walls become stiff and hard, a condition known as hardening of the arteries, or arteriosclerosis. This condition reduces the flow of blood.

Atheroma Blood flow along an artery is sometimes slowed or stopped altogether by a layer of fatty substance, called cholesterol, stuck to the artery walls (Fig. 6.13). This type of blockage is called an atheroma. Arteriosclerosis and atheroma are very dangerous when they occur in the heart or brain.

Heart attacks (heart failures) A heart attack, or heart failure, is the sudden slowing or stoppage of the heart beat. A heart attack occurs when a coronary artery is blocked by a thrombosis or atheroma. Coronary arteries supply heart muscle with food and oxygen. Consequently, when one is blocked a section of heart muscle stops working and eventually dies. If the whole heart is affected death is instantaneous.

Angina pectoris If one or both of the coronary arteries is partly blocked due to atheroma, heart muscle is unable to work properly during exercise. This causes pains in the chest known as angina pectoris. Angina curis is pain in the calf muscles of the legs caused by atheroma of arteries which deliver blood to these muscles.

Strokes A stroke, or cerebral thrombosis, is a blood clot in the brain. The blood clot suddenly blocks an artery inside the brain causing the region served by this vessel to stop working, and die. The results of a stroke depend on the area of the brain affected. Muscles may be paralysed, and speech or memory affected. Death occurs if the brain damage is extensive.

Prevention of circulatory diseases
Studies in America and Britain involving many thousands of people have shown that these diseases occur far less often among people:
1. Who never smoke cigarettes
2. Who take regular exercise (walking, cycling, swimming, active sports etc.)

A Healthy artery (cut lengthwise)

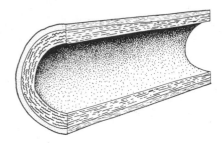

B Clogged artery

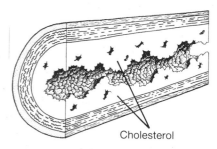

Cholesterol

Fig. 6.13 The hidden enemy. Cholesterol clogs arteries, slows blood flow, and may stop it altogether. Clogged arteries are more likely to occur in people who eat food rich in animal fats (e.g. dairy produce), who take little exercise, who smoke, and whose lives are full of stress (tension, anxiety, fear, etc.)

3. Who never drink alcohol, or drink only moderate amounts
4. Who have seven or eight hours sleep a night
5. Who eat balanced meals
6. Who never eat between meals.

Investigations

Topics in Safety (ASE, 1982) provides clear guidelines for blood sampling in the classroom.

A   *An investigation of blood cells*
1. *Red cells*   Obtain a blood sample by gently stabbing with a sterile lancet the area of skin shown in Fig. 6.14A.
   a) Smear the blood on a slide (Fig. 6.14B) and allow it to dry.
   b) Place a drop of glycerine and then a cover slip on the smear. Examine the red cells under a microscope.
2. Repeat with blood from a bird and a frog. Note any differences between red cells in these animals and those in mammals.

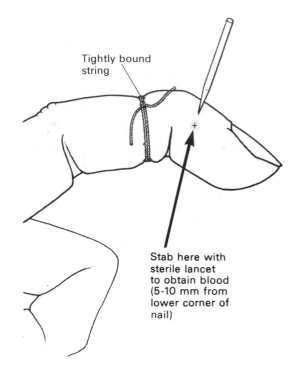

Tightly bound string

Stab here with sterile lancet to obtain blood (5-10 mm from lower corner of nail)

Fig. 6.14A   How to obtain a blood sample (exercise A)

3. *Stained red and white cells*   Obtain a blood smear as described in 1 above.
   a) When the smear is *nearly* dry pipette on to it about three drops of Leishman's (or Wright's) stain, and leave for 30 seconds.
   b) Pipette an equal number of drops of distilled water on to the stain, and rock the slide to mix the liquids. Leave for 10–15 minutes.
   c) Wash off the stain with water. Place a drop of glycerine and a cover slip on the slide.
   d) Observe the slide under low and high magnification. Red cells are pink, granulocytes have blue nuclei and pink granules, agranulocytes have blue nuclei and pale blue cytoplasm.

Fig. 6.14B   How to make a blood smear

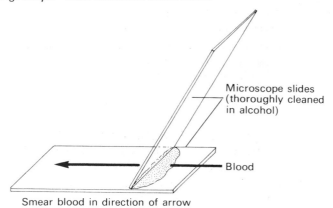

Microscope slides (thoroughly cleaned in alcohol)

Blood

Smear blood in direction of arrow

B *An investigation of blood clotting*

Place a drop of blood on a slide using the method given in A above, but do not smear it. Slowly pull the tip of a needle through the blood every 10–15 seconds and note the time taken before the needle draws a fine thread of fibrin from the blood.

C *An investigation of capillary circulation*

Capillary circulation is clearly visible in the tail fin of a fish. A goldfish, for example, can be wrapped in wet cotton wool and placed in a petri dish with half a microscope slide resting on its tail fin (Fig. 6.15). (Do not keep a fish under these conditions for more than 10 minutes; *remember* an investigator has a responsibility to act humanely towards animals.) The tail fin must be observed under high power magnification.

1. Observe the movement of red cells through the vessels and distinguish between arteries (arterioles), capillaries, and veins (venules), by comparing the relative sizes of the vessels.

Is the rate of blood-flow the same in each type of vessel? If not, which has the fastest flow and which the slowest?

2. Cool the specimen with ice or refrigerated water.

a) How does this affect the rate of blood-flow and capillary diameter?

b) What does this result reveal about the possible effects of cold weather upon capillary circulation?

3. Apply one or two drops of the following chemicals, in the order indicated, directly on to the tail fin either of different fish, or the same fish after thoroughly washing away all traces of the previous substance: adrenalin, alcohol, nicotine, lactic acid. Allow two minutes for each to diffuse in to the fin.

a) How does each chemical affect the diameter of blood vessels and the rate of blood flow?

b) Why does a drink of alcohol appear to make a person feel warmer in cold weather, and yet contributes to a rapid heat loss from his body?

c) Strenuous exercise results in the accumulation of lactic acid in muscles. How will this affect the circulation of blood through the muscles?

d) Why should an athlete not smoke before an event?

D *Observing capillaries in human skin*

Soak the area of skin between a finger-nail and first joint in cedarwood oil. This should clear the skin sufficiently to make capillaries visible, in strong light, under low-power magnification.

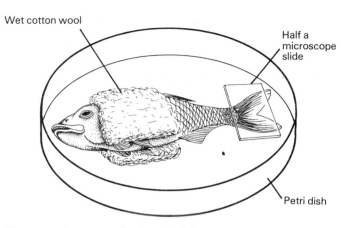

Wet cotton wool

Half a microscope slide

Petri dish

Fig. 6.15 Apparatus for Practical C

E *An investigation of pulse rate*

Pulse points occur in the wrist and in the throat (on either side of the windpipe). Find these and count the number of pulsations over a 30-second period: when sitting at rest in a chair, and after vigorous exercise such as press-ups. Record the time taken for the pulse to return to its original rate. Account for the changes in rate after exercise.

F *An investigation into the effect of temperature on heart-beat in cold-blooded animals*

Observe the heart-beat of a *Daphnia* (water-flea) by mounting one in water on a cavity slide under a microscope. Count the heart-beats by tapping a pencil point on paper in time with the pulsations for 5 or 10 seconds, then count the pencil dots. Do this several times and find the average. Compare heart-beat rates with the animal in water at various temperatures.

G *An investigation of gaseous uptake in blood*

Obtain fresh blood from a slaughter-house and immediately add about 5 cm³ per litre of 0.1% sodium oxalate to prevent it clotting. Alternatively, use an aqueous solution of crystalline haemoglobin.

1. Place equal amounts of blood or haemoglobin solution in three flasks.

a) Bubble oxygen through one flask and carbon dioxide through another. Note any colour changes compared with the blood in the third flask.

b) Reverse the procedure, i.e. bubble carbon dioxide through the flask which previously received oxygen and oxygen through that previously treated with carbon dioxide. Note any colour changes.

c) Do these experiments help to verify that haemoglobin both absorbs and releases oxygen? If so, explain why.

2. Use a vacuum pump to evacuate air from the flask containing oxyhaemoglobin. Note and explain any colour change.

Questions

1. The diagram below represents the structure of the heart.

a) Copy the diagram and use arrows to indicate the direction of blood flow through the heart.

b) Name the parts labelled A to I.

c) What are the functions of D and G?

d) Why do ventricles have thicker muscular walls than atria?

e) Describe what happens in one heart beat.

f) How does cardiac muscle differ from muscles which move the body.

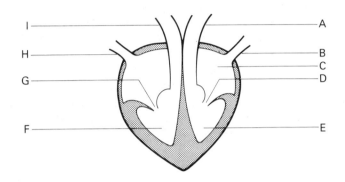

2. Blood contains red cells, granulocytes, lymphocytes, and platelets. Which of these four does each of the following sentences describe:

a) contain haemoglobin,

b) most are phagocytic,

c) transport oxygen,

d) made in bone marrow,

e) play a part in blood clotting,

f) move like an *Amoeba*,

g) produce antibodies.

3. Trace the path of a glucose molecule (naming only the larger blood vessels) from intestinal capillaries to the brain where it is respired, and then trace the path of respiratory carbon dioxide from the brain to the lungs.

4. Blood which flows through capillaries in the feet is at low pressure. Explain how it is returned to the heart against the force of gravity.

5. Unlike oxygen, carbon monoxide gas (present in car exhaust fumes and certain household gas supplies) forms a chemically stable compound with iron atoms in haemoglobin.

a) In view of this, explain why it is very dangerous to breathe this gas for more than a few minutes.

b) If a person's blood is damaged by carbon monoxide gas, what two processes may eventually return his red cell supply to normal?

6. Prolonged starvation reduces the amount of protein in the blood. One consequence of this is an increased amount of tissue fluid which tends to gather in the abdomen and lower limbs. What has this to do with B in Figure 6.11?

7. When papillary muscles (Fig. 6.7) contract they pull on the valve tendons and draw the edges of heart valve flaps together. At what point does this event occur in the sequence of atrial and ventricular contractions, and what function does it serve?

Summary and factual recall test

The liquid part of blood is called (1). It consists of 92 per cent (2), and contains dissolved substances such as (3–name twelve). When this liquid is forced through the (4) walls under pressure it forms (5) which bathes all body cells, supplying them with (6) and (7), and removing their (8) products.

Red cells are (9) in shape, are made in the (10), and have no (11). They are red because of a chemical called (12) which contains (13) atoms that combine loosely with (14) forming an unstable substance called (15).

Granulocytes are leucocytes which are (16) in shape, have a (17) nucleus, and contain (18) in their cytoplasm. They are said to be (19) which means they 'eat' cells. Their function is to (20). Agranulocytes have (21) nuclei, and no (22) in their cytoplasm. Most of them produce (23) which help prevent (24). Platelets help blood to (25) in wounds. Blood is pumped by the (26) into vessels called (27) which have (28) muscular walls. In the tissues these vessels branch to form very narrow vessels called (29), the walls of which are (30) to water and dissolved substances. These vessels join together and lead into (31) which have (32) at intervals along them which ensure that blood flows only towards the (33).

The blood of mammals circulates through two systems, the (34) and the (35). Deoxygenated blood fills the (36) at the (37)-hand side of the heart, and oxygenated blood fills a chamber with the same name at the (38)-hand side. Both these chambers contract, forcing blood to open the (39) and (40) valves and pass into the (41). These chambers contract forcing blood from the (42)-hand side to go to the lungs through the (43) artery, and blood from the (44)-hand side to go to the body cells through the (45).

7

Support and transport in plants

The roots of plants absorb water and minerals from the soil. These substances are then transported up the stem to the leaves where they take part in many different processes. Water is combined with carbon dioxide to form sugar during photosynthesis, and minerals are used in the manufacture of proteins and other complex chemicals. Later, these end-products are transported to other parts of the plant where they may be stored, or used in the growth of new tissues.

Substances move from one part of the plant to another through a system of narrow tubes known collectively as **vascular tissue**. Plants which have vascular tissue are known as the **vascular plants**, and these include the Pteridophytes (ferns), the Gymnosperms (mostly conifers), and the Angiosperms (flowering plants).

This chapter describes the structure of vascular tissue in flowering plants; the movement of water and minerals through this tissue from roots to leaves; and the movement of photosynthetic products from the leaves to storage areas and growing points.

7.1 Vascular tissue in flowering plants

There are two types of vascular tissue: **xylem** and **phloem**. In the roots of most flowering plants the xylem tissue is arranged in an X-shaped mass and the phloem is between the 'arms' of the X (Fig. 7.1). In most stems, however, the xylem and phloem are arranged together in compact masses called **vascular bundles** (Fig. 7.2). Usually the xylem is situated in the part of each bundle near the stem centre, with the phloem towards the outer surface of the stem. The arrangement of xylem and phloem in leaves is illustrated in Figures 5.2 and 5.3.

Xylem

Xylem is a tissue composed of tubes called **xylem vessels**, and long pointed **fibre cells**. These are best observed in vertical sections through a plant stem, as illustrated in Figure 7.2.

Xylem vessels develop from cylindrical cells arranged end to end (Fig. 2.6C). The cytoplasm dies and the cross-walls break down leaving a dead, empty tube. In trees these tubes may be over 30 metres long. Xylem vessels form a transport system through which water and dissolved minerals move from the roots, through the stem, and into the leaves.

Both the vessels and the fibre cells of xylem are very strong owing to the presence in their walls of a substance called **lignin**. In xylem vessels lignin appears either as rings, coils, or layers perforated at intervals by tiny holes called **pits**.

Phloem

Phloem tubes are also formed from cylindrical cells arranged end to end. Unlike xylem vessels, however, the cross-walls of the phloem tubes do not disappear; they develop perforations like a sieve. These cross-walls are called **sieve plates**, and the tubes are called **sieve tubes**. Sieve tubes form a transport system through which photosynthetic products, such as sugar, move from the leaves to the storage areas and growing points of a plant.

Phloem tissue also contains **companion cells**, so called because they grow side by side with sieve tubes and are apparently essential to their transport functions. Unlike xylem vessels and fibre cells, sieve tubes and their companion cells contain living cytoplasm.

7.2 Support in plants

The strength of xylem tissue provides a great deal of support for the softer tissues of roots, stems, and leaves against the force of gravity and the pressure exerted by strong winds. In other words, xylem is the hard woody 'skeleton' of plants. (The structure of this skeleton is best observed in trees, whose trunks consist of layers of xylem laid down year by year in concentric rings. The xylem is so strong that in gale force winds trees are more likely to be uprooted than suffer a broken trunk.)

In addition to the mechanical support provided by xylem and fibre cells (wood), plants are supported

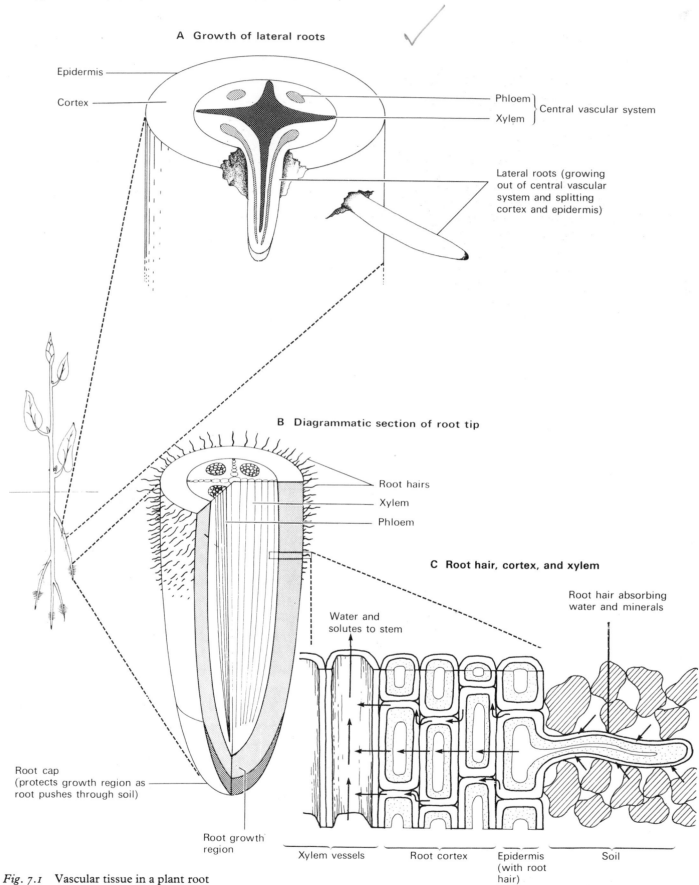

A Growth of lateral roots

Epidermis

Cortex

Phloem ⎱
Xylem ⎰ Central vascular system

Lateral roots (growing out of central vascular system and splitting cortex and epidermis)

B Diagrammatic section of root tip

Root hairs

Xylem

Phloem

C Root hair, cortex, and xylem

Root hair absorbing water and minerals

Water and solutes to stem

Root cap (protects growth region as root pushes through soil)

Root growth region

Xylem vessels

Root cortex

Epidermis (with root hair)

Soil

Fig. 7.1 Vascular tissue in a plant root

Cross-section of a vascular bundle

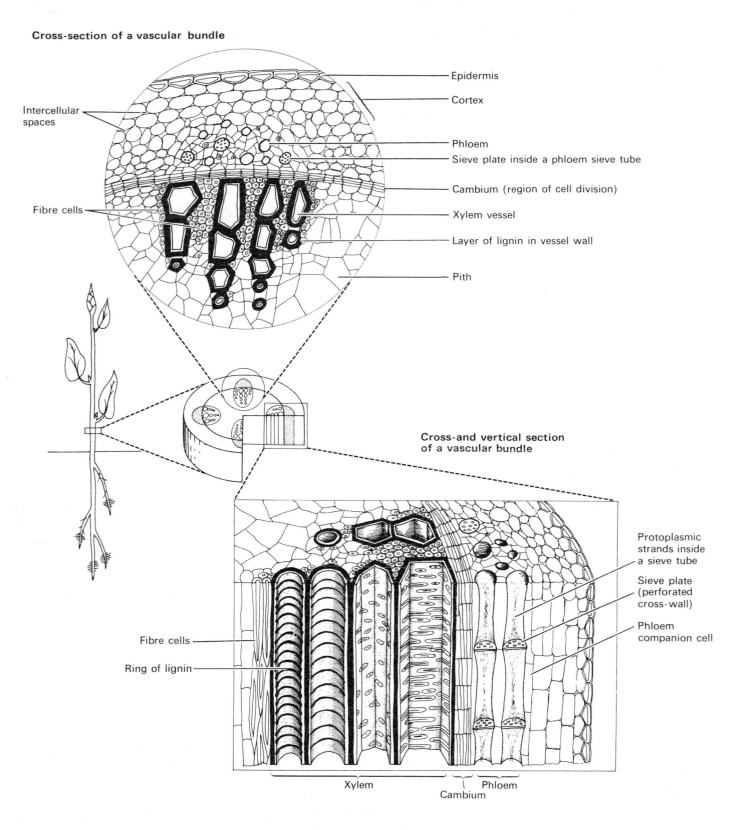

Epidermis

Cortex

Intercellular spaces

Phloem

Sieve plate inside a phloem sieve tube

Cambium (region of cell division)

Xylem vessel

Fibre cells

Layer of lignin in vessel wall

Pith

Cross-and vertical section of a vascular bundle

Protoplasmic strands inside a sieve tube

Sieve plate (perforated cross-wall)

Phloem companion cell

Fibre cells

Ring of lignin

Xylem

Cambium

Phloem

Fig. 7.2 Vascular tissue in a plant stem

by the water in their tissues. Seedlings, non-woody plants, and soft parts like leaves and flowers would be limp were it not for the fact that their cells are inflated with water which makes them firm, like a tyre inflated with air. Water enters cells by a process called osmosis (section 2.7).

7.3 Flow of water through plants

Water flows in a continuous stream through a plant. It enters through the roots and flows up the root and stem xylem vessels to the leaves, where it evaporates from the surface of spongy mesophyll cells and diffuses out of pores in the leaves called stomata (Figs. 5.2 and 7.3).

Evaporation of water from leaves generates the main force which moves water through a plant. The technical name for evaporation of water from leaves is **transpiration**, and the flow of water through a plant which this causes is called the **transpiration stream**.

Transpiration and transpiration stream

Water lost from a leaf by transpiration is replaced by more water which flows from leaf xylem vessels (veins). Water flows from leaf xylem vessels to the surface of mesophyll cells by two routes.

Most of the water flows directly through and around the porous cellulose walls of the mesophyll cells (Fig. 7.3). But a small amount of water is thought to flow by osmosis from cell to cell. Loss of water by evaporation from cells near stomata gives these mesophyll cells a lower osmotic potential than cells next to xylem vessels, where water is plentiful. In other words, a **gradient of osmotic potential** is created which is highest near the xylem and lowest in cells near stomata. Movement of water down this gradient is very slow because of the great resistance caused by the many cell membranes and layers of cytoplasm through which it must pass.

Flow of water up the xylem

Transpiration causes water to be continually removed from leaf xylem vessels. The result is a lower water pressure at these points than in root xylem vessels. This pressure difference causes water (and dissolved minerals) to be sucked up from the roots to the leaves, like lemonade being sucked up a drinking straw.

Flow of water from soil to root xylem

Water sucked up xylem vessels from the roots to the stem and leaves is replaced by water which flows from the soil through and around the porous cellulose walls of cells in the root cortex (Fig. 7.3). This flow of water continues as long as a plant transpires.

In spring, however, before leaf buds open, there is little or no transpiration. At this time water flows across the root to the xylem by osmosis. When there is no transpiration, mineral salts accumulate in root xylem. At certain concentrations these salts give xylem sap a lower osmotic potential than adjacent cells in the root cortex. As these cells lose water by osmosis to the xylem sap they develop a lower osmotic potential than root hairs and other root epidermal cells, since these are bathed in soil water. Thus, a gradient of osmotic potential is created which is highest in cells nearest the soil and lowest in cells next to the root xylem. Water flows from cell to cell by osmosis from the high to the low osmotic potential and then into the root xylem.

Root pressure

Water entering root xylem by osmosis causes a build-up of pressure in the xylem vessels. This is called **root pressure**, and is responsible for forcing water, and minerals, up into the root stem in early spring before the transpiration stream is established.

Root pressure can be observed if the stem of a healthy, well watered plant is cut off just above ground level. The flow of liquid is slow, but under considerable pressure, particularly if the plant had not been transpiring before the stem was cut.

For example, cut tomato stems exert a root pressure which would be sufficient to force liquid to a height of about 90 metres, or high enough to reach the top of most trees.

Root hairs

These are extremely narrow tubes which grow out from individual cells on the root surface (Fig. 7.1). They develop in a narrow zone of the root close to the root tips. There are millions of root hairs on even quite small plants, and they give the plant a far greater surface area over which it can absorb materials from the soil than if the root surface were smooth.

In addition, root hairs stick to the soil particles between which they grow, holding the soil firmly in place around the root and thereby helping to anchor the plant in the ground.

Individual root hairs usually function for a few days and are then replaced by new ones which grow nearer the root tip as it grows through the soil. In this way the root hair zone is always in contact with new regions of soil.

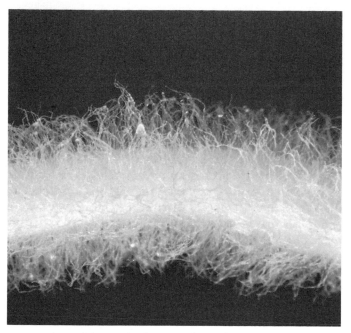

Root hairs (broad bean)

7.4 Absorption of minerals by roots

It was once thought that mineral salts were absorbed in solution along with water which plants take in through their roots. But there is now considerable evidence that water and mineral salt absorption are independent processes.

Plants absorb minerals from the soil against concentration gradients; that is, until their cell sap contains a far greater concentration of them than is present in surrounding soil water. This suggests that neither osmosis nor simple diffusion can be responsible for mineral absorption since, by definition, these two processes could do no more than produce the same concentration of minerals in plants as there is in soil water.

The actual mechanism of mineral absorption is very complex. All that need be said here is that evidence suggests a mechanism which requires the expenditure of energy on the part of plants, and for this reason mineral absorption is described as an **active transport** mechanism.

For instance, it has been shown that the rate of mineral absorption varies according to the rate at which energy is released by respiration in plants. This is demonstrated by growing plants in solutions of minerals. If oxygen is bubbled through the solutions mineral absorption increases, but it slows down when the oxygen is turned off, and almost stops when chemicals which inhibit respiration are added to the solution.

7.5 Transpiration rate

Transpiration rate depends upon the same factors that govern evaporation rate. Although these factors are listed seperately below it must be remembered that they affect transpiration simultaneously.

Humidity

In general, transpiration only occurs when there is a lower humidity level (concentration of water vapour) in the atmosphere than exists in the air spaces inside the leaves. Transpiration stops when the atmosphere is saturated with water vapour, and resumes when the air becomes drier. Anything which produces a change either in the humidity of the atmosphere or of the air spaces in leaves will alter transpiration rate.

Temperature

A rise in air temperature affects transpiration rate in two ways. First, it increases the capacity of air to absorb water from leaves. Second, it warms the water inside leaves making it evaporate more quickly. Direct sunlight has the same effect since it warms leaves to a higher temperature than the atmosphere. Transpiration is therefore generally faster on warm sunny days than on cold dull ones.

Wind

Air movements carry away water vapour from leaves and this prevents air around them becoming saturated with water vapour. Consequently, depending upon temperature and humidity, transpiration is faster on a windy day than in still air.

The best conditions for a high rate of transpiration are the same as those needed for drying washing on a line: a warm, dry, sunny, windy day.

The importance of transpiration

Transpiration is important for at least three reasons. First, it results in the transport of water and minerals from the soil to the leaves where they form the raw materials of photosynthesis and other types of food manufacture. Second, it ensures that the walls of spongy mesophyll cells are kept moist, which is essential for the efficient absorption of the carbon dioxide needed for photosynthesis. Third, evaporation of water from a leaf has a cooling effect which helps prevent hot direct sunlight from damaging delicate cells.

On the other hand, transpiration brings far more water to the leaves than is needed for photosynthesis and, during drought, loss of water by transpiration may result in wilting which can kill a plant.

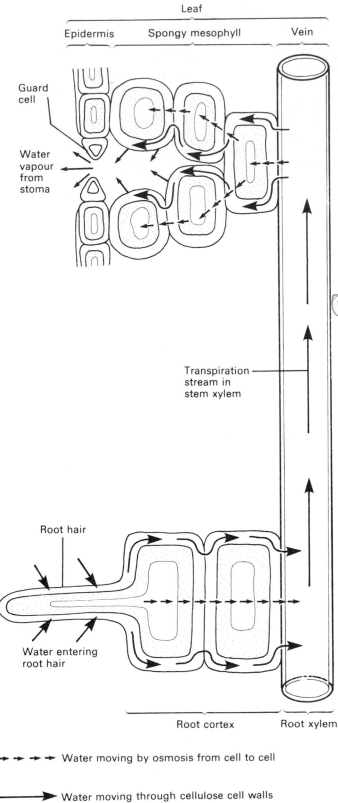

Leaf

Epidermis   Spongy mesophyll   Vein

Guard cell

Water vapour from stoma

Transpiration stream in stem xylem

Root hair

Water entering root hair

Root cortex   Root xylem

- ► - ► - ► Water moving by osmosis from cell to cell

———► Water moving through cellulose cell walls

Fig. 7.3 Diagram summarizing water absorption, transpiration, and the transpiration stream

Apart from plants with special adaptations which prevent excess water loss in dry environments (Fig. 7.8 and question I), it seems that most plants suffer from the unfortunate coincidence that most of those features which result in maximum photosynthetic efficiency (thin, flat leaf shape, large area of moist cell walls, many stomata, etc.) are also responsible for continuous rapid loss of water by evaporation from the leaf in warm, dry, windy conditions. However, there is evidence to suggest that two mechanisms protect the plant as it begins to wilt. First, the stomata close, which limits water loss to evaporation through the cuticle of the leaves and stem. Second, various chemical reactions increase the ability of roots to absorb water from the soil. In prolonged dry conditions, however, neither of these mechanisms can prevent the ultimate death of a plant.

7.6 Stomata

Stomata consist of two **guard cells** surrounding a central pore. Stomata are found all over the aerial parts of a plant, even on petals, anthers, and ovaries. In most flowering plants, however, they are concentrated on the under-surface of leaves (Fig. 7.4).

Generally speaking, stomata open in the light and close in the dark. It is almost certain that these movements are the result of variations in guard cell turgidity. It is known that each guard cell has a thick and relatively inelastic wall bordering the pore, and a thin elastic outer wall (Fig. 7.5). Experiments show that in the light guard cells take up water and become turgid, which causes their thin outer walls to bulge outwards, and the thicker inner walls to bend along with them. This gives the guard cells a curved banana-like shape, and opens the pore. In the dark the guard cells lose water, become flaccid, and the pore closes.

7.7 Translocation

Translocation is a term generally used to describe the transport of substances throughout the plant, but in particular it describes the movement of sugar and other manufactured materials from leaves to storage areas and growing points. Glucose, for example, moves from the mature leaves, where it is made during photosynthesis, upward through phloem sieve tubes in the stem to developing leaves, flowers, and fruits. It also moves downwards into the growing root tips. Consequently, there are regions of stem in which food moves up and down at the same time through adjacent, and possibly even the same sieve tubes.

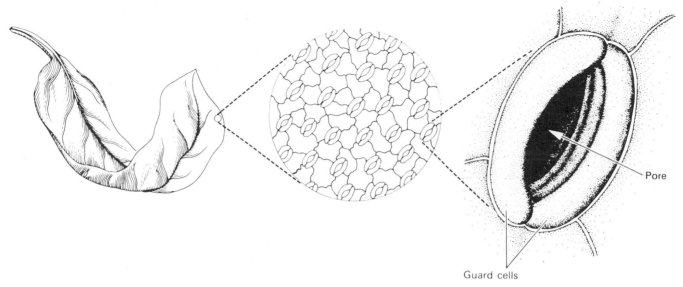

Fig. 7.4   Stomata

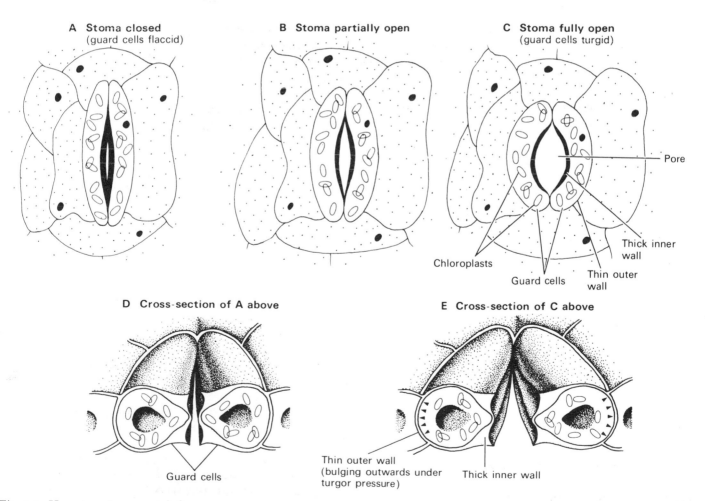

Fig. 7.5   How stomata open and close

Evidence of food transport through phloem
Food substances such as glucose are necessary for the development of new tissue at the growing points of a plant. If these foods are transported through sieve tubes, then damage to or removal of the phloem will affect growth. The following experiments demonstrate this in various ways.

Girdling experiments   A stem is girdled by completely removing a ring of bark and phloem from one or more places on the stem, leaving the xylem vessels unharmed (Fig. 7.6A).

1. When a girdle is cut around a stem just above ground level, root growth is suppressed and tissues immediately above the girdle swell, owing to the accumulation in them of food which would otherwise have gone to the root (Fig. 7.6B).

2. When a girdle is cut between the mature leaves of a plant and developing flowers, fruits, or leaves, further growth of these organs is suppressed.

Heat and cold treatments
1. The translocation of food from a leaf stops when its petiole is scalded with steam. Similarly, the girdling of a stem with steam gives the same result as the above-mentioned girdling experiments where tissue is removed.

Fig. 7.6   Girdling of a woody stem

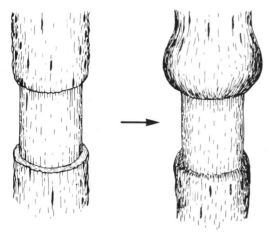

A A ring of bark is removed from a woody shoot, leaving the xylem undamaged

B After a few weeks the tissue above the girdle has swollen with accumulated substances which have moved down the phloem from the leaves

2. Translocation of food from a leaf is considerably reduced by subjecting the petiole to a temperature of $1°C$.

It is argued that heat and cold treatments adversely affect the living sieve tubes but not the dead xylem vessels. The results suggest that food passes through the sieve tubes since, in these experiments, there is no sign of it by-passing the girdle via xylem vessels.

Use of radio-active isotopes   It is possible to follow the translocation of sugar during photosynthesis by supplying a plant with carbon dioxide which contains radio-active carbon, i.e. $^{14}CO_2$. This procedure shows that sieve tubes are the major if not the only path for translocation of sugar.

Investigations

A   *An investigation of root and stem structure*

1. *Root hairs*
Germinate some broad bean seeds by trapping them against the sides of a jar with a cylinder of blotting paper filled with damp sand. Examine the seedlings for root hairs.

2. *Section cutting*
a) Put a bean seedling from 1 above into a beaker with its root immersed in water coloured red with eosin dye. After the dye becomes visible in the petioles and leaf veins, cut very thin slices of root and shoot. Examine these slices under a microscope looking for the structures shown in Figures 7.1 and 7.2, noting that xylem vessels are stained red with eosin.

b) Cut sections of various other plant tissues. Place the sections on a slide in a freshly made mixture of equal parts of phloroglucinol and dilute hydrochloric acid. This liquid will stain the xylem vessels bright red.

3. *Tissue maceration*
Warm small pieces of root, stem, and leaf separately in 10% potassium hydroxide solution for 5–10 minutes (handle this liquid carefully; it is very caustic). This procedure will separate the cells, and is known as maceration.

a) Put the macerated tissue in water on a slide and tease the cells from each other with mounted needles.

b) Stain the tissues as in 2(b) above.

c) Examine the shapes of all types of cell and vessel. Try to relate structure to function, using Figures 7.1 and 7.2 as a guide.

B *An investigation of transpiration rate*

1. Prepare the potometer apparatus shown in Figure 7.7. Be careful to keep the twig stem in water from the moment it is cut from the plant. This is necessary to prevent air entering its xylem vessels.

2. Allow a bubble to enter the capillary by:

a) removing the apparatus from the beaker of water;

b) pressing the rubber tube until a drop of water is squeezed from the capillary;

c) easing pressure on the rubber tube slightly to draw a little air into the capillary;

d) finally, replacing the apparatus in water. It is now ready for use.

Fig. 7.7 A potometer (exercise B)

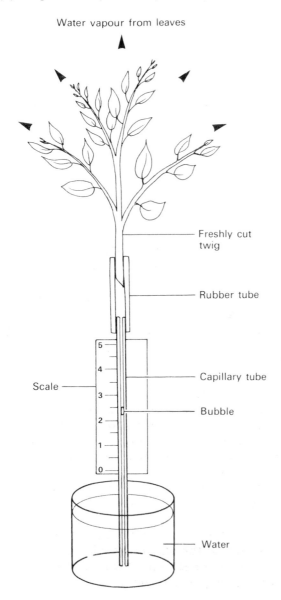

Water vapour from leaves

Freshly cut twig

Rubber tube

Scale

Capillary tube

Bubble

Water

3. Measure the time taken for a bubble to move along a specific length of scale when the potometer is:

a) indoors in the dark, and then in the light;

b) outdoors under different weather conditions;

c) next to an electric fan heater blowing hot, and then cool air (note that some unexpected results may be achieved);

d) arranged so that the leaves are covered with a plastic bag.

Between each of these operations return the bubble to the base of the capillary by: squeezing the rubber tube until the bubble is removed from the capillary; easing pressure on the rubber tube to draw water back into the capillary; and then repeating 2(*a*)-(*d*) above.

4. Explain how the results illustrate the explanation of transpiration given in the text.

C *An investigation of stomata*

1. *Stomatal structure*

Obtain a small piece of lower epidermis by roughly tearing leaves into pieces (e.g. privet, *Tradescantia*, or rhubarb). Mount the pieces in water on a slide under a cover slip. Observe under high magnification using Figure 7.4 as a guide.

2. *Epidermal 'prints'*

a) Smear a thin layer of latex rubber (e.g. 'Copydex' adhesive) on the lower epidermis of a leaf and allow it to dry completely.

b) Press a strip of clear self-adhesive tape (e.g. 'Sellotape') firmly on to the dry latex. When the tape is peeled off the leaf the latex comes with it, baring a perfect imprint of the leaf surface structure including stomata.

c) Press the tape on to a slide and observe the imprints under low and high magnification.

d) Make prints of several different leaves, and of different regions on the same plants. Estimate the number of stomata per square millimetre in each region examined.

3. *Stomatal behaviour*

a) Fill two dishes with water and on each dish float a plant leaf with its lower side uppermost.

b) Place one dish under a bench lamp and the other in the dark for 30 minutes, then take a strip of lower epidermis from each and plunge them into absolute alcohol. Alternatively, obtain a latex 'print' from each using method 2 above.

c) Look at 25 different stomata on each strip and estimate what percentage of stomata are open and what percentage are closed.

d) What do the results reveal about the effect of light on stomatal behaviour?

D *An investigation of water loss through stomata*

1. *Cobalt chloride method* Cobalt chloride is blue when dry, and turns pink in the presence of water.

a) Use a strip of 'Sellotape' to stick a small piece of dry cobalt chloride paper to the upper and lower surface of a leaf which is still attached to a plant.

b) How do the results help to confirm that more water is lost through the lower epidermis than the upper epidermis of a leaf?

c) What control is necessary in this experiment?

2. *Vaselined leaves* Vaseline smeared on the surface of a leaf seals the stomata and prevents water loss. Devise a controlled experiment using this technique to show that water is lost through only one particular leaf surface in most angiosperm leaves.

Questions

1. Study the illustrations of the two plants with adaptations which enable them to survive in conditions where water is usually in short supply.

a) List at least three ways (there are more) in which the *Oleander* leaf is adapted to restrict water loss by evaporation through its cuticle and transpiration through its stomata.

b) What is the advantage of a two-layered palisade in *Oleander*? (Clue: what effect will the thick

Cactus (*Opuntia*)

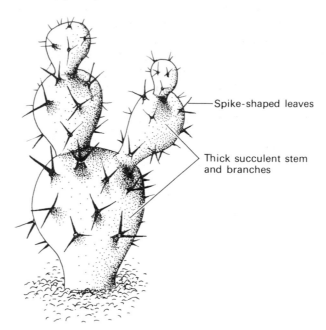

Spike-shaped leaves

Thick succulent stem and branches

cuticle and two-layered epidermis have on light penetration?)

c) In the course of evolution, the leaves of *Opuntia* have become reduced to non-photosynthetic spines. How does this help to restrict water loss by transpiration?

d) Since *Opuntia* has no leaves, what part of the plant carries out photosynthesis?

e) What other characteristics illustrated in the figure enable *Opuntia* to survive in dry conditions?

2. The diagram below shows the position of various tissues in a cross-section through a plant.

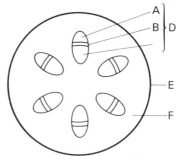

a) Which part of a plant has its tissues distributed as shown in this diagram?

b) Name the parts labelled A to F.

c) Give the label letter, and the name of the parts which:

are vascular tissues,
transport the products of photosynthesis,
transport water,
divide by mitosis,
support the plant.

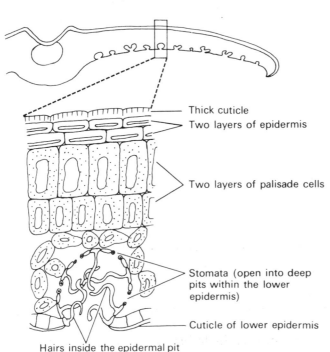

Thick cuticle
Two layers of epidermis

Two layers of palisade cells

Stomata (open into deep pits within the lower epidermis)

Cuticle of lower epidermis

Hairs inside the epidermal pit

Oleander leaf
(vertical section with a small area highly magnified)

3. A complete ring of phloem was removed from the stem of a plant (see Figure 7.6). At intervals during the next 24 hours liquid was removed from xylem and phloem above and below the ring. This liquid was analysed for its sugar content. The results are summarized in the graphs below, together with the results of similar tests on a normal (unringed) plant.

a) What do these results tell you about the functions of phloem and xylem in an unringed plant?

b) What do these results tell you about the effects of removing a ring of phloem?

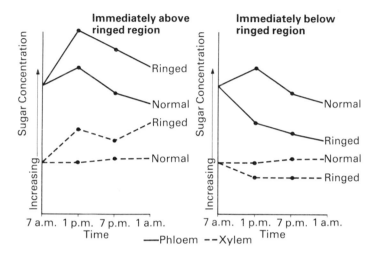

4. A potometer was assembled, as shown in Figure 7.7. The time for a bubble to travel 50 mm along the capillary tube was found under the following conditions:

plant leaves enclosed in a plastic bag (left for 30 minutes before readings taken),
potometer in wind,
potometer in shade,
potometer in darkness,
potometer in bright sunlight.

Five readings were taken under each of these conditions and average times recorded:

Condition	Average time taken (seconds) for bubble to travel 50 mm
Enclosed in plastic bag	70
Wind	42
Shade	57
Darkness	62
Bright sunlight	53

a) What conclusions can you draw from these results about the factors which control evaporation of water from leaves?

b) Why were five readings necessary for each condition?

c) Suggest reasons why evaporation rate varies in the way indicated by these results.

d) How could this apparatus be used to calculate the amount of water absorbed by the shoot?

Summary and factual recall test

Xylem (1) form a transport system in which (2) and (3) move from the roots to the (4). (5) tubes transport substances such as (6) from the (7), where they are made, to (8) and (9).

Transpiration is the loss of (10) by (11) from the (12) mesophyll of leaves. Water vapour diffuses into the atmosphere through (13) which occur on the (14) surface of most leaves. These consist of two (15) cells surrounding a pore. Transpiration rate varies according to the (16–list three factors). Transpiration is useful because (17–give three reasons), but can harm plants in conditions of (18).

Water lost from the leaves is replaced by water from (19) vessels, which produce a flow of water from roots to leaves called the (20) stream.

Translocation includes the movement of (21) from the (22) to the (23) areas and (24) points of a plant.

8
Respiration

The word respiration is derived from the Latin *respirare* which means to breathe. At first this term referred to the breathing movements which cause air to be drawn into and pushed out of the human lungs, but now, when defined with strict accuracy, respiration means something entirely different.

The modern definition of respiration is: the processes which lead to, and include, the chemical breakdown of materials to provide energy for life. These processes occur inside the living cells of every type of organism.

To avoid unnecessary confusion, it is strongly recommended that respiration be used only in this modern sense, so that it is clearly distinguished from the mechanism of breathing, which is concerned with the absorption of oxygen from the air. Breathing and related mechanisms are described in chapter 9. This chapter offers a brief introduction to the ways in which energy is released by the chemical breakdown of body materials.

8.1 The release of energy from food

The energy for life is released during respiration from substances known loosely as 'food'. There are many different foods, and they are taken into the body in many different ways, but in the majority of organisms all foods are converted into glucose before they are used as a source of energy. For the sake of simplicity, the following descriptions refer to the respiration of glucose.

In most organisms energy is released by a process called **aerobic respiration**, which requires a continuous supply of oxygen molecules obtained from the air or water surrounding the organism.

In certain circumstances, however, energy can be released without the use of oxygen molecules. This is known as **anaerobic respiration**. These two different but related types of respiration are described in the following sections.

8.2 Aerobic respiration

The aerobic respiration of glucose is summarized by the following chemical equation:

$$\underbrace{C_6H_{12}O_6}_{\text{Glucose}} + 6O_2 \longrightarrow 6CO_2 + 6H_2O + 2898\,kJ \text{ of energy}$$

In words this means:

$$\text{Glucose} + \text{Oxygen} \longrightarrow \text{Carbon dioxide} + \text{Water} + 2898\,kJ \text{ of energy}$$

Aerobic respiration releases all the available energy within each glucose molecule; that is, it produces the same amount of energy that is released when glucose is burnt in oxygen gas.

The chemical equation above gives the false impression that respiration involves only one chemical reaction, because it shows only the raw materials and end-products of respiration. The whole process involves a sequence of some fifty separate reactions, each catalysed by a different enzyme. The result is a controlled release of energy which is far more useful to the organism than a sudden explosive burst of energy.

Look again at the equation, and see what happens to all the hydrogen atoms contained within a glucose molecule. Eventually these atoms combine with oxygen atoms to form water. In fact, the bulk of respiratory energy becomes available to the organism as hydrogen atoms are removed from glucose during respiration. This process is catalysed mainly by **dehydrogenase enzymes**. In other words, the oxygen which an aerobic organism has absorbed combines with hydrogen atoms from glucose or other foods to produce water, which may be excreted from the body.

8.3 Anaerobic respiration

Anaerobic respiration differs from aerobic respiration in three important ways. First, anaerobic reactions break down glucose in the absence of oxygen. Second, anaerobic reactions do not completely break down glucose into carbon dioxide and water but into intermediate substances such as lactic acid or alcohol. Third, anaerobic respiration releases far less energy than aerobic respiration, because glucose is not completely broken down.

Organisms which respire anaerobically are called **anaerobes**. Certain bacteria are complete anaerobes. They live permanently in conditions where no oxygen exists and rely entirely upon anaerobic respiration for energy. Some of these bacteria are actually poisoned by oxygen, even in small quantities. Many organisms are partial anaerobes, in which case their cells are capable of carrying out both types of respiration, either separately or at the same time.

During periods of maximum effort the heart pumps blood at 34 litres per minute, and this delivers oxygen to the muscles at 4 litres per minute. But this is insufficient to meet the oxygen requirements of the muscles. They change over to anaerobic respiration and produce lactic acid instead of carbon dioxide and water, and begin to incur an oxygen debt

Anaerobic respiration in micro-organisms

Micro-organisms such as yeast and certain bacteria obtain most of their energy by a form of anaerobic respiration called **fermentation**. Typical products of fermentation are alcohol (ethanol) which is formed by yeast; and citric, oxalic, and butyric acids which are formed by certain bacteria. These chemicals are of great commercial value, and their production, with the help of micro-organisms, is now a major industry. Since these industries harness a biological process, their work is now described as **biotechnology**.

Many types of yeast are used in alcoholic fermentation. The equation for the fermentation of glucose is as follows:

$$C_6H_{12}O_6 \longrightarrow 2C_2H_5OH + 2CO_2 + 210\,kJ \text{ of energy}$$

Glucose $\qquad$ Ethanol $\quad$ Carbon dioxide

Compare this with the equation for aerobic respiration of glucose, and note two things. First, water molecules are not released in fermentation. This is because the reactions do not involve the removal of hydrogen atoms from glucose and their subsequent combination with oxygen. Second, very little energy is released in fermentation compared with aerobic respiration.

Yeast is not completely anaerobic, and in the production of alcoholic drinks aerobic conditions are maintained for some time so that yeast cells can carry out both types of respiration. In these conditions they grow rapidly and reproduce.

The type of alcoholic drink produced by fermentation depends largely upon the source of the sugar solution used. Fermentation of apple juice produces cider, grape juice produces wine, and malt extract from germinating barley produces beer. Distillation of certain fermentation products gives rise to much stronger alcoholic solutions called spirits. Brandy is a spirit produced by distilling wine.

The equation for anaerobic respiration shows that carbon dioxide is a product of alcoholic fermentation. In the making of bread, bakers' dough 'rises' because the yeast mixed into it produces carbon dioxide gas which fills the dough with bubbles as it escapes.

Anaerobic respiration in plants

Green plants can respire anaerobically for short periods. The time limit for this is determined by the rate at which alcohol accumulates in their tissues, since this substance is poisonous in high concentrations.

The ability of plants to live as temporary anaerobes allows them to survive in conditions where animals would quickly die of suffocation. When flooding occurs, for example, plants can survive for several days completely immersed in water, and for several weeks in waterlogged, airless soil. Anaerobic respiration is also necessary in the initial stages of germination, when the plant embryo is completely enclosed within an air-tight seed coat.

Anaerobic respiration in vertebrate muscle

The muscles of vertebrate animals can continue working for a minute or two without oxygen. This happens, for example, when an athlete runs in a race. For the first few yards his muscles respire aerobically, but soon they use up all their available oxygen. Despite his increased breathing rate and heart-beat, oxygen cannot be transported to his muscles fast enough to meet their requirements. Under these circumstances muscles obtain most of their energy anaerobically, but this type of respiration produces lactic acid instead of carbon dioxide and water:

$$C_6H_{12}O_6 \rightarrow 2CH_3CH(OH)COOH + 150\,kJ \text{ of energy}$$

Glucose   Lactic acid

If the athlete continues to run as fast as he can, lactic acid begins to accumulate in his muscles and it eventually reaches a critical level at which it prevents further muscular contraction. At this stage the athlete collapses, unable to run another step.

During the running period anaerobic respiration in the athlete's muscles is said to incur an **oxygen debt**; that is, the muscles have expended energy in excess of oxygen absorption. This 'debt' is 'paid' by rapid breathing in the recovery period after the race. For every 10 g of lactic acid accumulated in the body 1.7 litres of oxygen must be absorbed. This oxygen is used in aerobic respiration to break down about one-sixth of the lactic acid into carbon dioxide and water, and this releases enough energy to convert the remaining five-sixths back into glucose.

The largest amount of lactic acid which the body of a trained athlete can tolerate is about 127 g, and the largest tolerable oxygen debt is about 16 litres. However, months of continuous training are necessary before this level of fitness is reached.

8.4 How organisms use energy

Organisms use energy released by respiration to drive countless different processes, some of which are named in Figure 8.1. But cells cannot use energy as soon as it is released from respiration: the energy is first used to build up a temporary energy store, which takes the form of a chemical called **adenosine triphosphate**, or ATP for short.

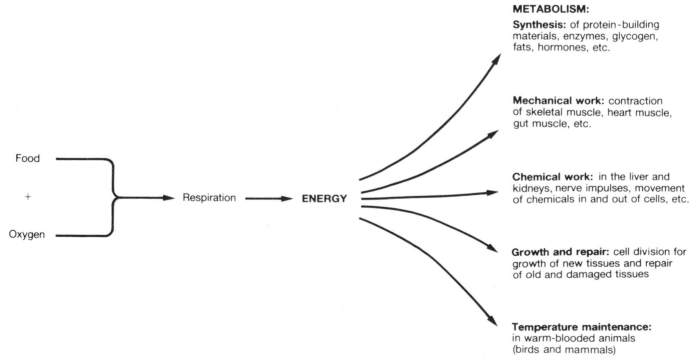

METABOLISM:

Synthesis: of protein-building materials, enzymes, glycogen, fats, hormones, etc.

Mechanical work: contraction of skeletal muscle, heart muscle, gut muscle, etc.

Chemical work: in the liver and kidneys, nerve impulses, movement of chemicals in and out of cells, etc.

Growth and repair: cell division for growth of new tissues and repair of old and damaged tissues

Temperature maintenance: in warm-blooded animals (birds and mammals)

Fig 8.1 Some of the ways in which respiratory energy is used in the body

Molecules of ATP are best thought of as 'packets' of energy, used to transfer energy from the chemical reactions which release it to the body processes which use it. Put very simply, respiration 'fills' these ATP packets with energy, and they are 'emptied' when energy is needed anywhere in the body (Fig. 8.2).

There are four main advantages to the ATP energy transfer system.

1. ATP takes up some energy which would otherwise have been lost as heat during the breakdown of glucose by respiratory enzymes.

2. Energy is released from ATP the instant it is required without cells having to go through the fifty-odd different reactions of respiration. This is important when sudden bursts of energy are required.

3. ATP delivers energy in precise amounts.

4. Energy can be transferred from ATP to other chemicals without energy loss. This converts relatively inert substances into highly reactive ones, so they can take part in processes which require energy input – like the synthesis of complex chemicals out of simple raw materials.

Investigations

A To verify that respiration produces heat

1. Heat production during germination
a) Obtain sufficient seeds (e.g. wheat, barley, or small peas) to fill two vacuum flasks, and soak them in water for 12 hours.

b) Divide the seeds into two equal portions. Boil half of them for two minutes to kill them. Wash both portions in 10% formalin to prevent the growth of bacteria and fungi.

c) Place the boiled seeds in one vacuum flask and the live seeds in the other. Plug the mouth of each flask firmly with cotton wool, then insert a thermometer through each plug deep into the flask. Finally, mount the flasks upside-down in clamps.

d) Record any temperature changes over two or three days.

e) Why are boiled seeds included in the experiment?

f) Why are the flasks inverted, and why is cotton wool preferable to a rubber bung? (Clue: heat rises and carbon dioxide is heavier than air.)

g) Why does the temperature rise in one flask and not in the other? What is the source of this heat?

h) Why would a growth of fungi on the seeds interfere with experimental results?

2. Heat production by animals
a) Obtain sufficient maggots (fly larvae) to fill two vacuum flasks. (Maggots are obtainable from shops which sell bait for fishermen.)

b) Kill half of them by boiling, then proceed as in 1(b) above omitting the formalin treatment and inversion of the flasks.

c) Results should be obtainable after 5 or 10 minutes.

B To verify that carbon dioxide is produced during respiration

1. To verify that plants and land animals produce carbon dioxide
a) Set up the apparatus shown in Figure 8.3.

b) Almost any small organisms can be placed in the specimen chamber: wood-lice, maggots, earthworms, germinating seeds, mushrooms or other large fungi, and small plants. (If plants are used place the apparatus in the dark.)

c) Draw air through the apparatus, noting the milky precipitate which develops in flask 4, owing to the effects of carbon dioxide gas on lime water.

d) Why is flask 2 necessary?

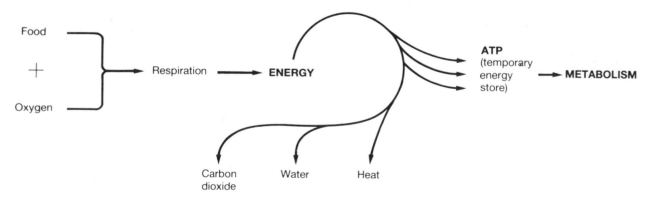

Fig. 8.2 ATP is a temporary energy store

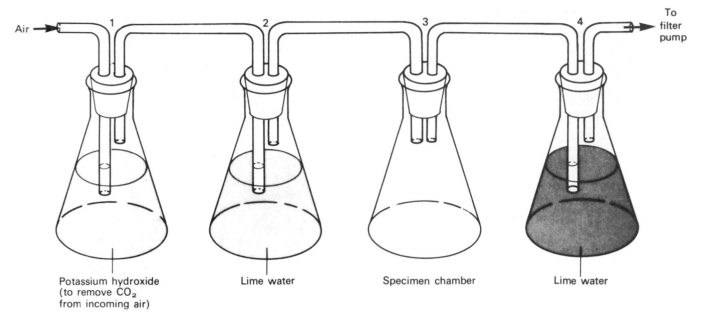

Fig. 8.3 To verify that carbon dioxide is produced during respiration (exercise B)

2. *Simplified alternatives to B1 above*

a) Prepare the apparatus in Figure 8.4 for use with small animals. Use an identical apparatus without a specimen as a control.

b) Prepare the apparatus in Figure 8.5 to find out whether all parts of a plant produce carbon dioxide. Wrap pieces of root, stem, leaves, and flowers separately in gauze, and hang them inside the apparatus in the manner illustrated. When testing green parts of plants place the flask in a dark cupboard. Prepare an identical flask without a specimen as a control.

C *To verify that organisms consume oxygen during respiration*

1. Prepare the apparatus in Figure 8.6.

Almost any small organism can be put into the flask (see B1 (b) above). In addition, try a thick suspension of yeast cells in sugar solution. Spread the suspension over crumpled filter paper in the flask.

a) Organisms take in oxygen and produce carbon dioxide at about the same rate. Carbon dioxide is absorbed by the potash (or soda lime) therefore oxygen consumption by the organism reduces air pressure inside the flask, which draws coloured water up the capillary.

b) Does this experiment actually *prove* that oxygen is consumed by the specimens?

c) How would the result differ if potash were omitted from the flask?

Fig. 8.4 Simplified version of apparatus in Figure 8.3 (exercise B2(a))

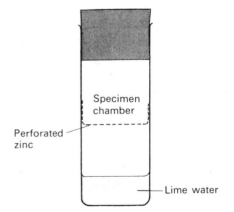

Fig. 8.5 To verify that plants produce carbon dioxide (exercise B2(b))

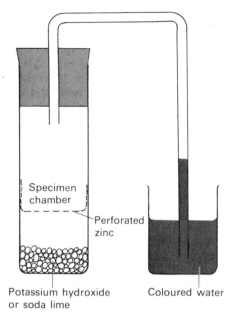

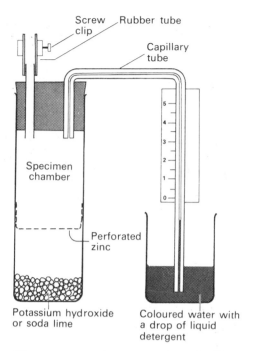

Fig. 8.6 To verify that organisms consume oxygen (exercise C)

Fig. 8.7 To measure respiratory rate (exercise D)

D *An investigation into the effect of temperature changes on respiratory rate*

1. Prepare the respirometer apparatus shown in Figure 8.7. This is a development of that shown in Figure 8.6, and works on the same principle (see C1(a) above).

a) Place different kinds of organisms (one type at a time) in the apparatus (see B1(b) above), and leave it, with the screw clip open, in a water bath at 20°C for 5 minutes. After this time close the screw clip and measure the time it takes for fluid to be drawn up a specific length of capillary tube. Repeat this procedure at the same temperature two or three times to obtain an average result. At the end of this stage, remove the rubber bung to let fresh air reach the organisms.

b) Repeat (a) above at 0°C, 10°C, 30°C, and 40°C, using the same organisms, and allowing them 10 minutes to adjust to each new temperature before closing the screw clip.

c) Graph the results, and if the capillary bore is known calculate the volume of oxygen consumed per hour per gram of the organism's body weight, at each temperature.

d) How do these results relate to findings from exercise C in chapter 6, concerning the effects of temperature on blood flow?

e) Why does temperature affect oxygen consumption in cold-blooded organisms? Is it likely to have the same effect on mammals? Explain your answer.

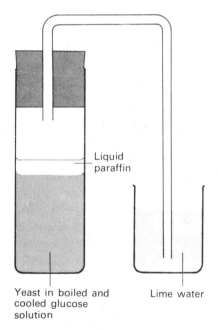

Fig. 8.8 To verify anaerobic respiration in yeast (exercise E1)

E *To verify that yeast and plant tissues respire under anaerobic conditions*

1. *Anaerobic respiration (fermentation) in yeast*

a) Prepare the apparatus shown in Figure 8.8 using 5% glucose solution which is boiled and cooled before adding a small amount of dried yeast. Finally, cover the yeast suspension with a layer of liquid paraffin.

b) How does boiling the glucose and then covering it and the yeast with oil produce, and maintain, anaerobic conditions?

c) Design an appropriate control for this experiment.

d) A larger scale version of this experiment (without the liquid paraffin) will produce enough alcohol for distillation.

2. *Anaerobic respiration in germinating seeds*

a) Soak 4 or 5 peas in water for 12 hours, put them in a test-tube, then fill the tube to its rim with mercury.

b) Hold a piece of card tightly over the tube mouth, then invert the tube into a deep dish of mercury. Remove the card once the tube mouth is submerged, then clamp the tube to hold it in a vertical position (Fig. 8.9). (*Note:* This experiment should be placed in a fume cupboard or under a large bell-jar.)

c) Over a period of 2 or 3 days a colourless gas appears at the top of the tube. When the tube is half full of this gas test it with lime water.

d) Design an appropriate control experiment.

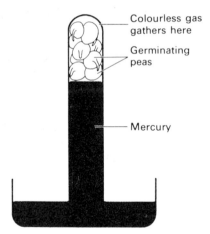

Fig. 8.9 To verify anaerobic respiration in green peas (exercise E2)

Labels on figure:
- Colourless gas gathers here
- Germinating peas
- Mercury

Questions

1. *a)* What are the differences between aerobic and anaerobic respiration?

b) The table below shows the number of times a boy could raise a 400 g weight with one finger during three, one-minute tests with no rest period in between.

	1st test	2nd test	3rd test
Number of times raised	95	81	62

Explain these results, using the words lactic acid, oxygen debt, and anaerobic respiration.

2. The following table shows the energy expended by a man depending on what he was doing:

Activity	kJ/minute
Sleeping	4.2
Sitting reading	5.8
Sawing wood	37.6
Heavy lifting	48.0

a) Name two body processes which require energy while the man is asleep.

b) Assuming the man sits still while reading, what activity is causing extra energy expenditure?

c) For what purpose would the majority of extra energy be used during sawing and lifting?

Summary and factual recall test

Respiration is the chemical (1) of materials to provide (2) for (3). Aerobic respiration requires a supply of (4) and results in the complete breakdown of glucose into (5) and (6) with the release of (7). Anaerobic respiration requires no (8), produces far less (9), and breaks down glucose into intermediate substances such as (10) and (11).

Micro-organisms such as (12) and (13) obtain energy by fermentation. Commercially useful prod-

ucts of fermentation are (14–name three). Baker's dough rises because (15).

Green plants can respire anaerobically for some time until (16) accumulates in their tissues. This is (17) to them and prevents further respiration. Vertebrate muscle can respire anaerobically until (18) prevents further (19). This condition is known as an oxygen (20).

9

Gas exchange and breathing

Organisms which respire aerobically must absorb oxygen into their bodies. At the same time they must remove carbon dioxide gas which is a waste product of respiration. In other words they must 'exchange gases' with the air or water around them. This process is called **gas exchange**.

9.1 Gas exchange in small organisms

Small organisms such as *Amoeba*, *Paramecium*, *Hydra*, and jelly fish have a large surface area relative to their small body volume. In these organisms gas exchange takes place over the whole body surface and, because of their small body volume, diffusion alone is sufficient to transport oxygen and carbon dioxide into, around, and out of their bodies. Diffusion in small organisms is explained further in section 6.1.

Higher organisms such as insects and vertebrates have a small surface area relative to their large body volume. In these organisms gas exchange takes place at a specialized region of the body called a **respiratory surface**. This surface is often part of an elaborate **respiratory organ**. Examples of respiratory organs are fish gills and human lungs.

9.2 Characteristics of respiratory surfaces

Below is a list of features common to all respiratory surfaces. They allow oxygen and carbon dioxide to be exchanged rapidly between an organism and the air or water which surrounds it.

1. Respiratory surfaces have a large surface area to ensure maximum contact with air or water. This is achieved in several ways. An insect's respiratory surface is the lining of thousands of finely branched tubes. A fish's respiratory surface is a folded skin over the gills. A mammal's respiratory surface consists of millions of tiny bubble-like air sacs.

2. All respiratory surfaces are moist. This is necessary because oxygen and carbon dioxide can only diffuse in solution across a respiratory surface.

3. A respiratory surface is extremely thin. In most organisms it is only one cell thick.

4. The inside of a respiratory surface is in contact with a network of capillary blood vessels. This allows gas exchange to take place between the blood and surrounding air or water.

5. There is usually a mechanism which ensures that a respiratory surface is well ventilated; that is, it receives a steady flow of air or water. Ventilation mechanisms, such as human breathing movements, increase the rate of gas exchange by continually removing carbon dioxide as it comes through the respiratory surface, and by renewing supplies of oxygen as fast as it is absorbed.

9.3 Respiratory organs of mammals

The lungs of a mammal, together with the heart and major blood vessels, are situated in the **thoracic cavity**, or thorax. The walls of the thorax are strengthened by the ribs, and its floor consists of a sheet of muscle called the **diaphragm**. A system of passageways leads from the mouth and nostrils into the lungs (Figs. 9.1 and 9.2).

Structure of the respiratory organs
The nasal passages Air entering through the nostrils is drawn into the nasal passages where it is warmed to body temperature and humidified by moisture which evaporates from the warm nasal membranes lining the walls of these passages. The

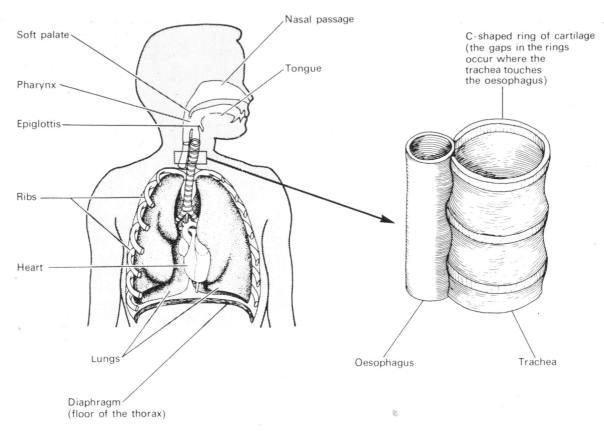

Fig. 9.1 The thoracic cavity

membranes covering the roof of the nasal passages contain the organs responsible for the sense of smell. The walls and base of the nasal passages are lined with a 'carpet' of microscopic hair-like structures called **cilia** (Fig. 9.2C). Between the cilia are **goblet cells** which produce a sticky fluid called **mucus**. Dust and germs inhaled from the atmosphere are trapped in the mucus and are carried by the rhythmic beating of the cilia towards the back of the mouth where they are swallowed. This mechanism helps to prevent germs and dirt from entering the lungs. Inhaled particles are trapped and passed out of the body through the digestive system. By the time air reaches the lungs it is relatively dust-free, warm, and moist.

Air is drawn out of the nasal passages into a channel called the **pharynx** at the back of the mouth. From here, air is drawn into the **trachea**, or wind-pipe. Food in the mouth is prevented from entering the trachea during swallowing by a mechanism described on page 45. Food accidentally entering the trachea touches a sensitive area and sets off a coughing reflex, which clears the trachea of any obstruction.

The larynx The larynx, or voice-box, is a cavity at the top of the trachea which contains the vocal cords. The vocal cords are two folds of membrane situated on opposite sides of the larynx. They are attached to muscles which vary the tension in the cords and the distance between them. When the muscles relax the cords are separated and slack so that air passes soundlessly between them; but when the muscles contract the cords become taut and close together so that air causes them to vibrate. This produces sound. The pitch of the sound varies according to the tension in the cords and the distance between them.

The trachea The trachea is a tube running from the pharynx (back of the mouth) to the lungs. It is held permanently open by C-shaped rings of cartilage (gristle) in its walls (Fig. 9.1). This feature prevents the trachea from being blocked by 'kinks' every time the neck bends during head movements. The cartilage rings also keep the trachea open when it develops a low internal air pressure during every intake of breath.

Cilia and goblet cells extend from the nasal passages for some distance into the trachea. They create an upward flow of mucus which removes dust and germs as explained above.

At its lower end, the trachea divides and

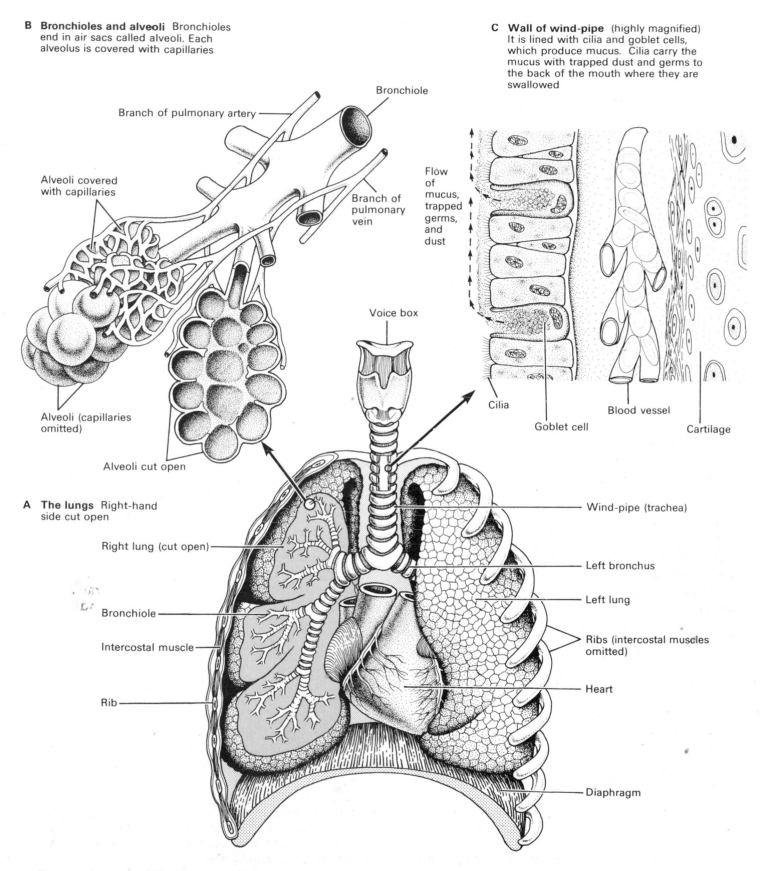

B Bronchioles and alveoli Bronchioles end in air sacs called alveoli. Each alveolus is covered with capillaries

Branch of pulmonary artery

Bronchiole

Alveoli covered with capillaries

Branch of pulmonary vein

Alveoli (capillaries omitted)

Alveoli cut open

Voice box

C Wall of wind-pipe (highly magnified) It is lined with cilia and goblet cells, which produce mucus. Cilia carry the mucus with trapped dust and germs to the back of the mouth where they are swallowed

Flow of mucus, trapped germs, and dust

Cilia

Goblet cell

Blood vessel

Cartilage

A The lungs Right-hand side cut open

Right lung (cut open)

Bronchiole

Intercostal muscle

Rib

Wind-pipe (trachea)

Left bronchus

Left lung

Ribs (intercostal muscles omitted)

Heart

Diaphragm

Fig. 9.2 Structure of the human respiratory system

subdivides to form the bronchial 'tree', which is made up of millions of **bronchial tubes**.

The bronchial tree   The main trunk of the bronchial tree is the trachea. This divides into two branches, the **bronchi** (singular bronchus), one leading to each lung. Inside the lungs each bronchus divides again and again to form a mass of very fine branches called **bronchioles**. Like the trachea, bronchi have C-shaped rings of cartilage in their walls, but these are only present to the point where the bronchi enter the lungs. From then onwards the rings are replaced by irregularly shaped plates of cartilage which perform the same function of keeping the tubes permanently open. Cartilage support of the bronchioles ceases altogether if they are less than 1 mm in diameter.

The bronchial tree terminates in air passages called **respiratory bronchioles**, which are about 0.5 mm in diameter. These branch into many short tubes of equal diameter called **alveolar ducts**, which end in tiny hollow bags called **air sacs**. The air sacs have many bubble-like pockets in their walls called **alveoli** (singular alveolus). The alveoli are the respiratory surface of a mammal (Fig. 9.2B).

Alveoli   There are about 300 million alveoli in one set of human lungs. Each alveolus is about 0.2 mm in diameter, and has walls made of membrane only 0.001 mm thick. It has been estimated that if the alveoli in both lungs could be spread out flat they would cover a surface area of 90 m². This is about the area of a singles tennis court.

This arrangement of bubble-like alveoli gives the lungs an appearance and texture similar to sponge rubber. It also gives the lungs a far greater internal surface area than if they consisted of two smooth-walled bags, like balloons.

The whole outer surface of each alveolus is covered by a dense network of capillary blood vessels (Fig. 9.2B). All of these capillaries originate from the pulmonary artery, and eventually drain into the pulmonary vein (Fig. 6.5). If all the capillaries which make up the network covering the entire 90 m² area of the lungs could be joined end to end, they would reach almost from London to New York, a distance of about 5000 km.

Blood flowing through this immense network of capillaries absorbs oxygen which diffuses through the alveoli walls from inside the lungs. At the same time the blood releases carbon dioxide which diffuses in the opposite direction into the alveoli. This two-way diffusion takes place through the alveoli and capillary walls which, in humans, have a combined thickness

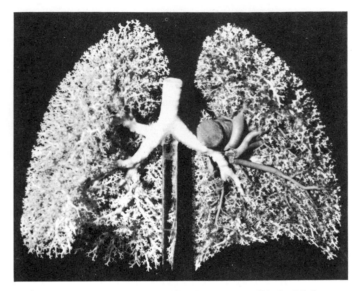

A latex cast of a human lung. The lungs were filled with latex rubber and all the tissues were digested away with enzymes

Part of the previous photograph enlarged. Identify the trachea, bronchus, and bronchioles

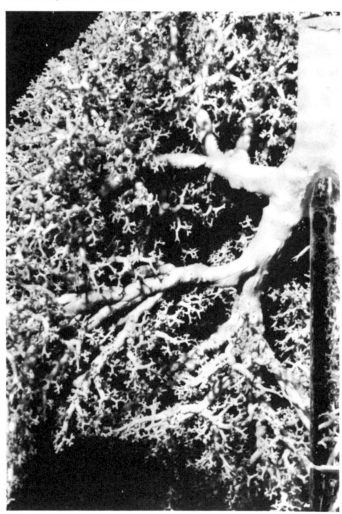

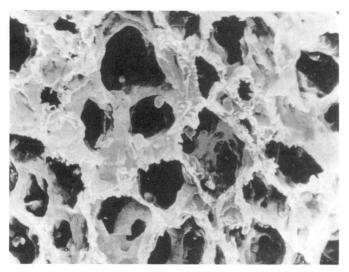

Healthy, spongy lung tissue

of only 0.005 mm. This is the distance which separates air from blood in the lungs, and across which gas exchange takes place.

9.4 Gas exchange in the lungs

Absorption of oxygen

In chapter 6 it was explained that blood entering the lungs is deoxygenated, because the haemoglobin in its red cells has given up all its oxygen to the body tissues.

The internal diameter of the lung capillaries is actually smaller than the diameter of the red cells which pass through them. The red cells are therefore squeezed out of shape as they are forced through the lungs by blood pressure, and the speed at which they move is considerably reduced by the resulting friction. This increases the rate of oxygen absorption in two ways. First, as the red cells squeeze through the narrow capillaries they expose more surface area to the capillary walls through which oxygen is diffusing, and thereby absorb more oxygen. Second, their slow rate of progress increases the time available for oxygen to diffuse into them and combine with haemoglobin.

The continuous removal of oxygen as fast as it diffuses into the lung capillaries, and the continuous arrival of oxygen in the alveoli owing to breathing movements, means that there is always a higher concentration of oxygen molecules in the alveoli than in the blood. This difference causes oxygen to diffuse from the alveoli into the lung capillaries since (as explained in chapter 2) diffusion continues so long as the molecules concerned are unequally distributed.

Release of carbon dioxide

It was explained in chapter 6 that blood entering the lungs is charged with carbon dioxide which it has absorbed from body tissues. This diffuses out of the blood into the alveoli for the same reason that oxygen diffuses in the opposite direction: the continuous arrival of carbon dioxide from the tissues, and its continuous removal from the alveoli by breathing movements, means that there is always a higher concentration of carbon dioxide in the blood than in the alveoli.

Figure 9.3 summarizes the process of gas exchange in the lungs and shows that it also occurs, in the reverse order, between body tissues and the blood in the capillaries which serve them.

Fig. 9.3 Diagram of gas exchange between an alveolus and the blood, and between blood and body cells

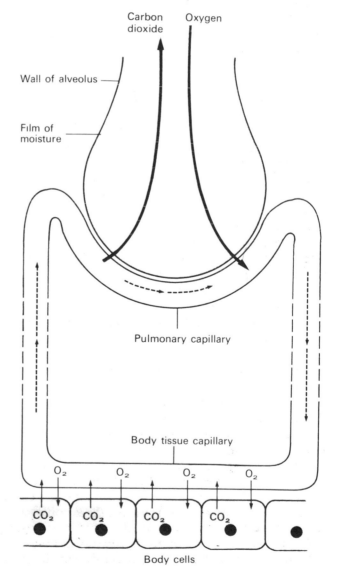

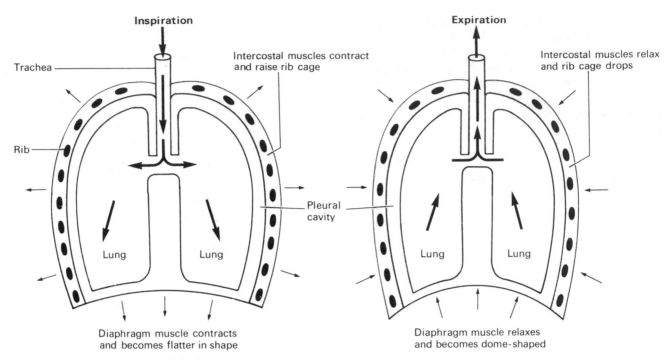

Fig. 9.4A Breathing (ventilation of the lungs)

9.5 Breathing (ventilation of the lungs)

The lungs are ventilated by muscular movements of the thorax wall which alter the volume of the thoracic cavity. When this volume is increased air is drawn into the lungs and they inflate, and when the volume is decreased air is pushed out and the lungs deflate. Before explaining how these volume changes are brought about it is necessary to describe the thorax in more detail.

Look at Figures 9.2 and 9.4 and note that the lungs hang down inside the thorax where they are surrounded by a very narrow space called the **pleural cavity**. This cavity is lined with a shiny, slippery skin called the **pleural membrane**, which produces an oily substance called **pleural fluid**. The pleural fluid acts as a lubricant which greatly reduces friction as the lungs rub against the thorax wall during breathing movements.

The pleural cavity is completely air-tight and contains a partial vacuum: its internal pressure is always less than the atmospheric pressure outside the body. On the other hand, the lungs are open to the atmosphere through the trachea and so there is always a higher pressure in the lungs than in the pleural cavity which surrounds them. This pressure difference is extremely important for two reasons. First, the higher pressure in the lungs than in the pleural cavity around them stretches the thin elastic alveoli walls so that the lungs as a whole almost fill the thorax.

Second, since this pressure difference is maintained during breathing movements, when the thoracic cavity increases in volume the lungs inflate to fill the extra space available.

The muscles which bring about these volume changes are the diaphragm and the intercostal muscles. The diaphragm is a dome-shaped sheet of muscle which forms the floor of the thorax, and the intercostal muscles consist of muscle fibres which cross the gap between each rib (Figs. 9.2 and 9.5).

Inspiration, or breathing in, results from an increase in the thoracic cavity volume brought about by the simultaneous contraction of the diaphragm and intercostal muscles (Figs. 9.4A and 9.5).

Diaphragm movements Immediately before inspiration the diaphragm is dome-shaped and its muscle is relaxed. Inspiration takes place when the diaphragm muscle contracts, making it flatter in shape. This contraction results in a downward movement of the central region of the diaphragm, and this increases the volume of the thoracic cavity.

Rib movements At the same time, contraction occurs in the external intercostal muscles between each rib. This raises the rib cage and increases its diameter and volume. Observe this movement by folding the arms across the chest and taking a deep breath. The arms move upwards and outwards along with the rib cage. Note from Figure 9.5 that this movement results from each rib pivoting at the point where it joins the backbone and sternum.

The increase in thoracic cavity volume which results from both these movements is automatically followed by an increase in lung volume. This temporarily reduces air pressure inside the lungs, and so air rushes into them from the atmosphere through the air passages.

Expiration, or breathing out, occurs when the diaphragm and external intercostal muscles relax. The rib cage drops under its own weight and the diaphragm returns to its original dome shape. Both of these movements reduce thoracic cavity volume and the lungs return to their original size, which pushes air out of the lungs. Air can be forced out of the lungs by contracting *internal* intercostal muscles (Fig. 9.5).

Control of breathing rate

A relaxed human adult breathes about 16–18 times per minute, using the diaphragm alone. During exercise this rate automatically rises, and the intercostal muscles come into operation so that the volume of each breath increases, as well as the breathing rate.

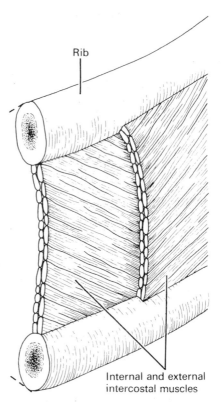

Fig. 9.5 External and internal intercostal muscles

Fig. 9.4B Side view of the rib cage showing rib movements during breathing

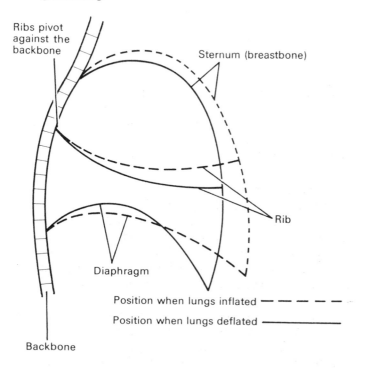

The rate and depth of breathing are partly under conscious control. In fact, they can be controlled with great precision, when playing a wind instrument such as a flute for example, or when singing. At other times, breathing is controlled by an unconscious (reflex) mechanism, which operates in the following way.

The walls of certain arteries contain nerve endings which are sensitive to changes in the levels of carbon dioxide and oxygen in the blood. During exercise there is a rise in the level of carbon dioxide in the blood owing to the increased production of this gas by the muscles. The sensory nerve endings detect this rise and send impulses to the medulla of the brain, which automatically increases the rate at which the diaphragm and intercostal muscles contract. As a result of this carbon dioxide does not accumulate in the body and extra oxygen is supplied to the muscles.

Experiments show that these nerve endings are relatively insensitive to reduced levels of oxygen in the blood. The oxygen level in inspired air can be artificially reduced by 50% (if the carbon dioxide level is held constant) before breathing rate alters. The body does not have to be equally sensitive to both gases since carbon dioxide variations are almost invariably related to corresponding changes in oxygen requirements. However, under conditions of low air pressure which occur, for example, at the top of a high mountain, the rate at which oxygen can be absorbed by the body is greatly reduced. In these circumstances breathing rate increases in response to a low oxygen level rather than a high carbon dioxide level in the blood.

Lung capacity

The total average capacity of adult human lungs is about 5 litres. But in normal breathing only 500 cm³ of air is breathed in and out. This is called **tidal air**. After a normal tidal inspiration a further 1500 cm³ of air can be drawn into the lungs. This is called **complemental air**. After a normal tidal expiration a further 1500 cm³ of air can be forced out of the lungs. This is called **supplemental air**. After a forced expiration a further 1500 cm³ of air remains in the lungs since the thorax cannot be completely collapsed. This is called **residual air**. Residual air is not stagnant since inspired air mixes with it at each breath.

9.6 Diseases of the respiratory system

Ailments of the respiratory system can be caused by inhaling disease organisms (germs), dust, chemicals, fumes, and vapours.

The best way to avoid lung infections is to maintain your physical fitness, eat nourishing balanced meals, and avoid overcrowded and badly ventilated places. Dust control in your place of work may also be necessary.

The common cold

There may be as many as 200 different viruses which cause the common cold.

Symptoms Sore throat, coughing, sneezing, and a 'runny' nose. Symptoms rarely persist for more than a week. If they last longer see a doctor because a secondary infection of bacteria may have occurred.

Spread of infection Mostly spread by inhaling viruses coughed or sneezed from infected people. Viruses remain infective for about three hours on a variety of surfaces including skin. So they may also be spread by touch.

Prevention Avoid over-crowded and badly ventilated places. People with colds should especially avoid such places.

Treatment At present no drug is effective because so many different viruses are involved. Drugs are available to treat the headache, sneezing, and coughing caused by colds. Large doses of vitamin C are no longer thought to prevent or cure colds, but they may reduce discomfort. Medicines which dry up a runny nose may cause unpleasant side-effects, especially in old people. Also they prevent the body getting rid of germs in mucus.

Influenza

Also caused by viruses. There are two types: Influenza A, which causes world-wide epidemics, and Influenza B, which causes local outbreaks.

Symptoms Sore throat, coughing, sneezing, headache, very high temperature, a weak 'shivery' feeling, sweating, and sickness.

Spread of infection Spread, like colds, by contact, and coughs and sneezes.

Prevention Patient should stay in bed to rest, and avoid contact with other people. Handkerchiefs, cutlery, cups etc., handled by the patient, should be thoroughly cleaned.

Treatment Vaccinations are now available which are quite effective.

Bronchitis

Bronchitis is inflammation of the lungs and bronchial tubes. It is not caused by one specific germ and can result from a number of causes. Untreated colds, influenza or whooping cough can develop into bronchitis. People who work in a hot dusty atmosphere, or where irritant gases and vapours are inhaled can develop bronchitis. Another major cause is cigarette smoke.

Symptoms A dry cough, followed by coughing with sputum.

Treatment There is no specific treatment. Bed rest is important. Coughing can be controlled by codeine and ephedrine cough mixture. A dry cough can be relieved by inhaling the steam from a bowl of hot water.

Emphysema

Emphysema is progressive thinning of lung tissue. Eventually the walls of the alveoli rupture, making the lungs far less efficient. Emphysema can result from chronic bronchitis, or coughs caused by dust, irritant gases and fumes, and cigarette smoke.

Symptoms Shortness of breath after even moderate exercise. The chest often appears to be enlarged, and breathing produces a 'wheezing' sound.

Treatment Exercises which strengthen the abdominal muscles and permit complete exhalation are recommended.

Pneumonia
Pneumonia is a build up of fluid in the lungs. There are several types of pneumonia. The commonest types are caused by *pneumonococci* and *streptococci* bacteria.

Symptoms Coughing, followed by fever, chest pains, and headache.

Spread of infection Pneumonia is spread by inhaling germs coughed or exhaled from the lungs of infected people.

Prevention Avoid overcrowded, badly ventilated places.

Treatment There are several antibiotic drugs which cure pneumonia. These drugs include penicillin, tetracycline, and erythromycin.

Investigations

An investigation of the mammalian respiratory system

1. *Gross structure*
Obtain the lungs of an animal from a butcher or slaughter-house.
   a) Identify and examine the trachea and bronchi.
   b) Slice open one lung and look for bronchioles.
   c) Note and account for the bright red colour of lungs and their spongy texture.

2. *Lung capacity*
   a) Squeeze all the air out of a large plastic bag, take a deep breath, and then exhale as much air as possible into the bag.
   b) Seal the bag, without letting air in or out, by tying a knot at its mouth.
   c) Measure the approximate volume of the bag by packing it into a large measuring cylinder, or other previously calibrated glass vessel.

3. *Analysis of inhaled and exhaled air*
   a) Prepare the apparatus illustrated in Figure 9.6.
   b) Measure the time taken for lime water to turn milky when breathing exhaled air down tube A; and again when sucking inhaled air through tube B (using fresh lime water for the second part of the experiment).
   c) Explain why one procedure takes longer to turn the lime water milky than the other. What exactly do the results prove?

4. *Breathing rate*
   a) Count the number of breaths per minute of a person at rest when he is unaware that he is being observed and then when he is aware.
   b) Count the breathing rate after one minute of vigorous exercise.
   c) Account for any differences in the three rates.

Questions

1. *a)* What are the characteristics of a respiratory surface?
   b) With reference to each of these features write a very brief description of the respiratory surface of an insect, a fish, a frog, and a mammal.

2. *a)* What is gas exchange?
   b) Explain briefly how gas exchange is accomplished in fish, and in frogs.

3. Put the following words in the correct order: bronchi, bronchioles, nasal passages, alveoli, trachea.

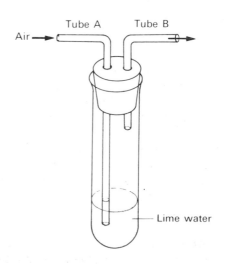

Fig. 9.6 Apparatus for analysing inspired and expired air

4. The following are some of the processes involved in breathing in mammals. Arrange then into the order in which they occur during breathing in, and during breathing out.

    a) External intercostal muscles contract.
    b) External intercostal muscles relax.
    c) Diaphragm muscle relaxes.
    d) Diaphragm muscle contracts.
    e) Volume of lungs decreases.
    f) Volume of lungs increases.
    g) Air pressure in lungs increases.
    h) Air pressure in lungs decreases.
    j) Air flows into lungs.
    k) Air flows out of lungs.

5. Study the chart below. What does it tell you about how air is changed inside the lungs, and about the effects of exercise on these changes?

Air	Unbreathed air	Breathed air from a sleeping man	Breathed air from a running man
Nitrogen	78%	78%	78%
Oxygen	21%	17%	12%
Carbon dioxide	0.03%	4%	9%

6. Explain the following experimental results:

a) An animal was placed in an apparatus which ensured that it re-breathed the same air continuously. Its breathing rate increased.

b) An animal was placed in the same apparatus but all carbon dioxide was removed from its expired air before it was breathed again. Its breathing rate remained normal for some time and then suddenly increased.

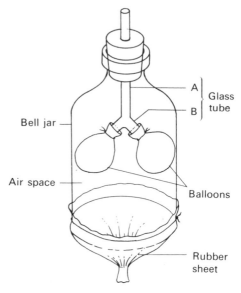

Fig. 9.7   Model of the lungs and thorax (question 7)

7. Study the apparatus illustrated in Figure 9.7, and construct it if possible. When the rubber sheet is pulled downwards the balloons inflate; when it is raised the balloons deflate.

a) To what parts of a mammal's body are the *labelled* parts of the apparatus equivalent?

b) How is each labelled part of the apparatus similar to, and yet different from, the part of a mammal's body to which it is equivalent?

c) How does the method of inflating the balloons resemble the mechanism of lung inflation? How does it differ?

d) Which mechanism of lung inflation is not illustrated by this apparatus?

Summary and factual recall test

Gas exchange is the absorption of (1) and the removal of (2) from organisms. In small organisms this can take place over the (3) because (4). In larger organisms gas exchange takes place at a special region of the body called a (5) surface. The features of such a surface are (6–list five).

In mammals the dome-shaped (7) muscle contracts and becomes (8) in shape. At the same time the (9) contract making the ribs pivot where they join the (10) and (11), so that the rib-(12) lifts upwards and outwards. All these movements (13) the volume of the (14) cavity, and this automatically expands the lungs because (15). This temporarily reduces air (16) in the lungs and they fill with air.

(17) and (18) in inhaled air are trapped in a layer of (19) produced by (20) cells. This substance is moved by the beating movement of (21) to the back of the (22) where it is (23).

The bronchial tree consists of the (24) or windpipe which divides to form two (25) which further sub-divide into (26). These end in millions of bubble-like (27), which are surrounded on the outside by a network of (28). This is where (29) exchange takes place.

Red cells are (30) in diameter than lung capillaries, and this increases the rate of oxygen absorption in two ways: (31) and (32).

10

Homeostasis: the maintenance of a constant internal environment

All the cells of the body are bathed in a liquid called tissue fluid. This fluid comes from the blood; it supplies cells with food and oxygen, and removes their waste products (section 6.4). Tissue fluid forms the conditions in which cells live or, in other words, it forms the **internal environment** of the body.

A number of organs work non-stop to adjust the temperature and contents of tissue fluid so that it is always as near perfect an environment as possible for health, growth, and efficient functioning of cells. The organs which perform this task are said to be involved in **homeostasis** or *the maintenance of a constant internal environment*.

The organs which carry out these constant adjustments act like a boy trying to balance a stick on the end of one finger. The stick is always falling sideways in one direction or another, but it does not fall to the ground if the boy moves his finger quickly enough in the right direction to bring the stick back to an upright position. The boy and the organs of homeostasis can never relax: the finger must be moved constantly, and the organs of homeostasis must work continuously, in an effort to achieve perfect balance.

10.1 Homeostasis and feed-back mechanisms

Stick-balancing and homeostasis are both controlled by **feed-back** mechanisms. The boy keeps his stick upright by attending to messages from his eye and other sense organs. The sense organs feed back information to his brain and from here messages are sent to muscles in his arm and hand which move accordingly. The results of these movements are detected by the same sense organs and this information is again fed back to the brain, which decides on subsequent responses.

The same kind of feed-back mechanism is involved in homeostasis. Sense organs detect a change somewhere in the body's internal environment. Information about the change is fed-back to a control centre, usually in the brain, which sends messages to an organ of homeostasis to make the necessary adjustments. The sense organs detect these adjustments and feed back this information to the control centre, which then modifies its message to the organ of homeostasis (Fig. 10.1).

There is an important difference between homeostatic feed-back and that used by the boy to balance a stick: in homeostasis, feed-back is largely an unconscious activity – that is, the person is unaware that it is taking place.

Organs concerned with homeostasis

Homeostasis is best developed in humans, other mammals, and birds. In these animals there are organs which keep the following features of blood and tissue fluid at a fairly constant level: temperature; dissolved substances such as carbon dioxide, oxygen, food, urea, and various poisonous substances; and osmotic potential.

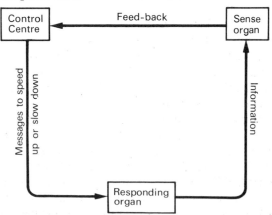

Fig. 10.1 The three parts of a typical feed-back mechanism

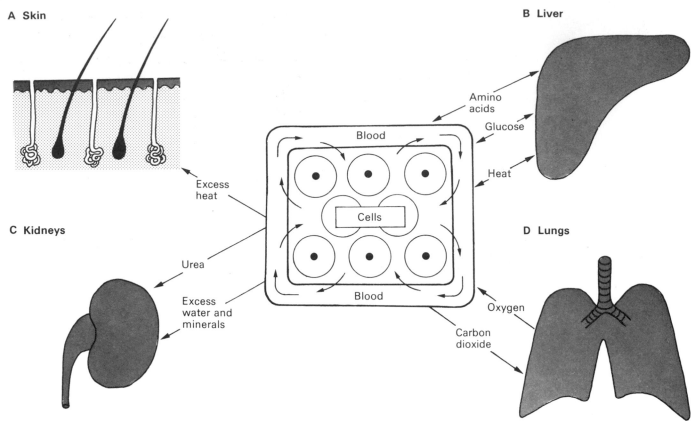

A Skin

B Liver

Amino acids

Glucose

Heat

Blood

Excess heat

Cells

C Kidneys

D Lungs

Urea

Excess water and minerals

Blood

Oxygen

Carbon dioxide

Fig. 10.2 Diagram of homeostasis   A number of organs work non-stop at maintaining the temperature and contents of tissue fluid at constant levels. This activity is called homeostasis or: the maintenance of a constant internal environment. Lungs control the levels of carbon dioxide and oxygen; the liver controls levels of amino acids and sugar; the skin helps keep temperature at about 37°C; and the kidneys remove waste substances such as urea, excess water, and mineral salts

The organs of homeostasis which constantly adjust these features are the lungs, liver, skin, and kidneys. Control of carbon dioxide and oxygen levels by the lungs is described in chapter 9. This chapter describes the part played by the liver, skin, and kidneys in maintaining a constant internal environment in the human body.

10.2   The liver

Before reading this section check the liver's position in the human body from Figure 4.13, and then its position in the circulatory system from Figure 6.6. The homeostatic functions of the liver are illustrated in Figure 10.3.

The liver of an adult man weighs about 6 kg, and is situated immediately below the diaphragm, where it extends from one side of the body to the other. It is dark red in colour owing to the large number of blood vessels which it contains. The liver is the largest gland in the body. It is an enormous chemical 'factory' which makes and releases into the body many useful substances. Obviously a chemical factory needs a large supply of raw materials, and the liver obtains its supplies from blood in the hepatic portal vein. In fact, the liver receives through this vein practically all of the digested food absorbed by the intestine.

The liver's main homeostatic function is to regulate the amount of food which reaches the blood and tissue fluid. It does this mainly by absorbing and storing the food which it receives, and then releasing it into the circulatory system at a rate which depends upon the body's current needs. This, and some additional functions, are described under the following headings.

Regulation of blood sugar

The liver, together with a set of glands in the pancreas, control with great accuracy the amount of glucose in the blood. It is very important that glucose is maintained at a certain constant level, first because this sugar is the body's main source of energy, and second because even slight changes in glucose concentration alter the blood's osmotic pressure and therefore alter the rate at which water moves in and out of the body cells by osmosis.

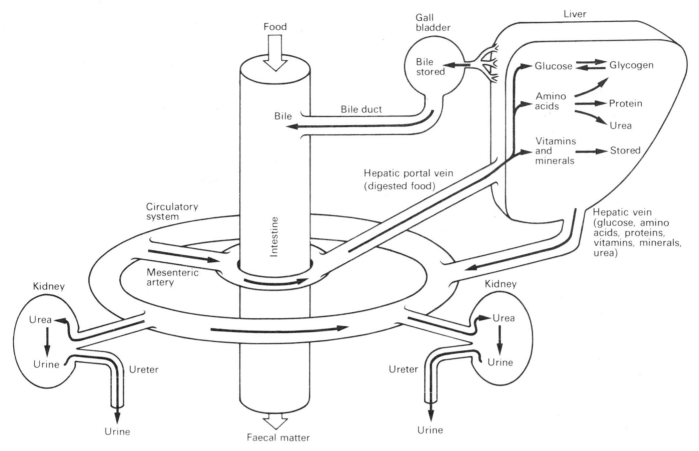

Fig. 10.3 Diagrammatic summary of the functions of the liver and kidneys

In man, the level of glucose in most arteries is normally $85 \, mg/100 \, cm^3$. After a heavy meal of carbohydrates it may temporarily rise to $180 \, mg/100 \, cm^3$. It never falls much below normal except during prolonged starvation.

Whenever there is an increase in the blood's glucose level, glands in the pancreas produce a substance called **insulin**. Insulin stimulates the liver cells to extract glucose from the blood. At first the liver converts this glucose into glycogen and stores it in its cells, but the liver can hold only $100 \, g$ of glycogen. When this limit is reached any remaining excess glucose in the blood is converted into fat and transferred to more permanent storage areas under the skin and around various body organs.

When there is less glucose in the blood than normal the pancreas slows down insulin production. This causes the liver to convert its stored glycogen into glucose, which then passes into the blood. When all the glycogen has been used up in this way stored fat is converted into glucose, and if after prolonged starvation there is no more fat in the body, protein is converted into glucose. In this way the liver keeps the body supplied with food for as long as possible when food is not available elsewhere.

Regulation of amino acids and proteins

The body can store only very small amounts of amino acids and proteins. When a meal supplies more of these than the body can use the liver gets rid of them by a process called **deamination**.

Deamination is the removal from each amino acid molecule of the part which contains nitrogen: that is, the **amino group** (which has the chemical formula NH_2). These amino groups would automatically change into ammonia (NH_3) which is very poisonous, were it not for the action of liver cells which immediately convert them into urea, which is far less poisonous.

Urea then passes from the liver into the blood and is eventually removed from the body by the kidneys. The remaining part of each amino acid molecule, which contains no nitrogen, is either converted by the liver into glucose and respired, or stored.

Storage of vitamins and minerals

The liver stores vitamins A, D, and B$_{12}$, together with minerals such as iron, copper, and potassium, until they are required by the body.

Production of fibrinogen

The liver manufactures an important blood protein called fibrinogen which is vital to the clotting of blood in wounds. In this way the liver indirectly helps to preserve the body's supplies of blood and tissue fluid.

Production of heat energy

The liver produces a great deal of heat as a by-product of the thousands of chemical reactions which take place within its cells. This heat warms the blood as it passes through the liver, which in turn warms body tissues as the blood circulates around the body.

Excretion of bile pigments

The liver excretes (removes from the body) waste substances called bile pigments which are produced during the breakdown of old 'worn-out' red blood cells in the spleen. Bile pigments are excreted into the intestine along with bile.

Purification of blood

Many poisons are produced by metabolism and by germs which get into the body. Other poisons, including certain drugs and alcohol, are deliberately taken into the body. Another group of poisons, such as some food preservatives, lead, mercury, DDT insecticide, are taken unavoidably into the body in food, drink, and when we breathe.

The liver can help rid the body of poisons in many ways. Lead, for example, is excreted in bile. Many poisons, including drugs and food preservatives, are modified by the liver by combining them with other chemicals before they are excreted by the kidneys.

Alcohol, when taken in small amounts, can be respired as a source of energy. When drunk in large amounts alcohol causes serious damage to liver cells. The cells may die and be replaced firstly with fatty tissue, and later by fibrous tissue. Alcohol may also cause cancer cells to develop (see section 21.10).

10.3   Excretion

Excretion is the removal from the body of waste substances which are produced by metabolism, and of substances of which the body has an excess. In other words, excretion is the removal of unwanted substances from the body. Excretion is an extremely important part of homeostasis because if these substances were not removed, they would poison cells or slow down metabolism. The lungs and kidneys are the main organs of excretion.

It is important to distinguish between excretion and two other processes with which it may be confused: defecation and secretion. Fig. 10.4 illustrates the differences. **Defecation** is the removal from the body of indigestible substances; that is, the passage of faecal matter through the anus. Defecation differs from excretion in that faecal matter is *not* produced by metabolism. **Secretion** is the production by metabolism of *useful* substances such as enzymes and hormones.

The most poisonous of all the waste by-products of metabolism is ammonia. Ammonia is formed during the breakdown of excess amino acids in the liver. Ammonia is very soluble and kills cells if its concentration in the blood rises above 1 part in 25 000. The liver immediately converts ammonia into a relatively harmless substance called **urea**, which is released into the blood. The kidneys extract urea from the blood and excrete it from the body as part of a liquid called **urine**. Ammonia and urea contain nitrogen, therefore removal of urea from the body by the kidneys is known as **nitrogenous excretion**. The kidneys are part of a set of organs known as the **urinary system**.

10.4   Excretion and the kidneys

The urinary system

The parts of the urinary system and their position in the human body are shown in Figure 10.5.

Human kidneys are about 12 cm long by 7 cm wide, and are bean-shaped. A thin tube, the **ureter**, comes out of the concave side of each kidney and extends downwards to a single large bag called the **bladder**. The bladder has only one exit, a tube called the **urethra**, which leads to the body surface. The bladder end of the urethra is normally held closed by means of a ring of muscle (a sphincter), which controls the release of urine from the bladder.

Urine drains continuously out of the kidneys into the ureters where it is forced downwards into the bladder by wave-like contractions of the ureter walls. The bladder stretches and expands in volume as it fills with urine, and when it is nearly full the stretching stimulates sensory nerve endings in its walls so that nerve impulses are sent to the brain. This is how a person knows when his bladder must be emptied. The sphincter muscle around the urethra is then voluntarily (i.e. consciously) relaxed to let urine drain

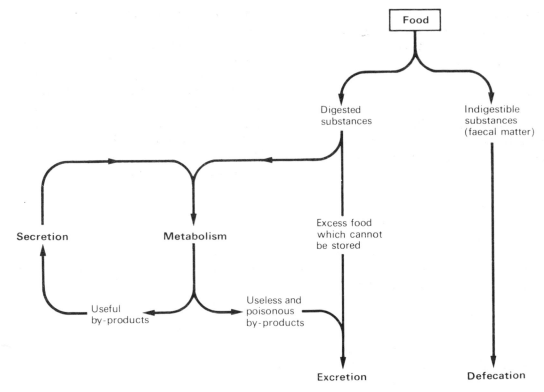

Fig. 10.4 The differences between excretion, defecation, and secretion

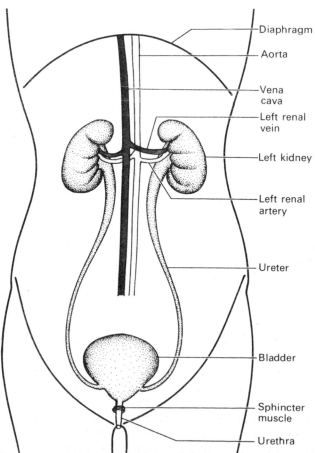

Fig. 10.5 The human urinary system (diagrammatic)

from the bladder, through the urethra, and out of the body. This is called **urination**.

Kidney structure

A human kidney contains about 160 km of blood vessels, and more than a million separate lengths of extremely fine tubes known as kidney tubules, which have a combined length of about 60 km. The arrangement of blood vessels and kidney tubules is best seen by studying, at higher and higher magnification, a kidney which has been cut in half. This is what Figures 10.6 and 10.7 represent.

A kidney has a dark-coloured outer zone called the **cortex**, and a paler-coloured inner zone called the **medulla**. The medulla consists of several cone-shaped areas called the **pyramids**. Urine drains continuously from the tips of the pyramids into funnel-shaped spaces formed by the top of the ureter.

Kidney tubules

In humans, each kidney contains about one million kidney tubules, each of which is about 3 cm in length. Figure 10.6 shows that they begin in the cortex of the kidney, where each one is expanded into a round, cup-shaped object called a

Bowman's capsule, which is about 0.2 mm in diameter. Each Bowman's capsule almost entirely encloses a ball of finely divided and inter-twined blood capillaries known as a **glomerulus** (plural glomeruli). Glomeruli were discovered by Marcello Malpighi, a seventeenth-century biologist who said that they looked like a coil of worms.

Each kidney tubule emerges from the Bowman's capsule on the side opposite to the glomerulus and, after a complicated series of coils and loops (Fig. 10.7), it joins a wider **collecting duct**. Collecting ducts collect urine from the kidney tubules and transport it straight through the medulla of the kidney to the tips of the pyramids.

Blood supply to the kidney tubules   Each kidney receives oxygenated blood at high pressure through a renal artery direct from the main aorta. Inside each kidney the renal artery divides into arterioles which convey blood to glomeruli capillaries with only a small loss in pressure. The blood vessel which leaves each glomerulus branches into a capillary network around the coiled and looped portion of each kidney tubule, before eventually joining the renal vein.

Formation of urine

Urine is formed in two stages. First, blood is filtered into the kidney tubules to form a clear fluid (i.e. a filtrate) which contains the waste substance urea and many useful substances which the body cannot afford to lose. Second, the useful substances are reabsorbed from the filtrate back into the blood leaving only urea and other substances in the kidney tubules which are of no use to the body. The technical words for these two stages are **filtration**, and **reabsorption**.

Filtration   In the kidney the liquid to be filtered is the blood. The blood is filtered by the cup-shaped Bowman's capsules, of which there are more than a million in humans. The blood is filtered through two layers of living membrane: the capillary wall of each glomerulus and the inner wall of each Bowman's capsule. Blood is filtered as it passes through these membranes on its way to the kidney tubule. In humans these membranes have a total surface area of one square metre.

However, there are two important points about filtration in the kidney which should be noted. First, blood is forced through the two membranes at a very high pressure. Second, not all the blood is filtered. In fact, the remaining unfiltered blood has an important function to perform after leaving the glomeruli on its way to the renal vein.

The filtering process in kidneys is very similar to the formation of tissue fluid described in section 6.4. Both processes consist of blood being forced at high pressure through capillary walls, with the result that blood cells and large protein molecules are filtered out (i.e. held back) leaving a clear liquid. In kidneys, this liquid is called **glomerular filtrate**. It enters the cavity of the Bowman's capsules and then drains into the kidney tubules.

A high blood pressure in the glomeruli is essential to the filtering process, and this pressure is achieved in three ways. First, blood entering the glomeruli is already at high pressure because the renal artery takes blood from the main aorta at a point close to the heart. Second, this pressure is further increased by the pressure build-up which results from the fact that the vessel which leaves a glomerulus is narrower than the one which enters it. Third, there is a tiny ring of muscle around the vessel which leaves each glomerulus. These ring muscles automatically alter in diameter to maintain a constant pressure in the glomerular capillaries despite changes in blood pressure in the rest of the body.

In humans, the kidneys filter about 60 litres of blood an hour, and it takes them only five minutes to filter an amount which is equal to the body's entire blood supply. The filtering process produces about 7.5 litres of glomerular filtrate an hour, and this liquid contains not only urea, but many useful substances such as glucose, amino acids, mineral salts, and vitamins, dissolved in a large amount of water. If all of this were excreted the body would lose most of its water and soluble food supplies in a few hours. This does not happen because 99% of the glomerular filtrate is reabsorbed.

Reabsorption   Reabsorption occurs as glomerular filtrate passes out of the Bowman's capsules and down the kidney tubules towards the collecting ducts. All along this route, cells in the kidney tubule walls extract useful substances from the glomerular filtrate so that by the time it reaches the collecting ducts a liquid called **urine** has been formed. Urine contains only waste excretory substances. Humans produce about 1.5 litres of urine daily depending on the amount of liquid they drink. (The full extent of reabsorption is shown in Table 1.)

The useful reabsorbed materials pass back into the blood through the capillaries which surround the coiled part of each kidney tubule (Fig. 10.7).

Reabsorption is not yet fully understood, but it is known to require a lot of energy. For this reason blood which leaves the kidneys has not only lost most of its waste material but some oxygen and food as

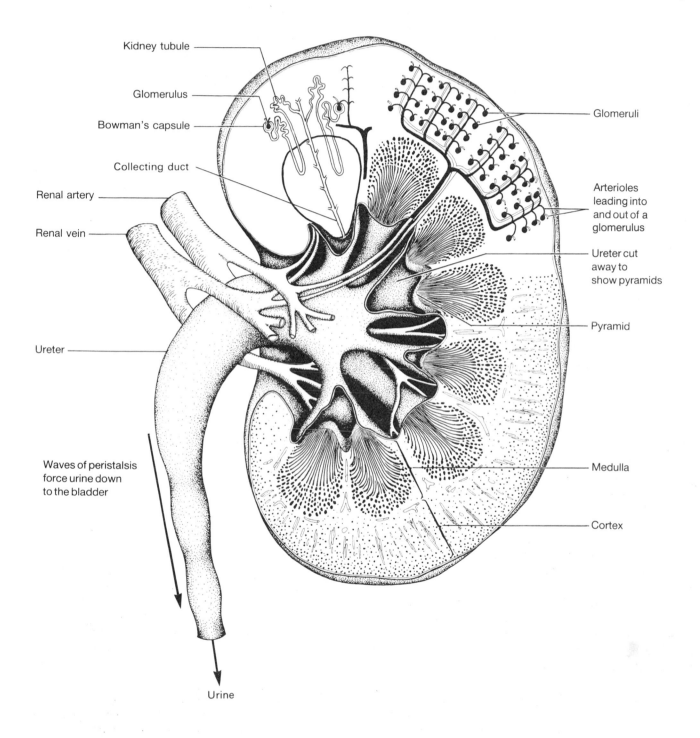

Kidney tubule

Glomerulus

Bowman's capsule

Collecting duct

Renal artery

Renal vein

Ureter

Waves of peristalsis force urine down to the bladder

Urine

Glomeruli

Arterioles leading into and out of a glomerulus

Ureter cut away to show pyramids

Pyramid

Medulla

Cortex

Fig. 10.6 The kidney of a mammal (human) cut in half to show internal structure. (The top half is purely diagrammatic whereas the bottom half gives a more realistic impression of the kidney.)

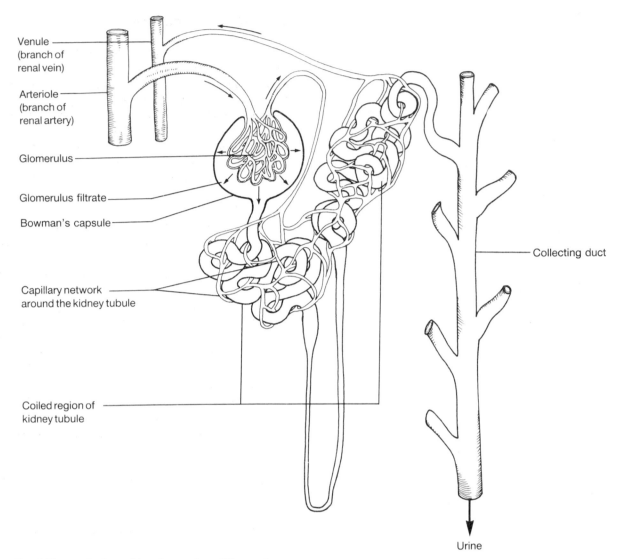

Venule
(branch of
renal vein)

Arteriole
(branch of
renal artery)

Glomerulus

Glomerulus filtrate

Bowman's capsule

Capillary network
around the kidney tubule

Coiled region of
kidney tubule

Collecting duct

Urine

Fig. 10.7 One kidney tubule and its glomerulus. (Fig. 10.8 gives a simplified version)

Table 1 Summary of filtration and reabsorption in humans

| Substance | Daily Output | |
	Glomerular Filtrate	Urine
Glucose	200 g	Trace
Sodium	600 g	6 g
Potassium	35 g	2 g
Calcium	5 g	0.2 g
Urea	60 g	35 g
Water	180 litres	1.5 litres

well, which have been taken up by the kidneys for respiration.

Reabsorption is not merely a rescue operation to save useful materials in danger of being lost from the body. It is also a means of controlling the level of water and dissolved substances in blood and tissue fluid so that the osmotic potential of these liquids remains at a constant level.

The technical name for the process which controls osmotic potential is **osmoregulation**.

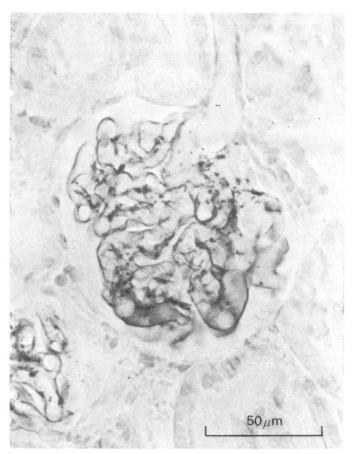

Vertical section through a glomerulus from a rat kidney. The space around the glomerulus is the cavity of the Bowman's capsule

10.5 Osmoregulation

Blood and tissue fluid must be kept at a constant osmotic pressure to avoid unnecessary movements of water in and out of cells by osmosis. If, for example, the osmotic potential of these liquids is too high, cells lose water by osmosis and the body becomes dehydrated.

In chapter 7 it was explained that the osmotic potential of a solution depends upon its strength, that is, the amount of dissolved substance which it contains. For instance, the osmotic potential of a sugar solution depends upon the amount of sugar dissolved in it. If more sugar is added, its osmotic potential decreases. If more water is added, its osmotic potential increases. Therefore, osmotic potential can be controlled in two ways: either by altering the amount of dissolved substance in a liquid, or by altering its water content. It is by these methods that kidneys control the osmotic potential of blood and tissue fluid. Quite simply, they vary the amount of water and dissolved substances which are re-absorbed back into the blood.

Sometimes the osmotic potential of blood drops because it contains too much dissolved food. This can happen after a very heavy meal, or because of failure in the liver-insulin mechanism described in section 10.2, which controls the level of blood sugar. In these circumstances the kidneys do not reabsorb all of the food material from glomerular filtrate, so that the excess amounts pass from the body in urine.

Fig. 10.8 Diagrammatic summary of nitrogenous excretion

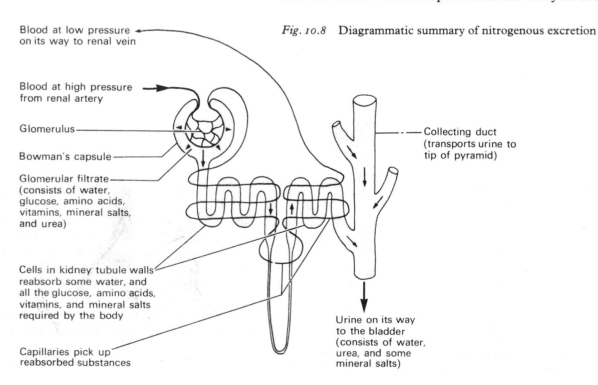

Blood at low pressure on its way to renal vein

Blood at high pressure from renal artery

Glomerulus

Bowman's capsule

Glomerular filtrate (consists of water, glucose, amino acids, vitamins, mineral salts, and urea)

Cells in kidney tubule walls reabsorb some water, and all the glucose, amino acids, vitamins, and mineral salts required by the body

Capillaries pick up reabsorbed substances

Collecting duct (transports urine to tip of pyramid)

Urine on its way to the bladder (consists of water, urea, and some mineral salts)

The osmotic potential of blood also varies according to its water content. If blood contains too much water, as happens when a lot of liquid is drunk, less water is reabsorbed by the kidney tubules. The result is a large amount of dilute urine. On the other hand, if blood osmotic potential drops because too little fluid is drunk then the kidney tubules reabsorb a maximum amount of water, and the result is a small quantity of very concentrated urine. In hot weather this same mechanism ensures that more water is available for cooling the body through perspiration.

To summarize: **the main function of the kidneys is nitrogenous excretion. They also help conserve the body's water supply so that more is available for perspiration in hot weather. In addition, they make fine adjustments to the blood's dissolved contents and this, among other things, maintains a constant osmotic potential in the blood and tissue fluid.**

10.6 Kidney machines and transplants

There are several diseases which affect kidneys. The commonest infections cause inflammation of the kidneys, obstruction of urine flow, or reduction of blood flow into the kidneys. These disorders reduce kidney efficiency and can cause them to fail altogether. When this happens urea and other wastes accumulate in the blood to dangerous levels, which can lead to death.

Treatment of serious cases include the use of a kidney machine to 'clean' the blood, or a kidney transplant.

Kidney machines

A kidney machine receives blood through a tube connected to an artery. Inside the machine, blood flows through **dialysis tubing** which allows small molecules, including urea, to pass through its walls. The 'cleaned' blood is returned to the patient through a tube connected to a vein.

The dialysis tubing is bathed in a liquid similar to blood plasma, except that it lacks the plasma waste substances. Consequently wastes, but not useful substances, diffuse out of the blood and are carried away by the machine.

Kidney machines allow patients with kidney failure to remain healthy provided dialysis is carried out every few days. But this is time-consuming and unpleasant. An alternative treatment is a kidney transplant.

Kidney transplants

Of all the organ transplant techniques, kidney transplants have been the most successful. The best results occur when a kidney is moved from one identical twin to another: their tissue and body chemistry are identical – so the patient's body accepts the new kidney as if it were its own.

Most problems occur when the donor is unrelated to the patient. When this happens the patients body may treat the new kidney as if it were a disease organism, and produce antibodies and white blood cells to destroy it. This is called **tissue rejection**. To avoid this happening the patient is injected with **anti-rejection drugs**. But these weaken the body's ability to fight off infections, so the patient must be kept in a germ-free atmosphere until tissue rejection ceases to be a significant risk.

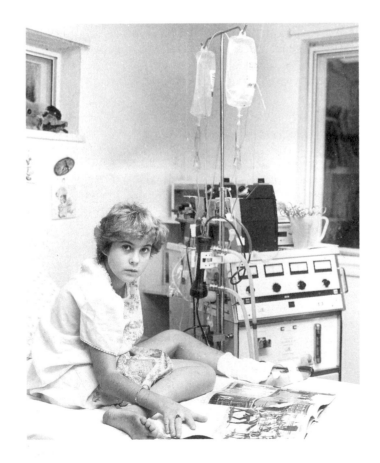

A kidney machine purifies blood by passing it through dialysis tubing. This allows wastes to diffuse out of the blood. But regular dialysis is expensive, time-consuming, and can cause psychological problems. An alternative is kidney transplantation

10.7 Temperature control and the skin

Mammals and birds have a more or less constant body temperature. That is, their body temperature remains about the same despite variations in the temperature of their surroundings. For this reason, mammals and birds are often called 'warm blooded', while all other animals which cannot maintain a constant temperature are called 'cold blooded'. Unfortunately these phrases are very misleading, because there are occasions when so-called warm-blooded animals are cooler than their surroundings, and cold-blooded animals are often warmer than their surroundings. As always, it is better to use accurate technical terms.

Birds and mammals are called **homoiothermic**, which means they have mechanisms of homeostasis that keep their body temperature at a constant level despite changes in their surroundings. All other animals are called **poikilothermic**, which means they have no temperature control mechanism and so their temperature is approximately the same as that of their surroundings. Reptiles and amphibia are examples.

A homoiothermic animal has several mechanisms which work non-stop to balance heat production in its body against heat lost through its skin. This balance is achieved by a temperature control centre in the brain attached to sense organs which are very sensitive to temperature changes in the blood. Whenever such a change occurs this control centre adjusts many different processes concerned with heat production and heat loss, so that a balance is restored and body temperature returns to normal.

Control of over-heating

The body makes adjustments which prevent over-heating under at least two different circumstances: first, whenever conditions outside the body are near to, or hotter than, normal body temperature (37°C in humans); second, whenever there is an increase in heat production by the body which may occur during vigorous exercise, or when the body is fighting a disease. Some of the adjustments which help prevent over-heating in humans are described below and illustrated in Figure 10.9.

1. *Sweating* Sweating is the production of a watery fluid containing dissolved salt from sweat glands in the skin. As sweat evaporates from the skin it has a cooling effect because the evaporating liquid carries away body heat. The evaporation of sweat is an extremely efficient cooling mechanism, and in climates where air temperature approaches body temperature sweating is the only mechanism which can effectively cool the body.

The rate at which sweat evaporates from the body, and therefore its effectiveness in cooling the body, depends on two things: humidity (i.e. the amount of water vapour in the air), and air movements (e.g. winds or fans). Sweat evaporates and cools the body very rapidly in hot, dry, windy conditions. In climates of this type people can tolerate temperatures near to that of their own bodies, and can even take part in vigorous physical exercise without much discomfort. But in hot, humid conditions, especially in still air, sweat evaporates and cools the body very slowly. In climates of this type temperatures near to that of the body can be intolerable, and physical exercise will generate heat that may not be removed from the body. In this case the body temperature rises. The critical body temperature seems to be 41°C, above which sudden collapse and unconsciousness are likely and death may result.

The rapid sweating which results from strenuous exercise in hot climates may cause the loss of up to 30 litres of water per day and 30 g of salt. Loss of water at this rate soon causes the blood to become thick and concentrated so that it no longer circulates properly. Loss of salt causes muscle pains (heat cramp). In hot climates a person must not only drink a lot of fluids; he must also increase the amount of salt in his diet. This is why salt tablets are manufactured for use in the tropics.

When a person is forced to work for long periods in a hot climate her ability to sweat may suddenly fail altogether. This is called **heat stroke**. If the victim is not taken to a cool place immediately her temperature will rise uncontrollably with possibly fatal results.

2. *Panting* Animals of the dog family have sweat glands only in the pads on their paws, therefore evaporation and cooling by sweating is very limited. Their main method of losing heat is to pant rapidly with the tongue hanging out. This causes evaporation from the mouth and lungs and cools the body.

3. *Vasodilation* Vasodilation is the expansion of blood vessels; that is, an increase in their diameter so that more blood flows through them. Whenever the body gets too hot vasodilation occurs in the dense network of capillaries which lie just beneath the epidermis of the skin, i.e. in the **superficial capillaries** shown in Figure 10.9. These capillaries open up and let a large volume of over-heated blood flow very close to the body surface. Here the blood rapidly loses heat by radiation through the skin, which cools the body. This is why a person's skin becomes a flushed pink colour and feels hot to the touch when he is over-heated.

4. *Relaxation of hair erector muscles* The hair erector muscles are shown in Figure 10.9. Whenever heat must be lost from the body, hair erector muscles relax and so the hairs lie more or less flat against the skin. In this position they offer the least possible obstruction to heat loss by radiation and convection. The reasons for this will be clearer after reading what happens to hair in cold weather.

Control of over-cooling

There are similarities between the way in which homoiothermic animals keep warm in cold weather, and the way in which people keep their houses warm in winter. A house can be kept warm by turning on a central heating system, by fitting double glazing, and by laying down felt or other insulating material above the ceilings. More heat is produced inside the house, while at the same time the loss of heat from the house is reduced to a minimum. This is exactly what happens in the bodies of mammals and birds when there is a danger of their body temperature dropping blow normal.

1. *Increased heat production* The 'central heating system' in a homoiothermic animal is the heat generated by metabolism mainly in the liver and muscles, and in particular the heat which comes from the breakdown of food by respiration in these organs. Vigorous exercise warms the body, because it increases the rate of respiration and heat production in the muscles. However, when a person tries to rest in cold conditions his muscles begin jerky or rhythmic movements against his will. This is called shivering, and it is a mechanism which helps to keep the body warm by automatically causing the muscles to generate heat whenever necessary.

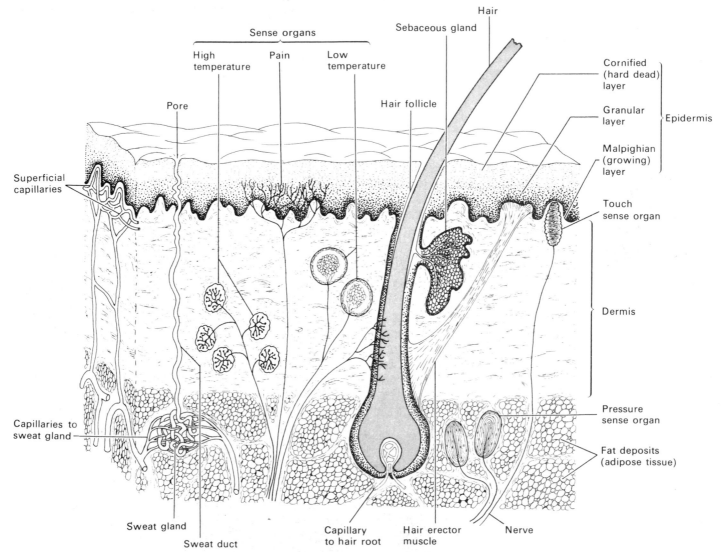

Fig. 10.9 Skin of mammal highly magnified

In addition, there is a general increase in the rate of metabolism during cold weather which brings about an increased appetite for food. This further increases heat output, and helps maintain a constant body temperature.

2. *Reduction of heat loss* The equivalent of double glazing and roof insulation in mammals is hair, and the layer of fat beneath the skin.

a) In cold weather the hair erector muscles contract, which raises the hair shafts to an almost vertical position: i.e. the hairs 'stand on end'. This helps prevent heat loss in two ways. First, the upright hairs prevent cold winds from reaching the skin where they would rob the body of heat. Second, the upright hairs cause a layer of still air to develop around the body. This air is slowly warmed by body heat, and helps to insulate the body against heat loss, since air is a very poor conductor of heat (i.e. heat passes very slowly into it from the body). This mechanism does not work very well in humans owing to their lack of body hair. In fact, without clothes the human body can maintain its characteristic 37°C only at outside temperatures no lower than 27°C. It is only by putting on clothing that humans are able to survive in colder conditions. Putting on a warm winter overcoat is man's equivalent of the contraction of the hair erector muscles in other mammals. Birds achieve a similar effect by means of muscles which make their feathers fluff out.

b) Animals which live in cold climates, such as seals and polar bears, have a very thick layer of fat beneath the skin called **adipose tissue** (Fig. 10.9). This fat is quite effective as a layer of insulation and so helps prevent heat loss from the body. It is also a store of food.

c) The sweat glands cease to operate in cold weather. This reduces heat loss by evaporation, but a small amount of water still evaporates through the epidermis from moist underlying tissues.

d) In cold weather, **vasoconstriction** occurs in the skin's superficial capillaries. The capillaries become smaller in diameter, which restricts blood flow near the body surface and so reduces heat loss by radiation through the skin to a mimimum. This explains why the skin looks pale in cold weather.

A vertical section through a human scalp. Using Figure 10.9 as a guide identify the epidermis, dermis, hair roots, and hair follicles

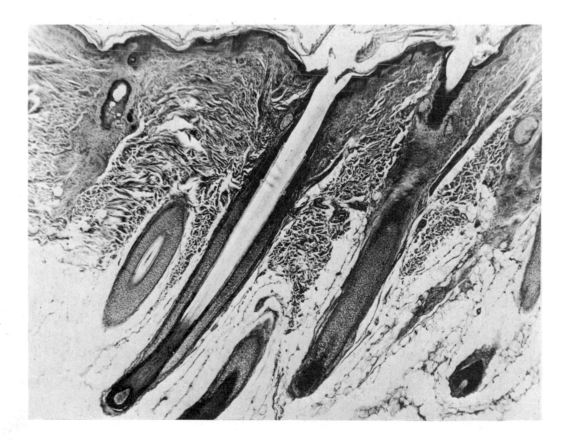

10.8 Hypothermia

Hypothermia is a gradual cooling of the body until, even deep inside, its temperature is well below the normal 37°C.

At 2°C below normal, movements and speech become slow, the victim becomes drowsy, and then unconscious. If the temperature drops much further death can result.

Old people are especially vulnerable to hypothermia if their blood circulation is poor and they are underweight. The main problem is that, because hypothermia slows mental processes, they usually do not realize what is happening to them.

Treatment Victims of hypothermia feel cold all over, even under the arms. They should be wrapped in blankets, given a wrapped hot water bottle, a warm (not hot) sweet drink, and placed in a warm room. Severe cases need hospital treatment.

Avoiding hypothermia People at risk from hypothermia should try to keep one room warm at all times. Windows and doors should be kept shut and drafts excluded. Several layers of clothing are better than one thick layer. The bed and main arm chair should be placed against an inside wall if possible.

Sitting in one place for long periods is dangerous. It is a good plan to spread jobs throughout the day so that circulation is restored at regular intervals.

Investigations

Investigating dialysis

1. Warm a beaker full of water to approximately 37°C.

2. Tie a tight knot at one end of a length of dialysis (Visking) tubing. Use a syringe to fill the tubing with a mixture of starch and glucose solution. Close the other end of the tubing with a paper clip.

3. Rinse the outside of the dialysis tubing with tap water to remove all traces of starch and glucose. Place the filled tubing in a large test tube of warm water (Fig. 10.11). Withdraw some of the water in contact with the tubing with a pipette. Test this water for starch and glucose (as explained at the end of chapter 3).

4. After 20 minutes withdraw some more water in contact with the tubing and test it for starch and glucose. If the results are negative repeat the tests again after 10 minutes.

5. The dialysis tubing in this experiment represents the tubing used in a kidney machine to purify blood.

a) What does the glucose/starch mixture represent?

b) What part of a kidney does the dialysis tubing represent?

c) What do the results of this experiment tell you about the part played by dialysis tubing in purifying blood?

Fig. 10.11 Apparatus for dialysis investigation

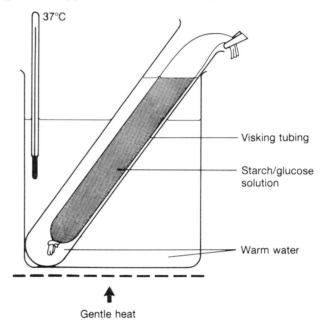

Questions

1. Explain the meaning of 'homeostasis'. Describe two functions of the liver which are examples of homeostasis in action.

2. The graphs opposite show the effects of exercise and immersion in cold water on human body temperature.

a) What was body temperature after 10 minutes exercise, and after 10 minutes immersion in cold water?

b) How long did it take for body temperature to return to normal after exercise, and after the cold bath?

c) What causes body temperature to rise during exercise?

d) Describe all the temperature control processes which ensure that body temperature returns to normal after exercise.

e) Describe the temperature control processes which returned body temperature to normal after the cold bath.

f) What are the advantages to mammals of having temperature control mechanisms?

3. *a*) The human body produces between 75 and 140 litres of glomerular filtrate in 24 hours. The quantity of urine formed in the same period is between 1 and 1.5 litres. Account for the difference.

b) Explain why an increase in an adult's daily intake of protein causes a corresponding rise in the amount of urea in the adult's urine.

c) Why does this happen to a lesser extent if a child eats more protein?

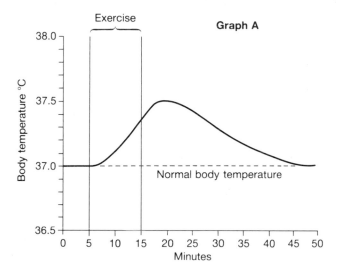

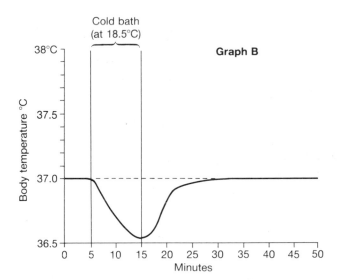

Summary and factual recall test

Homeostasis is the (1) of a constant (2). It is achieved mainly by (3)-back mechanisms. Homeostasis is best developed in (4), but is found in most (5), and in (6). In these animals there are homeostatic organs which keep the following features of (7) and (8) fluid at a constant level: (9–list seven features).

The main homeostatic functions of the liver include the regulation of (10), (11), and (12) in the blood. The liver helps to achieve homeostasis indirectly by storing vitamins (13–list three), and minerals such as (14–list three). In addition, the liver purifies the blood by (15); helps the (16) of blood in wounds by producing (17); produces (18) energy which is distributed by the blood and so (19) body tissues; and excretes (20) pigments.

Excretion is the removal of (21) substances produced by (22), and the removal of substances of which the body has an (23). The kidneys extract a waste substance called (24) from blood and excrete it in a liquid called (25). This is called nitrogenous excretion because (26). Urine is formed in two stages. Blood is

filtered by passing through the walls of capillaries arranged in tiny balls called (27). It then passes into cup-shaped objects called (28). From here the filtrate passes along the kidney (29) where useful substances such as (30–list four) are removed from it and passed back into the blood. This process is called (31).

A homoiothermic animal is one which (32). Examples are (33–name two). All other animals are said to be (34), which means (35). It is misleading to use the phrases 'warm-blooded' and 'cold-blooded' because (36).

When the body is over-heated (37) glands in the skin produce a liquid which (38) and so carries away (39). In addition, the (40) capillaries of the skin undergo expansion, or (41), so that a large volume of (42) blood flows (43) to the body surface where it loses heat by (44). Over-cooling is controlled mainly by an involuntary (45) of the muscles which generates extra (46), and by contraction of the hair (47) muscles which reduces heat loss in two ways: (48) and (49).

11

Sensitivity and movement in plants

When a farmer scatters seeds over the soil he does not worry which way up they land. He knows that even if they land on their sides, or upside-down, they will all automatically send their roots down into the soil, and their stems and leaves up towards the light and air. If a potted geranium is placed on a windowsill it is not necessary to worry which way its leaves are facing. The plant will automatically turn its leaves towards the light, and it will also arrange them so that those nearest the light do not overlap those further back. As a result no leaves are in the shade and the plant receives maximum illumination.

Both these examples and a great deal of experimental evidence show that plants respond to certain kinds of stimulation. In biological terms, they respond to a **stimulus**. Plant roots grow downwards in response to the pull of gravity, and they also grow

A geranium on a window-sill turns its leaves to the light, an example of phototropism

towards water. Plant shoots, on the other hand, respond to gravity in the opposite way by growing upwards, and they respond to light by growing towards it. In short, plants are stimulated by, and can respond to, gravity, water, and light.

The response of a plant to a stimulus is known as a **tropism**, or a **tropic response**. The tropic response to light is called **phototropism**; the response to gravity is called **geotropism**; and the response to water is called **hydrotropism**. This chapter describes tropisms in general with special reference to phototropism and geotropism. A detailed description of hydrotropism is beyond the scope of this book.

11.1 Tropisms

Tropisms are defined as growth movements. This is because the 'movement' involved in the response is produced by a plant's growing points, such as those immediately behind the root and shoot tips. What happens is that the direction of growth alters according to the direction of the stimulus received.

A potted geranium on a windowsill, for instance, detects light coming from only one direction – through the glass. Its stem and leaf stalks respond by growing towards the light, i.e. they bend so that the leaves face the light. But the tropic response is different when a geranium is planted out of doors. Here, it is illuminated from above and it responds by growing straight upwards.

Tropic responses can be either positive or negative. A root, for example, is behaving in a **positively geotropic** manner when it grows downwards in the same direction as the pull of gravity. But shoots are **negatively geotropic** when they grow upwards in the opposite direction from the pull of gravity.

Tropisms are essential to a plant's survival. They enable seedlings to become established with their roots in the soil where they can obtain water and minerals, and their leaves displayed in the air for maximum illumination. Furthermore, tropisms enable plants to alter their pattern of growth at any time throughout their lives to suit changing circumstances. For example, if a plant survives being blown over in a strong wind, new growth curves upwards under the influence of negative geotropism. Similarly, if the water supply from one direction dries up, and the roots can detect water in another direction, new root growth curves in that direction under the influence of positive hydrotropism.

11.2 Phototropism

Phototropism is a growth movement in response to the direction of light; that is, the direction of the growth movement depends on the direction from which the light is coming. Normally, plant stems grow *towards* a source of light, i.e. they are positively phototropic.

Phototropic responses are not normally obvious in plants grown out of doors because they are evenly illuminated from above by the sun and so they grow straight upwards. It is quite possible that the way in which light affects plant growth was first discovered accidentally by farmers who observed some seeds which had grown in the dark, and others which had begun to grow in the shade where they received light from only one direction. If this happened, the farmers would have seen plants which look like those in Figure 11.1. A plant grown out of doors, and evenly illuminated from above, has a short straight vertical stem, and well-developed leaves held horizontally. One grown in the dark has a long thin spindly stem and poorly developed yellow leaves which lack chlorophyll. A plant which is illuminated from only one side has a stem which curves towards the light, and has leaves held at right angles to the light rays.

The responses of plants B and C in Figure 11.1 show how plants are equipped to survive in difficult conditions. If, for example, a seed falls into a shady place the seedling which it produces does not merely die off from lack of light. It uses food reserves stored in the seed to make a long thin stem, and does not waste energy by producing properly developed leaves. Even if it did grow proper leaves they would contain no chlorophyll, because light is necessary for its formation. The light-sensitive stem is able to grow rapidly over or around obstructions towards any source of light until all the food reserves are exhausted. If this race against time is successful and the light is reached before the food reserves run out, the plant produces green leaves and then restores its depleted food reserves by photosynthesis. Phototropism is therefore part of a plant's 'survival kit' in a dangerous world.

Fig. 11.1 The effect of light on the growth of broad bean plants. **A** was grown out of doors; **B** was grown in the dark; **C** was illuminated from one side

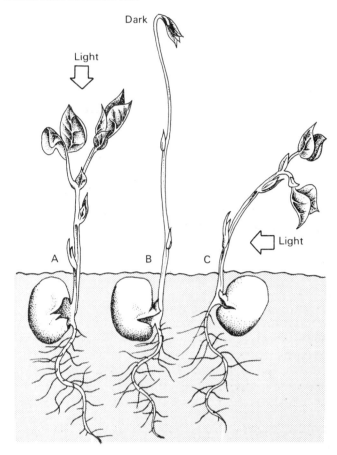

Early experiments on phototropism

One of the first people to make a scientific study of how plant growth is affected by light was Charles Darwin, the English biologist famous for his study of evolution. Late in the nineteenth century Darwin experimented with a type of grass seedling to discover which part of a plant detects light.

Darwin's method, illustrated in Figure 11.2, was to grow three sets of seedlings. The first set had their shoot tips covered with black paper caps; the second set had everything *except* their shoot tips covered with black paper; and the third set was left to grow uncovered, as a control. Darwin then placed all the seedlings near a window where they received light from one direction only.

The completely uncovered (control) seedlings, and those with their shoot tips uncovered, grew towards the light. The seedlings whose shoot tips were covered with black paper caps grew straight up. Darwin concluded that the shoot tip of a plant is sensitive to light. In addition, he assumed that some 'influence' passes from the tip down the stem causing it to bend.

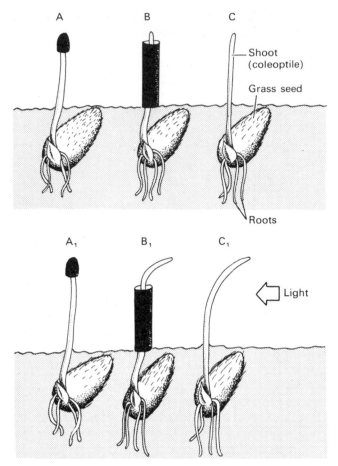

A   B   C

Shoot (coleoptile)

Grass seed

Roots

A₁   B₁   C₁

Light

Fig. 11.2 Darwin's experiment to discover which part of a plant responds to light

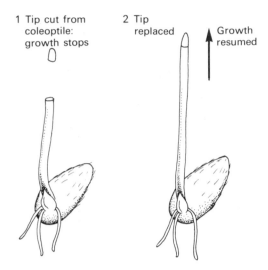

1 Tip cut from coleoptile: growth stops

2 Tip replaced

Growth resumed

Fig. 11.3 Experiment to demonstrate the growth-promoting properties of a coleoptile tip

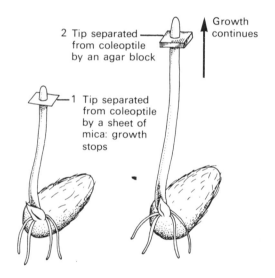

2 Tip separated from coleoptile by an agar block

Growth continues

1 Tip separated from coleoptile by a sheet of mica: growth stops

Fig. 11.4 Experiment to show that a growth-promoting substance diffuses from the coleoptile tip

This experiment is a fine illustration of Darwin's scientific genius. The method is extremely simple and yet it gives clear evidence that shoot tips are sensitive to light. What is more his use of grass seedlings is important because plants in this group germinate (begin to grow) by producing a hollow sheath-like tube known as a **coleoptile**, which encloses and protects the developing leaves until they are free of the soil. Thus, coleoptiles are experimental material which is entirely unencumbered by leaves, side-shoots, and buds. For these reasons. Darwin's technique was borrowed and greatly developed early in the twentieth century.

In 1910 it was discovered that an oat coleoptile stops growing when the tip is cut off. But when the severed tip is replaced a few hours later, growth is resumed (Fig. 11.3). This result gave added support to Darwin's theory that the tip of a plant in some way influences the lower part of a shoot. Could this 'influence' be a chemical, or chemicals?

Figure 11.4 illustrates an experiment which suggests that this is true. Growth stops when the tip of a coleoptile is separated from the rest of the plant by a thin sheet of mica (a substance which does not let chemicals pass through). On the other hand, growth does not stop when a thin piece of agar jelly is placed between the tip and the rest of a plant (agar does let chemicals pass through). Subsequent experiments suggested that the mysterious chemical, or chemicals, which influence plant growth can be collected from a shoot tip. This was done by putting the severed tip of a coleoptile on a block of agar for an hour or so, then discarding the tip and placing the agar block on a coleoptile stump (Fig. 11.5). After this the coleoptile resumed normal growth.

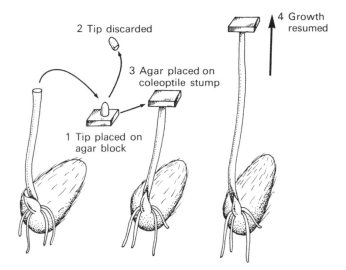

Fig. 11.5 Experiment to show that a growth-promoting substance can be collected from the coleoptile tip

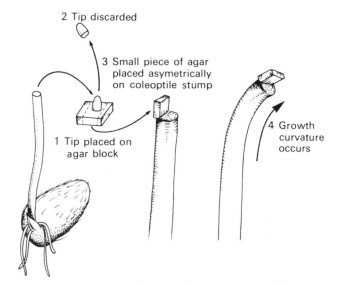

Fig. 11.6 Experiment to show that the growth-promoting substance from a coleoptile tip can cause a growth curvature

These experiments were the beginning of a long line of investigations which eventually led to the discovery that growth in a coleoptile is controlled by a plant hormone of a type known as **auxin**.

There is now strong evidence that auxin is produced by the region of dividing cells which exist at the tip of a coleoptile, and it is now clear that auxin is produced in the growing tips of most other plants. The auxin passes downwards where it stimulates the growth of cells lower in the shoot. Under the influence of auxin cells grow lengthwise (i.e. they elongate) which results in a corresponding increase in the shoot's length. In other words, auxin promotes growth in plant shoots by increasing the rate of cell elongation. Auxin is the mysterious 'influence' discovered by Darwin more than a hundred years ago.

Auxin and phototropism

What is the relationship between the growth-promoting properties of auxin, and the behaviour of a plant undergoing a phototropic response? A possible answer is illustrated in Figure 11.6. This shows that, under certain conditions, auxin will cause a stem to grow in a curve just as if it were responding to light from one direction.

The tip of a coleoptile is cut off and placed on a block of agar for an hour or so. The tip is then discarded and the block placed on a coleoptile stump a little to one side of centre, i.e. asymmetrically. In time the coleoptile grows in a curve as shown. The most popular explanation of this result is as follows. Auxin diffuses from the block down only one side of the shoot. Since auxin promotes cell growth, there will be a faster rate of growth on this side of the shoot than on the other side which presumably has less auxin diffusing through it. If this is so, then growth curvature results from an unequal rate of growth (i.e. cell elongation) under the influence of auxin.

Figure 11.7 illustrates an experiment which suggests that light may bring about a growth curvature by causing more auxin to pass down one side of the stem than the other. The method is to expose a coleoptile tip to light from only one direction. The tip is then placed in the dark on two agar blocks which are separated from each other by a strip of mica. When, still in the dark, these two blocks are placed on coleoptile stumps, the block from the side of the tip which was furthest from the light causes a greater curvature than the block from the side nearest the light. A generally accepted conclusion is that light somehow causes more auxin to gather on the side of the coleoptile tip furthest from the light, i.e. the shaded side. Consequently, more auxin passes into the block under the shaded side of the tip, and it is this block which causes the greatest amount of curvature when placed on a coleoptile stump.

To summarize phototropism so far: **it is thought that light affects the direction of growth in most plants by somehow controlling the distribution of auxin in the shoot tip, so that more auxin gathers on the side furthest from the light. Therefore more auxin passes down the shaded side of the shoot where it increases the rate of cell elongation and causes a growth curvature towards the light.**

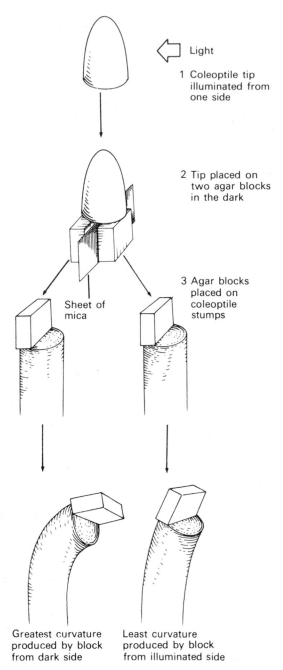

Light

1 Coleoptile tip illuminated from one side

2 Tip placed on two agar blocks in the dark

Sheet of mica

3 Agar blocks placed on coleoptile stumps

Greatest curvature produced by block from dark side

Least curvature produced by block from illuminated side

Fig. 11.7 Experiment to show that exposure to light from one direction causes the redistribution of growth-promoting substances in a coleoptile

11.3 Geotropism

Geotropism is a growth movement in response to the pull of gravity. When a young bean seedling, for example, is placed in a horizontal position its shoot curves upwards and its root curves downwards. Therefore the root is said to be positively geotropic and the shoot negatively geotropic.

This experiment is more informative if the seedling's root and shoot are marked with lines at 1 mm intervals, as illustrated in Figure 11.8. The spacing of these lines after curvature has taken place shows that the actual bending occurs within only a small region behind the root and shoot tip. There is a way of showing that this region is, in fact, the only part of a plant which can form a growth curvature. The method is simply to turn a seedling from the above experiment through 90° (drawing 3) which brings the root and shoot once more into a horizontal position. The plant does not respond by straightening out existing growth curvatures, it produces new curves and these are again in the region behind the root and shoot tips. The explanation is that curvatures can form only in these regions because they are the only parts where the majority of cells are growing by elongation. Elsewhere cells are either dividing, e.g. at the extreme tips of the root and shoot, or are fully grown and often specialized for a particular function.

Roots grow downward no matter which way up a seed is planted

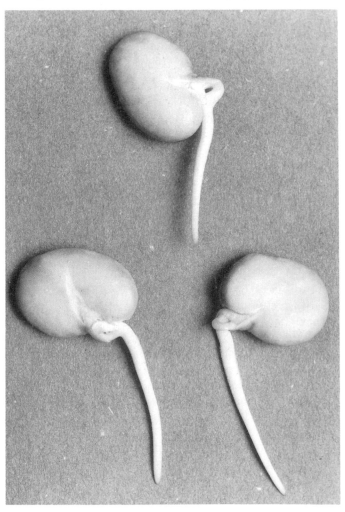

Fig. 11.8 Experiment to demonstrate geotropism in root and shoot

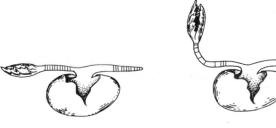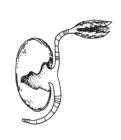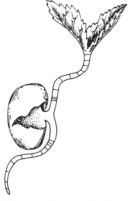

1 Bean seedling placed in a horizontal position in the dark, with markings at 1 mm intervals

2 The shoot develops an upward (negatively geotropic) curvature, while the root develops a downward (positively geotropic) curvature. The markings indicate where cell elongation has occurred

3 Seedling rotated through 90°

4 Old growth curvatures remain unaltered because they are established (differentiated) tissue. New curvatures develop only behind the root and shoot tips

The seedlings described above seem to be responding to the downward force of gravity, and yet scientific method demands a control experiment to give actual evidence that this is so. A control must somehow be devised which provides a plant with a force equivalent to gravity which pulls equally in all directions rather than just downwards. This is done by using a **clinostat** (Fig. 11.9). A clinostat consists of a cork disc which can be rotated at various speeds by a motor. This machine can be set up as shown to rotate a seedling in a horizontal position. When the speed of rotation is set at about once per hour, the force of gravity acts equally on all parts of the plant. Furthermore, the seedling does not remain in any one position long enough to make a geotropic response. Hence, its shoot and root grow horizontally. At slower speeds of rotation the plant does have time to begin a geotropic response but the direction of the stimulus changes while this is happening. The result is that the root and shoot develop twisted curvatures. like a corkscrew.

There is evidence that auxin plays a part in geotropic responses. One theory is that when a plant is placed in a horizontal position the force of gravity causes auxin to gather in the lower half of the root and shoot (Fig. 11.10). Presumably the auxin accelerates cell elongation in the lower half of the shoot thereby causing an upward curvature. On the other hand, if auxin becomes distributed by gravity as illustrated then it must have an opposite effect on root cells: in order to produce a downward curvature auxin must retard cell elongation in the lower half of the root. Recent investigations suggest that this is an oversimplified explanation of geotropism. Apparently auxin is only one of several plant hormones at work in geotropic responses.

Fig. 11.9 A clinostat in operation

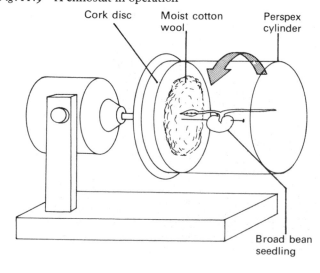

Cork disc   Moist cotton wool   Perspex cylinder

Broad bean seedling

11.4 Practical uses for plant hormones

Biochemists are now able to make a range of synthetic plant hormones which are useful to gardeners, and in agriculture.

Fig. 11.10 Gravity is thought to cause a redistribution of auxin in root and shoot

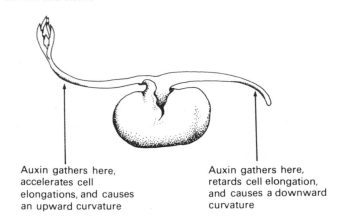

Auxin gathers here, accelerates cell elongations, and causes an upward curvature

Auxin gathers here, retards cell elongation, and causes a downward curvature

The action of selective weed killer

Weedkillers

Synthetic hormone weedkillers work by making a plant grow so fast that it quickly exhausts itself and dies. Most hormone weedkillers are **selective**, which means that they kill certain plants but leave others unharmed.

One of the best-known selective hormone weed-killers is used on lawns. It kills broad-leaved weeds, such as daisy, plantain, and dandelion, and not grasses. Selective weedkillers are also available to kill grasses but not broad-leaved plants.

Selective weedkillers act at very low concentrations, so it is important to wash out the watering can used to dispense these chemicals, before using it again. You could damage or kill valuable plants in this way.

Rooting compounds

There are a number of synthetic hormone compounds which promote rapid root growth in stem cuttings. A length of stem is cut from a plant and its lower leaves are removed. The cut end of the stem is dipped into the hormone compound and then planted in damp compost. The hormone stimulates root growth so that they appear much sooner than in untreated cuttings.

Growth inhibitors

Biochemists have extracted chemicals from plants which slow down, or inhibit, growth. These have several commercial uses. They can be sprayed onto potatoes to prevent sprouting during transport and storage. They can also be sprayed onto hedges like privet; this slows down leaf growth so the hedge does not have to be trimmed so often.

Investigations

A *To verify that light affects stem growth*

1. Germinate four sets of mustard seeds in small dishes on moist cotton wool in the dark.

a) Keep one dish of seedlings permanently in the dark.

b) Put another dish of seedlings in a cardboard box with a circular hole about 3 cm in diameter cut in one end. Position the box so that light shines into the box through the hole.

c) Arrange a clinostat so that its disc is in a horizontal position and set it to rotate about once an hour. Place a dish of seedlings on the rotating disc, then put the whole apparatus in a cardboard box with a hole cut in one end, as in (b).

d) Put another dish of seedlings out of doors.

e) After a few days compare the shape of plants in each of the four situations, and explain the results.

f) From the position of the leaves in experiment 1(b) explain why leaves are sometimes described as 'dia-phototropic'?

B *An investigation of phototropic responses*

1. *An 'obstacle course' for plants*

a) Prepare three cardboard shoe-boxes as shown in Figure 11.11.

b) Put a dish of germinating mustard seeds at the end of one box furthest from the hole.

c) Put a pot of germinating broad beans at the same end of another box.

d) Put a pot containing a sprouting potato at the same end of another box.

e) Compare the ability of each type of plant to reach the light. Why are some more successful than others?

Fig. 11.11 An obstacle course for plants (exercise B1)

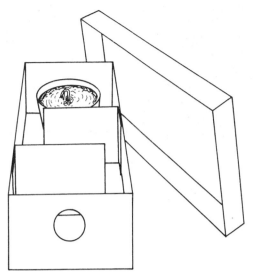

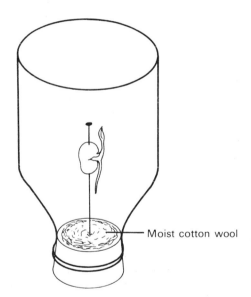

— Moist cotton wool

Fig. 11.12 Apparatus to demonstrate positive and negative phototropism (exercise B2)

f) Examine and explain the shape of each type of plant at the conclusion of the experiment.

g) Devise other types of 'obstacle course' to test the ability of plants to reach light.

2. *Positive and negative phototropism*

a) Prepare a jar with a cork bung as shown in Figure 11.12 and then cover the whole jar with black paper except for a narrow vertical slit down one side. (Alternatively cover the jar with black paint except for a slit down one side.)

b) Place the jar in a position where light will shine through the slit.

c) Explain the growth curvatures which occur in the bean root and shoot.

C   *To verify that gravity affects the growth of plants*

1. *a*) Prepare a clinostat as shown in Figure 11.9, and place it in the dark.

b) Using different seedlings each time run the clinostat for two or three days at different speeds. Repeat the experiment without rotating the clinostat.

2. *a*) Explain the results obtained in each experiment.

b) Why must the clinostat be placed in the dark?

c) The moist cotton wool provides the seedlings with a uniformly humid atmosphere. Give at least two reasons why this is a necessary part of the experiment.

D   *To verify that auxin affects growth curvatures*

1. Germinate three sets of oat or wheat grains in small dishes on moist cotton wool in the dark. Put about five grains in each dish. Obtain some commercially prepared lanolin paste containing the auxin indoleacetic acid (IAA for short), and some plain lanolin. *All* of the following operations must be carried out in red light, after which the plants must be kept in the dark.

a) After about five days, when the coleoptiles are about 2 cm long, select at least two of the straightest plants in each dish and cut down the rest. Do not use any plant in which the leaves have broken through the coleoptile sheath.

b) Leave one set of plants untouched as the control group. Smear a small quantity of warm lanolin with IAA down *one* side of another set of plant coleoptiles. Repeat this operation on the third set of plants using the plain lanolin.

c) Observe the plants again (in red light) after about three hours and explain any growth curvatures which have taken place.

d) Coleoptiles are insensitive to red light. Why is it necessary to observe the plants only in light of this colour?

Questions

1. Oat coleoptiles were used in the experiments illustrated below. The experiments were conducted in the dark.

   a) What can you conclude from the results of experiment P?

   b) Why was plain agar used on coleoptiles C of experiment Q?

   c) What can you conclude from the results of experiment Q?

   d) From the results of experiment R what can you conclude about the effects of auxin on coleoptiles?

2. Auxin is known to reduce the plasticity of cell walls in a coleoptile. That is, under the influence of auxin the cell walls are more easily stretched, and they do not spring back to their original length (like elastic) when the stretching ceases.

   Light from only one side causes auxin to gather on the opposite (shaded) side of a coleoptile.

   Cells continue taking in water by osmosis until their walls can stretch no further.

   Use these three facts to *explain* the following:

   a) the elongation of a coleoptile;

   b) the fact that elongation ceases when the tip is removed;

   c) the curvature of a coleoptile when illuminated from one side;

   d) the fact that once such a curvature has formed it becomes a permanent feature of the plant;

   e) the fact that no bending occurs when a coleoptile is illuminated from one side provided its tip is covered by a black paper cap;

   f) the fact that shading the tip does not prevent elongation.

3. Why is statement (a) below a more accurate, and more scientific description of tropisms than statement (b)?

   a) Shoots bend towards the light because light causes an unequal distribution of auxin in the stem, which in turn causes an unequal rate of cell elongation.

   b) Shoots bend towards the light in order to receive more light for photosynthesis.

Experiment	Start of experiment	After 10 hours
P	Tip cut and replaced / Tip cut off — A, B, C	A, B, C
Q	Agar with auxin / Plain agar — A, B, C	A, B, C
R	Agar with auxin — A, B, C	A, B, C

Summary and factual recall test

Tropisms are defined as (1), because (2). The tropic response to light is called (3). Three sets of cress seedlings were treated as follows: set A was placed in the dark; set B was placed in a box with a slit cut out at one end; set C was evenly illuminated from all sides. The tallest set was (4). Set (5) had yellow leaves. The stems of set (6) were curved. Set (7) was the control group and had (8) leaves. These plants were necessary to show that (9). It is concluded that shoots are (10) phototropic, and that light is required for (11) formation.

Response to gravity is called (12). A bean seedling was placed in a horizontal position and its shoot curved (13), showing a (14) response, while its root curved (15) showing a (16) response.

There is evidence that tropisms are controlled by a hormone called (17). Light, for example, is thought to control the (18) of plant growth by affecting the distribution of this hormone at the shoot (19), so that more of it passes down the shaded side of the shoot where it (20) the rate of cell (21) thereby causing growth curvature (22) the light.

12

Support and movement in animals

Without support of some kind, the bodies of all but the smallest creatures would collapse. Jelly fish, sea anemonies, and other soft-bodied aquatic creatures are almost entirely supported by the water in which they live. In other animals the support which they require comes from a **skeleton** of some kind.

12.1 Types of skeleton

Soft-bodied creatures

Soft-bodied land animals, like earthworms, slugs, and caterpillars, are supported by liquid contained in their cells and in spaces inside their bodies. Animals supported in this way are said to have a **hydrostatic skeleton**.

If, for example, a caterpillar's skin is punctured, its supporting liquid runs out and it deflates like a punctured tyre.

External skeletons (exoskeletons)

Arthropods such as crabs, lobsters, flies, and beetles are supported by a skeleton of hard dead material which forms on the outside of their bodies like a suit of armour. An outer skeleton is known as an **exoskeleton**.

In crustaceans (e.g. crabs and lobsters) the exoskeleton is hard and heavy because it contains lime; whereas in insects the exoskeleton is made mostly of tough but comparatively light material called **chitin**.

All exoskeletons are made up of variously shaped plates and tubes held together at the joints by flexible membranes. Therefore, despite its hard exterior, the animal has considerable freedom of movement. The muscles and other living tissues of the animal are firmly attached to the inside of the skeleton.

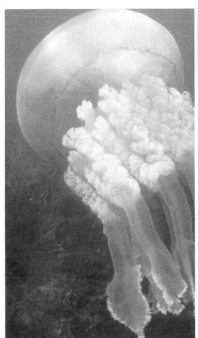

A jellyfish supported by water

A slug with a hydrostatic skeleton

A crab with a hard exoskeleton

An insect moulting

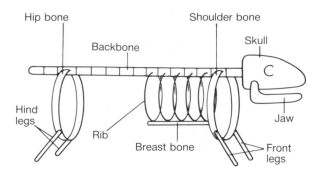

Fig. 12.1   General plan of an internal skeleton   All vertebrates have a backbone (vertebral column) and a skull. Land vertebrates also have limbs. Shoulder and hip bones hold the limbs in place and take the weight of the body

Living permanently inside a suit of armour causes problems during growth. Arthropods overcome this problem by **moulting**, or **ecdysis**. When they become too big for their exoskeleton they break it open, climb out of it, expand in size and grow a completely new one. This happens several times as they grow to adult size.

Internal skeletons (endoskeletons)

All animals with backbones (mammals, birds, reptile, amphibians, and fish) are supported by a hard skeleton of bones which forms inside their bodies. An internal skeleton is called an **endoskeleton**.

There are many different sizes and shapes of bones in an endoskeleton. All vertebrates (animals with backbones) have a **skull**, and a backbone or **vertebral column** made up of small bones called **vertebrae**. In addition, land vertebrates usually have a **rib cage**, **limb bones**, and **shoulder** and **hip bones** which hold the limbs in place (Fig. 12.1).

12.2   The mammal skeleton

Mammals and all other vertebrates have a skeleton based on the following plan.

General plan of skeleton

The skull, backbone (vertebral column), and rib cage are called the **axial skeleton** because they form the main axis of the body. The shoulder blades (pectoral girdles), hips (pelvic girdles), and the arm and leg bones form the **appendicular skeleton** (Figs. 12.2A and 12.2B).

Functions of the skeleton

The skeleton forms a rigid framework which supports all the soft parts of the body and maintains body shape. Some soft tissues are protected by the skeleton: the skull protects the brain, inner and middle ears, and nasal organs. The eyes are partly protected by sockets in the skull called **orbits**. The heart, major blood vessels, and lungs are protected by the rib cage. Marrow tissue inside the long bones and ribs manufactures red and white blood cells (Fig. 12.3).

One of the main functions of the skeleton is to act as a system of rods and levers which are moved by muscles.

Bones

Bones are alive. Or, to be more precise, bones consist of living cells surrounded by hard, dead mineral substances. These minerals are the building materials of the skeleton which gives bones their strength and shape.

The mineral part of the bone is mainly calcium with some carbonate and phosphate. These minerals are not permanently fixed in bones like the minerals in a piece of rock. They are replaced all the time so that bone structure is constantly changing. If an animal is deprived of food calcium and phosphate are removed from the bones and transported to other parts of the body where they are more urgently needed. Bones can lose up to a third of their mineral

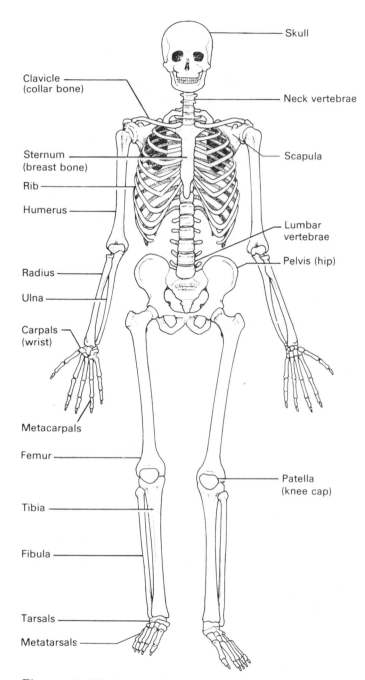

Fig. 12.2A The human skeleton

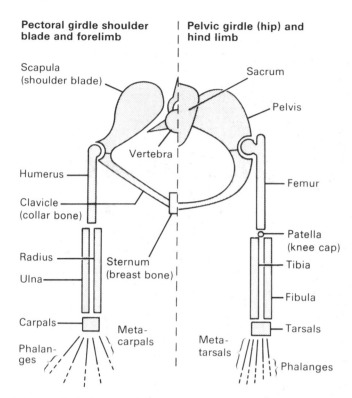

Fig. 12.2B Diagram of the appendicular skeleton

content during periods of starvation, which makes them soft and easily broken. In well-fed, healthy young mammals, however, the minerals are added to, and rearranged in such a way that bone structure slowly adapts to the constant stress it receives from gravity and the pull of muscles. As a result, the bones become perfectly shaped to perform special functions. Bone structure can change even in adults. For example, bones in the right arm of a right-handed tennis champion become stronger than those in his left arm.

The mechanism which constantly changes bone structure also has the power to repair bones when they break or fracture. Breakage is unlikely under normal conditions because bone is rigid but not brittle. Indeed, in some respects bone is almost as strong, weight for weight, as mild steel. The tremendous strength of certain regions of the human skeleton is illustrated by the capabilities of world champion weight lifters. Using a technique known as 'back lift', they can lift weights of up to 3000 kg, which is equivalent to three average-sized family cars!

At the same time bone is very light. This is important because less energy is required to move a light skeleton, and less material is required to make it. The large bones, like the thigh bone (femur), are light for their size partly because they have hollow **shafts**, and also because at their ends, or **heads**, the bone is full of holes like a sponge (Fig. 12.3). This hollowness does not weaken a bone very much. In fact bone absorbs most of the stresses to which it is subjected over its surface regions, so a hollow bone does almost as well as a solid one. The hollow region inside a bone is occupied by tissue which manufactures red and white blood cells.

A Human femur
(thigh bone)

Head (articulates
with pelvis)

Ridges
where
muscles are
attached

Shaft

Ridges (articulate
with tibia at
the knee joint)

B Longitudinal section

Articular cartilage

Light 'spongy'
bone

Hollow shaft
(contains bone
marrow)

Thick compact
bone

The 'spongy' bone is
concentrated in areas which
are subjected to greatest stress,
and is specially arranged in
patterns which will absorb
stress

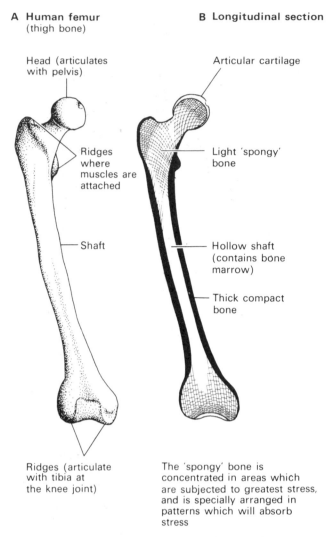

Fig. 12.3 Structure of a bone (human thigh bone)

Longitudinal section through the upper end of a human femur.
Identify the parts using Figure 12.3 as a guide

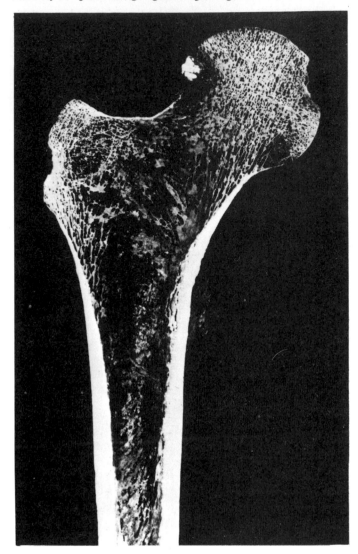

12.3 Joints

Joints occur wherever two or more bones touch. In some cases the bones are joined firmly together by fibrous tissue and may even have their edges dove-tailed into one another; this is the case with the flatt-ened bones which make up the roof of the skull. Other joints have a pad of flexible gristle (cartilage) bound between the bones, and slight movement is possible, e.g. the places where ribs pivot against the sternum (breastbone) during deep breathing (Fig. 9.4B), and between adjacent vertebrae in the backbone. In the latter case the cartilage pads or **intervertebral discs**, act as shock absorbers. That is, they absorb shocks and jolts transmitted to them through the limbs during running, jumping, and similar activities.

Without these discs, shocks would clatter along the line of vertebrae like the shock waves which pass down a line of shunted railway carriages.

There are also about seventy freely moveable or **synovial joints** in the mammalian skeleton. Synovial joints are so called because the region where one bone rubs against the next is enclosed within a capsule filled with **synovial fluid** (Fig. 12.4A). This fluid lubricates the joint when the bones move. Another feature of synovial joints which reduces friction is the layer of slippery **articular cartilage** which covers the surfaces of the bones that rub together, i.e. the articular surfaces. The whole joint is enclosed within a layer of tough fibres, the **ligament**, which holds the bones firmly in place and yet allows free movement of the joint.

Types of movement at the joints

Imagine two players engaged in a game of tennis. They are using all of their seventy-odd synovial joints and more than six hundred muscles to achieve the agility required to hit and recover the tennis ball. By analysing body movements in such a situation it is possible to learn a great deal about the marvellous flexibility of the human bone and muscle machinery. Study Figure 12.2 in conjunction with the following notes.

Hinge joints The elbows, knees, and knuckle joints of the finger are examples of hinge joints. They move, like the hinge of a door, in one plane only.

Test the flexibility of the hinge joints by exercising your elbows, knees, and knuckles. Think about the advantages and disadvantages of this particular joint.

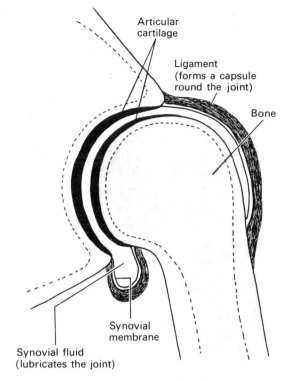

Fig. 12.4A Structure of a synovial joint (human shoulder joint)

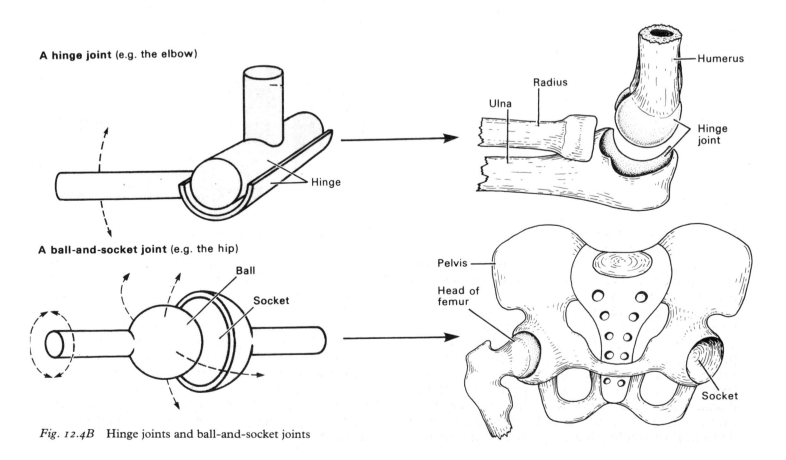

Fig. 12.4B Hinge joints and ball-and-socket joints

Ball-and-socket joints The shoulders and hips are examples of ball-and-socket joints. Here the rounded head of one bone fits into a cup-shaped socket in another. These joints give the greatest flexibility of movement of all joints (Fig. 12.4B).

Other types of joint There is an intricate arrangement of small sliding surfaces between adjacent vertebrae. These permit the backbone to bend at the waist and neck in all directions, but particularly forwards, together with a certain amount of twisting. Movements between vertebrae add considerably to the body's flexibility. To verify this point think how restricted one's movements would be if the backbone were a solid rod. The greatest amount of flexibility in the backbone exists where the skull articulates with (i.e. moves against) the topmost vertebrae. Here, the **atlas vertebra** takes the weight of the skull (like the Greek god Atlas, who according to mythology balanced the world on his shoulders). This articulation permits nodding movements of the head in all directions. Below the atlas, the **axis vertebra** permits swivelling or pivoting movements of the head, which occur, for example, when a person shakes his head to say no.

Another pivot joint occurs at the head of the radius bone near the elbow (Fig. 12.4B). This permits the forearm to twist, as it does when turning a screwdriver.

12.4 Muscles and movement

Muscles make up 40–50% of body weight in most mammals. They are the 'meat' of the body, and consist largely of protein, which is why they are a valuable food.

The muscles which move the body are referred to as **voluntary**, or **skeletal muscles**. They are under conscious control, and are attached to bone at both ends. The point of attachment between a muscle and a bone is called a **tendon** (Fig. 12.5). Tendons consist of very strong inelastic (i.e. non-stretchable) fibres which begin inside the bone and penetrate deep into the muscle tissue, attaching it firmly to the bone. In freshly killed meat, tendons appear as glistening silver-grey strands between bone and muscle.

When a muscle contracts it exerts tension between its two points of attachment. One of these points remains fixed. This is the anchorage point, or **origin**, of the muscle. The other end moves as a result of this tension, and is called the **insertion** of the muscle. The origins and insertion of the biceps muscle are illustrated in Figure 12.5.

In mammals, as with all other animals described in this chapter, muscles work opposite each other in **antagonistic systems**. In each of these systems one set of muscles causes the bending of a joint. These are the **flexor** muscles. The **extensor** muscles are those which work in the opposite direction to straighten the joint. Flexor and extensor muscles are illustrated in Figure 12.6A and B.

Sometimes both sets of muscles in an antagonistic system may contract at the same time. This is done to lock the joint at a particular angle. An example of this is the locking of arm muscles to hold something firmly in one particular position so that it doesn't move in any direction.

In locomotion the muscles do not work at random but in a precisely ordered sequence controlled by the nervous system. This sequence generally has to be learned, which is one reason why babies cannot walk at birth.

Even when a person is sitting perfectly still many of her muscles are at work to counteract the force of gravity on the body. These muscles are responsible for body posture, and if they were to stop working suddenly a person would collapse to the floor like a rag doll. The main posture muscles are those which run from the hip bone (pelvis) to the back of the lumbar vertebrae. These are the muscles which may become fatigued and cause backache if the body has to stoop forwards for prolonged periods.

Fig. 12.5 Muscles and bones of the elbow joint

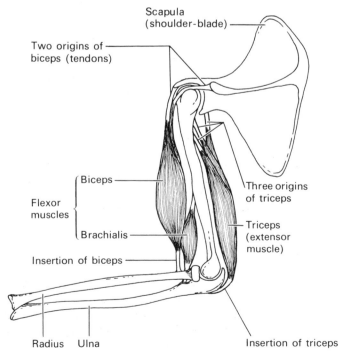

A Action of a flexor muscle

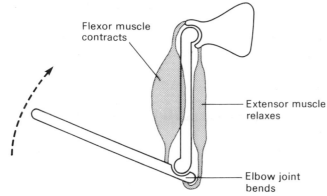

Flexor muscle contracts

Extensor muscle relaxes

Elbow joint bends

B Action of an extensor muscle

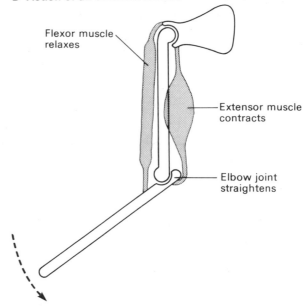

Flexor muscle relaxes

Extensor muscle contracts

Elbow joint straightens

Fig. 12.6 Muscles pull against the skeleton, bending or straightening it at the joints   Muscles which bend joints are called flexor muscles, and muscles which straighten joints are called extensor muscles. Flexor and extensor muscles pull in opposite directions and are said to form an antagonistic system

Investigations

A *An investigation of exoskeletons and movement*
  1. Obtain a live crab from the sea-shore. Keep it in shallow sea-water in an aquarium, and provide it with a pile of stones under which it can hide. Feed it on small pieces of fish or meat, but do not leave uneaten food in the tank; it will decay and foul the water.
  a) Watch the animal's movements, noting the range of movements at each leg joint.
  b) Always return these animals to the sea-shore when the study is finished.
  2. Obtain a dead crab or lobster.
  a) Examine all its leg joints, noting the range of movements which they possess.
  b) Detach the legs, open the 'body' from the top by sawing through the skeleton with a sharp knife, and remove all the internal organs. Boil the skeleton in several changes of water until only the hard parts remain. Study the *internal* structure of the skeleton and try to explain its functions. Is it strictly accurate to describe a crab's skeleton as 'external'.
  3. Obtain a wide variety of insect specimens (or pictures), e.g. water-beetle, 'mole' beetle, grasshopper, praying mantis etc.
  a) How do the various leg structures suit the insects' particular ways of life?

B *An investigation of endoskeletons by the transparency method*
  It is possible to study complete intact skeletons of small animals by making their soft tissues transparent, and staining their bones deep purple. The method is lengthy, but the results are very informative and extremely beautiful.

Amphibia and small mammals
  a) Skin the animal, remove its digestive system and liver, then soak it for 24 hours in 70% alcohol with a few crystals of iodine.
  b) Soak in pure alcohol for another 24 hours.
  c) Soak for 30 minutes in each of the following: 75% alcohol, 50% alcohol, then 25% alcohol, and lastly water.
  d) Mix yellow alizarin in pure alcohol to make a $\frac{1}{2}$% solution. Mix this with 1% aqueous solution of potassium hydroxide to make a deep purple colour. Stain the animal for 12 hours in this liquid.
  e) Put the animal in 1% potassium hydroxide until no more dye colour comes away. This may take 6 to 12 weeks, but must be watched carefully because the animal may disintegrate.
  f) Mix equal parts of 1% potassium hydroxide with 880 ammonia (take care, the fumes of this liquid are extremely unpleasant). Brief immersion in this liquid will remove any brown colour from the specimen.
  g) Transfer to equal parts of 1% potassium hydroxide and glycerine, and to pure glycerine.

C *Investigating the nature of bone*
  1. Obtain three small pieces of bone (e.g. a rib bone sawn into pieces about 2 cm long).
  2. Put one piece in a beaker of water, another in solium hydroxide solution, and another in dilute hydrochloric acid.
  3. After one day, wash and dry each piece of bone. What is the difference between the piece soaked in acid and the other pieces? What does this tell you about the nature of bone?

Questions

1. Sort the following into parts of the axial skeleton and parts of the appendicular skeleton:

hand, skull, ribs, vertebral column, leg, cranium, intervertebral discs, shoulder blade, arm, sternum, pelvis.

2. What are the functions of:
a) orbits,
b) cranium,
c) intevertebral discs,
d) transverse processes,
e) bone marrow,
f) neural canal,
g) rib cage?

3. The diagram opposite shows some of the muscles and bones of a human arm.

a) Name the bones labelled A, C, E, and F.

b) Describe the types of movement possible at joints D and H.

c) How is the shape of the bones at joints D and H related to the movements which they can perform?

d) Name the flexor and extensor muscles which move joint D.

e) Complete the sentences opposite by placing a √ under one of the headings provided.

4. Why would movements such as walking be impossible without a hard skeleton?

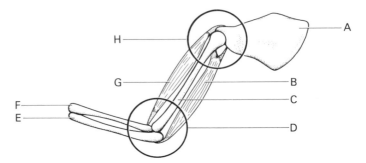

Fig. 12.7   Diagram for question 3

	Muscle G contracts	Muscle B contracts	Both muscles contract	Both muscles relax
When lifting a weight in the hand				
When pressing down on a desk top				
When the arm hangs loosely at your side				

Summary and factual recall test

Soft-bodied animals such as (1–name two) are supported many by (2) in their cells and in spaces in their bodies. These animals are said to have a (3) skeleton.

Arthropods have a skeleton like a suit of armour called an (4). In crabs it is made mainly of (5), but in insects it is made of (6). During growth this skeleton is replaced by ecdysis, which is (7–describe the process).

Vertebrates have an external skeleton, called an (8), made of (9). The axial skeleton is made up of (10–name three parts) and the appendicular skeleton is made up of (11–name three parts). The main functions of the skeleton are (12–list five).

Bones are made of living cells surrounded by minerals such as (13–name two). Large bones such as the

(14–name two) have a hollow inner region where (15) are made.

In mammals, movement occurs at the freely moveable, or (16) joints. Examples of hinge joints are (17–name three). These are called hinge joints because (18). Examples of ball-and-socket joints are (19–name two). These are called ball-and-socket joints because (20); Joints are held together by fibres called the (21), and muscles are attached to bones by fibres called the (22). The origin of a muscle is the end which (23), whereas the insertion is the end which (24). Every joint is moved by two opposing sets of muscles which are called an (25) system. The muscles which bend a joint are called the (26), while the (27) straighten the joint.

13

Co-ordination in animals

13.1 The need for co-ordination

The millions of cells and scores of different tissues and organs in the body of an animal do not work independently of each other. Their activities are **co-ordinated**. This means they work together performing their many tasks at times and rates that depend on the needs of the whole body.

Co-ordination is necessary so that an organism is ready to respond to happenings in the world around it.

One of the most familiar examples of co-ordinated behaviour is the way in which muscles work together during movement. When a boy runs to catch a ball, for example, he uses hundreds of muscles to move joints in his arms, legs, and back. Using information from his sense organs, the boy's nervous system co-ordinates these muscles so that they contract in the correct sequence, with the correct degree of power, and for precisely the correct length of time needed to get him to the spot where he can catch the ball. But this is not all. Muscular activities like running to catch a ball involve many other forms of co-ordination, such as those which increase the rate of breathing and heart-beat; adjust blood pressure; remove extra heat from the body; and maintain sugar and salt levels in the blood. Furthermore, all this co-ordination occurs without a single thought from the boy; it is an unconscious process.

13.2 The components of co-ordinated behaviour

Behaviour has five main components (Fig. 13.1):

stimulus⟶receptor⟶co-ordinator⟶ effector ⟶response

A stimulus
This is any change, inside or outside the body, which provokes a change in behaviour. It could be a sound, a sight, a smell, a pain, or a touch.

Receptors
Receptors are sense organs which detect a stimulus. The ears, eyes, nose, skin, and tongue are receptors.

Co-ordinators
These are organs which receive information from receptors, and use it to co-ordinate behaviour. The brain and spinal cord are co-ordinators which receive information in the form of 'messages', called **nerve impulses**, from sense organs.

Effectors
These are the parts of the body which are controlled by co-ordinators. Muscles are effectors; controlled by the brain and spinal cord.

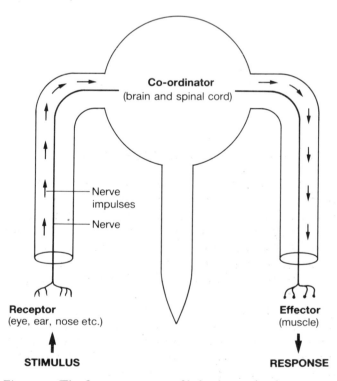

Fig. 13.1 The five components of behaviour: stimulus, receptor, co-ordinator, effector, response

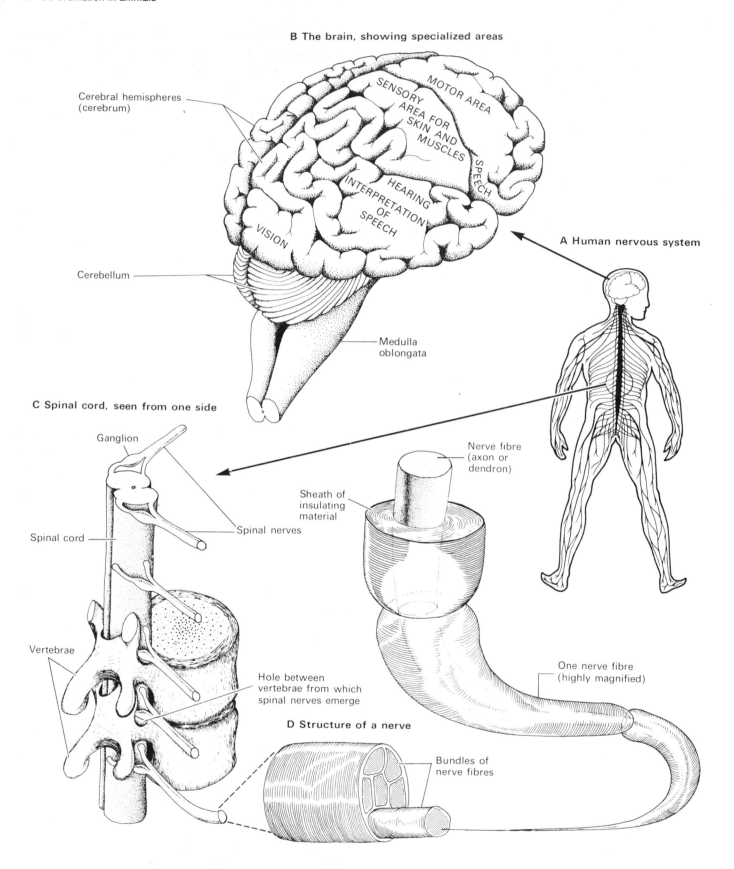

B The brain, showing specialized areas

Cerebral hemispheres (cerebrum)

MOTOR AREA

SENSORY AREA FOR SKIN AND MUSCLES

SPEECH

HEARING

INTERPRETATION OF SPEECH

VISION

A Human nervous system

Cerebellum

Medulla oblongata

C Spinal cord, seen from one side

Ganglion

Nerve fibre (axon or dendron)

Sheath of insulating material

Spinal nerves

Spinal cord

Vertebrae

One nerve fibre (highly magnified)

Hole between vertebrae from which spinal nerves emerge

D Structure of a nerve

Bundles of nerve fibres

Fig. 13.2 Structure of the human nervous system

Responses

A response is the behaviour provoked by the original stimulus. For example, pulling your hand away from something very hot or shouting to a friend seen across the street.

Using the human body as an example, this chapter describes how co-ordination is achieved in mammals through the activities of the **nervous system** and the **endocrine system**. Very simply, the nervous system consists of tissue which conducts 'messages', called **nerve impulses**, at high speed to and from all parts of the body. The endocrine system consists of glands which produce chemicals called **hormones**. These are released into the blood-stream and transported around the body. Unlike nerve impulses, hormones produce effects which are usually slow to appear, and which are often long-lasting.

13.3 The parts of the nervous system

The nervous system of mammals, and all other vertebrates, consists of a **brain** and a **spinal cord**, which together form a **central nervous system**. This system is connected to all parts of the body by **nerves**, which are made up of thousands of long thin **nerve fibres**. Figure 13.2 illustrates the arrangement of these structures in the human body. Like all other organ systems the nervous system consists of different types of cells. The cells which conduct the nerve impulses are called **neurones**.

Nerve cells (neurones)

Figure 13.3 illustrates the shapes of the main types of neurones as they appear when separated from various types of nervous tissue. Like all other cells, neurones have a nucleus surrounded by cytoplasm. The region of a neurone where the nucleus is located is called the **cell body**. Here, the cell is about a thousandth of a centimetre in diameter, which is slightly larger than most other animal cells. However, neurones differ from other animal cells in having cytoplasm which extends outwards from the cell body forming long fine threads as thin as 0.005 mm in diameter and, in humans, up to 1 metre in length. These threads are the nerve fibres along which travel the 'messages' made up of nerve impulses to and from all parts of the body.

Nerve impulses pass along nerve fibres in only *one* direction. They pass into the central nervous system along the fibres of **sensory neurones**, and out of the central nervous system along fibres called **motor neurones**.

Sensory neurones Sensory neurones, illustrated in Figure 13.3A, conduct impulses from the sense organs. That is, they conduct impulses from **receptors** in the body such as eyes, ears, nose, taste buds, and touch receptors. Sensory neurones have two long fibres: a **dendron** which conducts impulses from a sense organ to the cell body of the neurone, and an **axon** which conducts impulses from the cell body into the central nervous system.

Motor neurones Motor neurones, illustrated in Figure 13.3B, conduct impulses from the central nervous system to the **effector organs**, such as muscles and glands. The cell body of a motor neurone is embedded in the central nervous system. The cell body collects impulses from other neurones through its hundreds of tiny fibres, called **dendrites**. A long, single axon carries these impulses from the cell body to a muscle fibre or a gland. At the point where an axon of a motor neurone enters a muscle fibre there is a structure called a **motor end-plate**. When impulses reach an end-plate they set off chemical reactions which result in muscular contraction.

Most nerve fibres, whether dendrons or axons, that lie outside the central nervous system are encased in a sheath of fatty material which insulates them from one another. The main nerves of the body are, in fact, bundles of these insulated fibres wrapped together in a layer of connective tissue, something like a bundle of insulated wires which make up a large electric cable (Fig. 13.2D).

There are insulated fibres in the outer regions of the spinal cord and inner region of the brain. These fibres run parallel to each other in thick solid layers forming the **white matter** of these regions (Fig. 13.6). The **grey matter** which forms the inner core of the spinal cord and outer layers (cortex) of the brain is formed mainly of neurone cell bodies, and uninsulated nerve fibres (Fig. 13.3C).

13.4 Nerve impulses

Some nerve impulses originate inside the central nervous system; others come from the sense organs. Indeed, the sole function of a sense organ is to change various forms of stimulus, such as light or sound, into nerve impulses which pass along sensory neurone fibres to the brain. Figure 13.4 illustrates a simple way of explaining what happens during the conduction of an impulse.

Imagine blocks of wood, such as dominoes, arranged in a row. The first domino is knocked over and falls against the next, which falls against the next,

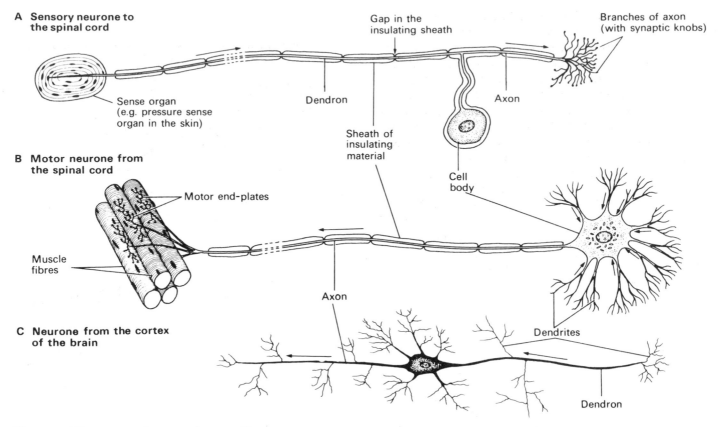

A Sensory neurone to the spinal cord

Gap in the insulating sheath

Branches of axon (with synaptic knobs)

Sense organ (e.g. pressure sense organ in the skin)

Dendron

Axon

Sheath of insulating material

Cell body

B Motor neurone from the spinal cord

Motor end-plates

Muscle fibres

Axon

Dendrites

C Neurone from the cortex of the brain

Dendron

Fig. 13.3 Three types of neurone (nerve cell)

and so on to the end of the line. Note two important facts. Nothing, except a certain amount of energy, has moved along the line of dominoes, and this cannot happen again until the dominoes have been stood on end. These events can be compared with a nerve impulse in the following way.

An impulse begins as a change in the arrangement of chemicals in a small area at one end of a nerve fibre. Like one domino falling against the next and knocking it over, this changed area of nerve fibre excites an identical change in the area adjacent to it, which excites the next area, and so on to the end of the fibre. No material object has moved along the fibre, only a wave of chemical rearrangement. Just as the dominoes must be stood on end before they can be knocked down again, so a nerve fibre must recover before it can conduct another impulse. But this recovery period is only a few thousandths of a second.

There is another similarity between falling dominoes and nerve impulses. If the first domino in the line is touched very lightly it may rock back and forth, but will not knock over the next and trigger off the chain of events described above. The first domino must be pushed with a specific amount of force before it falls against the next. Similarly, there are levels of

sound and intensities of light which are so weak that they do not stimulate the ears or eyes enough to trigger off impulses to the brain. Stimulation of a sense organ must reach what is called the **threshold level** before the organ sends a nerve impulse to the brain.

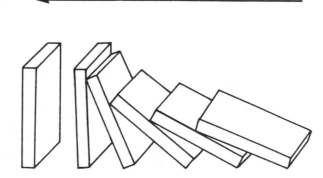

Fig. 13.4 A simple analogy of a nerve impulse

Another feature of nerve impulses is that there is no such thing as a weak or strong impulse. They either occur or they don't, and all are exactly alike no matter where they originate. This is called the 'all-or-none' principle. The only feature of impulses which ever varies is the number of them which pass along a nerve fibre per second. The **frequency** of the impulses depends on the strength of the stimulus. Thus, a strong stimulus, such as a flash of bright light, results in the eyes sending hundreds of impulses per second to the brain, whereas a dim light produces only a few impulses per second, and a very dim one, below the threshold level, produces none at all.

Impulses passing along a nerve fibre eventually reach the end of it, where they encounter an obstacle to their progress: a microscopic gap called a **synapse** between the tip of one fibre and the beginning of the next.

Synapses

Neurones are not continuous with one another. Nerve impulses must cross a synapse where the axon of one neurone meets the dendrites or cell body of another. The dendrites and cell body of a motor neurone, for example, have synapses with hundreds, even thousands, of other neurones (Fig. 13.5).

Synapses, as well as sense organs, each have a certain threshold level. The threshold of a synapse is the number of impulses per second at which the synaptic gap is 'bridged' and impulses begin to flow in the next neurone.

The threshold level of a sense organ, plus the thresholds of every synapse in the chain of neurones connected to it, form a barrier to the movement of impulses between that sense organ and the brain. This barrier is only penetrated by the high frequency impulses which result from strong stimuli. The low frequency impulses resulting from weaker stimuli cannot cross the synapse and do not reach the brain.

To summarize: **the nervous system is a mass of interconnected nerve fibres which conduct impulses from receptors to effectors throughout the whole body.**

13.5 Reflex actions

A reflex action is behaviour in which a stimulus results in a response which does not have to be learned, and which occurs very quickly without conscious thought. For example, a person does not have to learn or even think what to do when his hand accidentally touches a very hot object, he automatically pulls his hand away. Such responses are built in to the nervous system from birth.

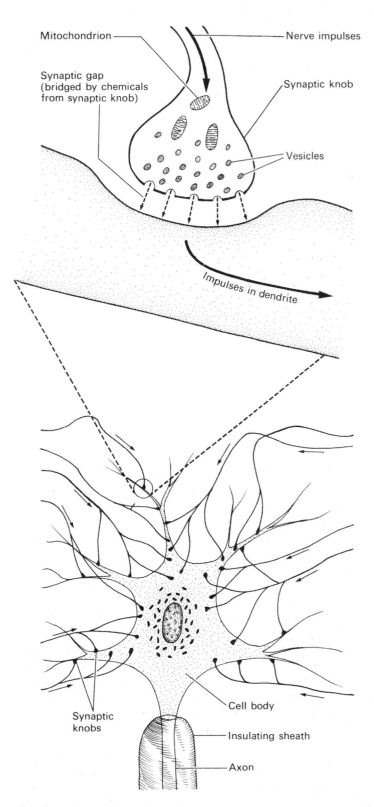

Fig. 13.5 Arrangement and structure of synapses on the cell body and dendrites of a motor neurone. (It is thought that impulses cause the release of a chemical from vesicles in the synaptic knobs which start new impulses in the next neurone.)

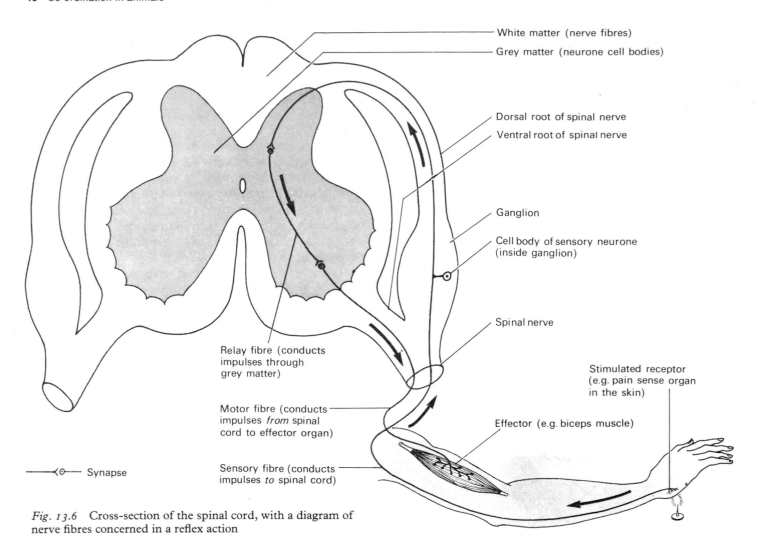

White matter (nerve fibres)

Grey matter (neurone cell bodies)

Dorsal root of spinal nerve

Ventral root of spinal nerve

Ganglion

Cell body of sensory neurone (inside ganglion)

Spinal nerve

Stimulated receptor (e.g. pain sense organ in the skin)

Effector (e.g. biceps muscle)

Relay fibre (conducts impulses through grey matter)

Motor fibre (conducts impulses *from* spinal cord to effector organ)

Sensory fibre (conducts impulses *to* spinal cord)

—<○— Synapse

Fig. 13.6 Cross-section of the spinal cord, with a diagram of nerve fibres concerned in a reflex action

Withdrawal from a painful stimulus is an example of a **spinal reflex**. The nerve impulses involved in it pass through the spinal cord along **spinal nerves**. The pathway of these impulses is illustrated very simply in Figure 13.6. From this illustration note the arrangement of sensory and motor neurones, and the synaptic connections between them which form a pathway for impulses from sense organs to a motor end-plate. Such nervous pathways are often called **reflex arcs**, perhaps because of their curved shape.

It is very important to realize that Figure 13.6 is a greatly simplified picture of a reflex action. In fact, there are not three but hundreds of neurones involved in even comparatively simple reflexes like pulling a hand from a hot object. These neurones activate more than a score of muscles to raise the arm and flex the fingers. But even simple arm movements require elaborate muscular co-ordination, the pattern of which differs according to the speed and direction of arm movements. Apparently simple reflex actions are in fact very complex events.

Figure 13.7 indicates nerve fibres which conduct impulses from the sensory side of a reflex arc to the brain, and other fibres which conduct impulses from the brain to the opposite side of the spinal cord. These connections with the brain enable a person to be aware of certain spinal reflexes and, up to a point, exert control over them.

Think what might happen when a father picks up a pan of boiling liquid from the cooker to find that the pan handle is very hot. Reflex action may make him drop the pan and spill the liquid. However, if his young child is standing nearby he can deliberately prevent, or **inhibit**, this reflex (using the downward nerve pathways from his brain described in Figure 13.7) and put the pan down safely even though its handle is burning his fingers.

In general, reflex responses to painful stimuli protect the body from injury. Other examples of reflex responses and their functions are dealt with in question 1.

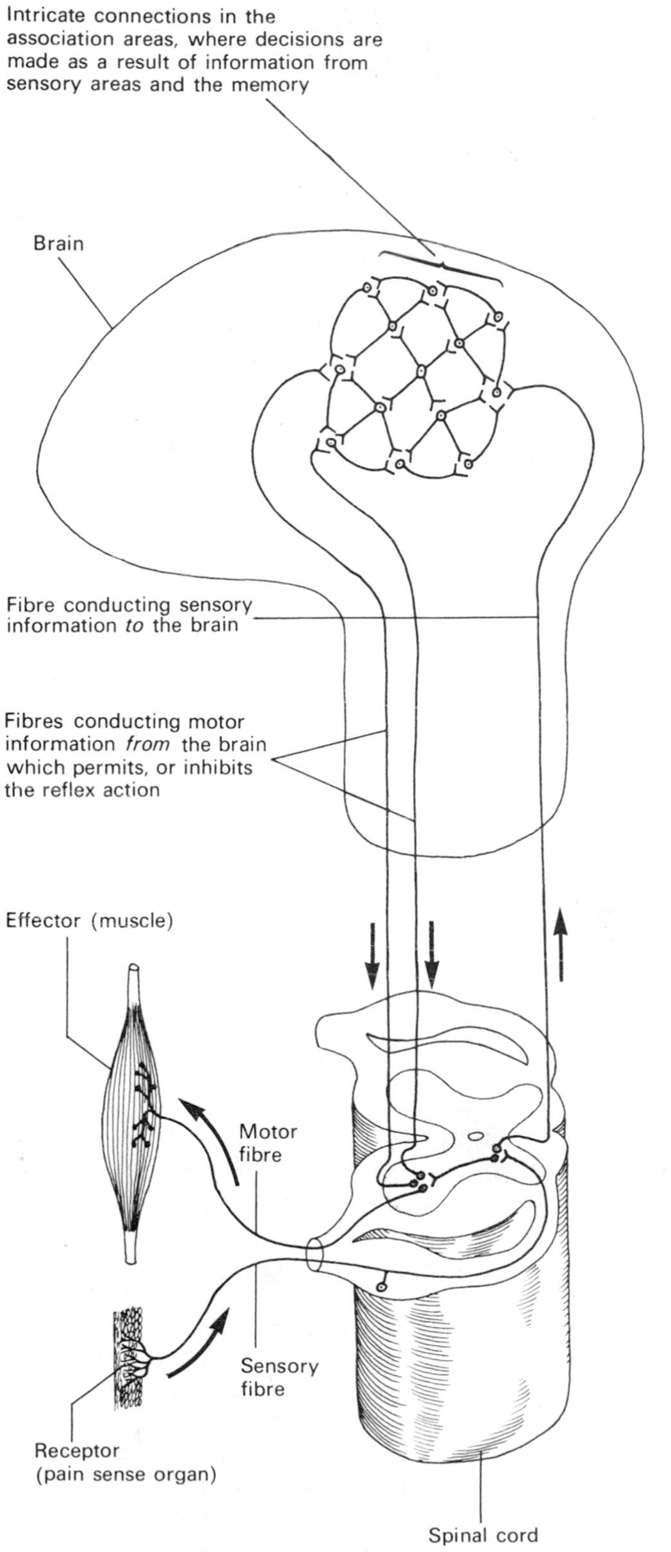

The nerve fibres which connect reflex arcs with the brain enable reflex actions to be modified with experience. Conditional reflexes are examples of such modifications.

13.6 The human brain

The human brain weighs about 1.5 kg and contains thousands of millions of neurones. Each neurone has synapses with thousands or more other neurones, making an immensely complex network of cells and nerve fibres. The 'lowest' part of the brain, that is, the part which merges with the spinal cord, is called the **medulla oblongata** (Fig. 13.8).

Medulla oblongata This region is concerned with many unconscious processes including the regulation of blood pressure, body temperature, and rates of heart-beat and breathing. The medulla also contains the mass of nerve fibres which connect the brain and spinal cord. Above the medulla is the **cerebellum**.

Cerebellum This receives impulses from sense organs concerned with balance and from stretch receptors in the muscles and joints. It uses information from these sources to maintain muscle tone and, thereby, a balanced posture, and it also co-ordinates muscles during activities like walking, running, dancing, and riding a bicycle. Above the cerebellum is the **cerebrum**.

Cerebrum This the largest part of the human brain. It is a dome-shaped mass of nervous tissue made up of two halves called the cerebral hemispheres.

The cerebrum consists of an outer layer of 1 mm thick of neurone cell bodies (grey matter) and a much thicker inner layer of nerve fibres (white matter). The outer layer is called the **cerebral cortex**, or **cortex** for short. This is the region where the main functions of the cerebrum are carried out.

The cortex is concerned with all forms of conscious activity. It is here that sensations of touch, taste, hearing, vision, and smell are generated. Reasoning takes place, decisions are taken, emotions are felt and memories are stored in the cortex. While the cerebellum co-ordinates the muscles, it is the cerebrum which *directs* them, so that movements have a definite purpose.

Intricate connections in the association areas, where decisions are made as a result of information from sensory areas and the memory

Brain

Fibre conducting sensory information *to* the brain

Fibres conducting motor information *from* the brain which permits, or inhibits the reflex action

Effector (muscle)

Motor fibre

Sensory fibre

Receptor (pain sense organ)

Spinal cord

Fig. 13.7 Diagram of the nerve fibres through which the brain becomes aware of, and can control, a reflex action

B The brain and spinal cord

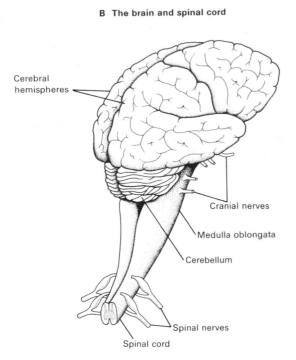

Cerebral
hemispheres

Cranial nerves

Medulla oblongata

Cerebellum

Spinal nerves

Spinal cord

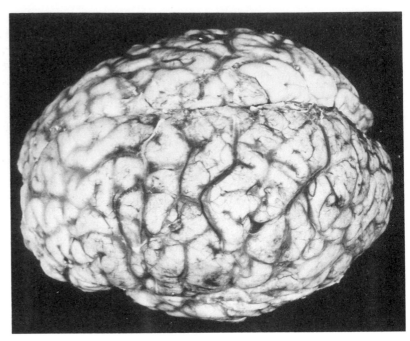

A human brain

Fig. 13.8 The human brain and spinal cord

13.7 The endocrine system

The endocrine system consists of glands, and like all other glands they produce and release (secrete) useful substances. Some non-endocrine glands, like the pancreas and salivary glands, secrete their products through tubes, or ducts, which lead directly to the place where these substances carry out their functions. Endocrine glands, however, have no ducts. They secrete their products – chemicals called **hormones** – directly into the blood-stream which carries them all round the body, where they affect organs sensitive to them.

Both hormones and nerve impulses co-ordinate the body's activities, but they do this in different ways. The difference between the way in which nerve impulses and hormones operate is like the difference between a telephone message and a message broadcast by radio. A telephone message goes along a wire to one person; and nerve impulses go along a nerve fibre to one particular muscle or gland. A radio announcement, however, is broadcast to everybody with a radio set, but only those actually concerned with the message respond to it. Similarly, hormones are 'broadcast' by the blood-stream to every cell in the body, but only certain cells respond to them.

Hormones are produced in minute quantities, but their influence on the body is often profound and long-lasting. For example, hormones affect the size to which the body grows, the development of sexual characteristics, and to a certain extent the development of mental powers and personality. In general, both physical and mental health depend on the endocrine glands producing the right amounts of hormones at the right times.

It is not known exactly how hormones carry out their functions. They may influence the rate at which chemicals become available for metabolic reactions, and they may also speed up or slow down the reactions. Equally important is the ability of some hormones to control the types of chemicals which cells take in or lose through their cell membranes.

The following notes describe only the major endocrine glands, the hormones they produce, and their effects on the body. Other chapters describe the actual mechanisms, such as reproduction, which are controlled by hormones. Figure 13.9 illustrates the position of endocrine organs in the human body.

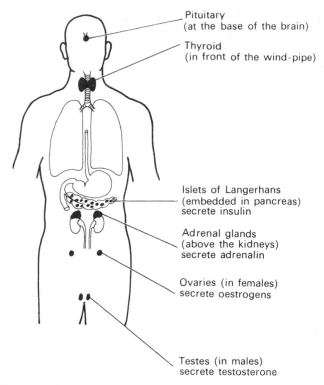

Fig. 13.9 Position of the major endocrine organs in the human body

Pituitary
(at the base of the brain)

Thyroid
(in front of the wind-pipe)

Islets of Langerhans
(embedded in pancreas)
secrete insulin

Adrenal glands
(above the kidneys)
secrete adrenalin

Ovaries (in females)
secrete oestrogens

Testes (in males)
secrete testosterone

The pituitary gland

The pituitary gland is situated in a small cavity of the skull 13 mm wide, beneath the brain and above the roof of the mouth. The pituitary secretes several hormones. Some of these affect the body generally, and some have a controlling influence over the other endocrine glands. For this last reason the pituitary has been called the 'master gland', and even 'the conductor of the endocrine orchestra'.

The pituitary strongly influences growth, both directly through its own growth hormone, and indirectly through its influence on other endocrine glands. Pituitary growth hormone controls the size of the bones. If the pituitary produces too much growth hormone the result is abnormal growth, or giantism. Humans, for instance, may grow to two metres or more in height. Too little growth hormone results in delayed or permanently retarded growth.

The thyroid gland

The thyroid is a butterfly-shaped gland situated in the neck, in front of the wind-pipe. There are several thyroid hormones, including thyroxin, and they control the rate at which sugar is consumed in cellular respiration. Through these hormones the thyroid controls the rate of metabolism in the whole body.

The thyroid has a major influence upon physical and mental development from birth to old age. In the young, failure of the thyroid to produce its hormones can result in stunted growth and severe mental retardation, a condition known as **cretinism**.

In cases of over-active thyroid, the rate of metabolism increases, the body gets thinner, the victim becomes very restless, over-excited, and even mentally unstable. A person's whole physical and mental well-being can depend therefore on the balanced activity of this one tiny gland.

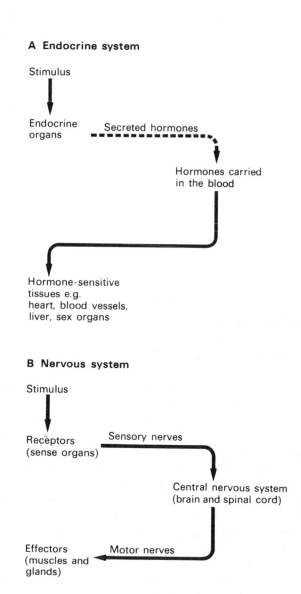

A Endocrine system

Stimulus

Endocrine organs — Secreted hormones → Hormones carried in the blood

Hormone-sensitive tissues e.g. heart, blood vessels, liver, sex organs

B Nervous system

Stimulus

Receptors (sense organs) — Sensory nerves → Central nervous system (brain and spinal cord)

Effectors (muscles and glands) ← Motor nerves

Fig. 13.10 A diagram summarizing the differences between endocrine organs and the nervous system

147

Adrenalin

Adrenalin is a hormone produced by the inner region (medulla) of the adrenal glands. There is one of these glands above each kidney in mammals. The sudden and dramatic effects of adrenalin must be well-known to everyone, for they are experienced when ever a person is faced with a dangerous or very exciting situation.

Imagine a man walking slowly down a street feeling tired after a hard day's work. A car appears from around a corner and suddenly skids out of control towards him. Almost instantaneously the man's tiredness is forgotten and his body becomes super-efficient. He runs faster than he has for years, then jumps effortlessly over a wall which would normally have seemed impossibly high. What has caused this amazing transformation?

The sight of the car hurtling towards him causes a very rapid outpouring of adrenalin from the man's adrenal glands. Within seconds this hormone stimulates his heart to beat much faster; it makes the bronchioles of his lungs increase in diameter, and his breathing rate increases. This charges his blood with extra oxygen and sends it, along with a massive dose of glucose which adrenalin releases from his liver, rushing to his muscles and brain. The flow of blood to these regions is increased in two ways. First, adrenalin constricts blood vessels in his skin and gut, thus diverting blood from these regions. Second, the hormone enlarges blood vessels to his muscles, so that blood flow is greatest where it is most needed.

In short, adrenalin prepares the body for sudden, possibly violent effort. In wild animals, adrenalin is released in situations such as fights between predators and their prey. In humans, situations that require sudden bursts of physical activity are quite rare. Nevertheless, there are many other occasions in which adrenalin is released, such as during emotional excitement, stress, fear, or anger. Familiar symptoms of adrenalin's effects at these times are: a dry mouth, a pounding heart, and an unpleasant 'sinking' sensation in the stomach. (X-rays have shown that the stomach can actually drop several centimetres within the abdomen during emotional stress.)

Insulin

Within the pancreas, a gland which secretes digestive juices (Fig 4.13), there are patches of tissue with no digestive function. These patches, poetically named the **Islets of Langerhans**, produce a hormone called **insulin**.

Insulin controls the level of glucose in the blood by increasing the rate at which the liver converts glucose into glycogen. Insulin also enables body cells (except neurones and muscles) to absorb glucose, which is their main source of energy. In addition, insulin stimulates the production of fat from glucose, and the synthesis of proteins.

Diabetes is a disease caused by the slowing down of insulin production. This sets off a complex chain of events in the body. The major symptom of diabetes is a massive increase in glucose level in the blood. First, glucose from digested carbohydrate is no longer turned into glycogen or fat. Second, without insulin both fat and protein in the body cells tend to be broken down, which eventually yields even more glucose. Fat breakdown also produces fatty acids, and these are eventually further broken down into poisons such as acetone and acetic acid. Some of the excess glucose in the blood is excreted from the body by the kidneys as part of urine, but this requires large amounts of water and so the diabetic develops an almost insatiable thirst. An added complication is that many body cells can no longer absorb glucose and so their respiration rate slows down. Unless the victim of diabetes is treated with insulin (extracted from cow or pig pancreas) he may become unconscious as a result of dehydration, acetic acid poisoning, and slow cellular respiration.

Sex hormones

The sex hormones are produced by the ovaries of female mammals, and the testes of males (Fig. 13.9).

Female sex hormones The ovaries produce a number of hormones known collectively as **oestrogens**. These have three main functions. First, they control the development of secondary sexual characteristics; that is, the external features of females. In humans these include breasts, wide hips, and a high-pitched voice. Second these hormones prepare the uterus to receive a ripe, fertilized ovum. Third, they maintain the uterus in a state whereby it can nourish and protect the developing embryo.

Male sex hormones The testes produce the hormone **testosterone**. This promotes the development of secondary sexual characteristics, which in humans include a deeper voice, more body hair, and more powerful muscles than in females.

Hormones and the nervous system affect the whole body and therefore each other in an enormous number of different ways. In fact, the interrelationship between the two is so close that a disturbance in any endocrine organ or any part of the nervous system can have far-reaching effects all over the body, sometimes with disastrous consequences.

Hormones and nerves affect a person's behaviour, but it must also be remembered that a person's behaviour can affect his hormones and nerves. Any behaviour which causes excessive stress, either physical or emotional can, if prolonged, cause damage to the endocrine and nervous systems, and this damage may cause a chain-reaction of events so that symptoms appear which have no apparent connection with the original cause. The original cause may be very difficult to trace. But worst of all the damage may be permanent.

Investigation

How fast is your reflex response to a stimulus?

1. Work in pairs. One student holds a ruler between thumb and forefinger so that the ruler hangs with its zero mark at the bottom.

2. The other student waits with thumb and forefinger of one hand about 2 cm apart and level with the zero mark of the ruler.

3. The student holding the ruler says 'ready', then drops the ruler within five seconds without further warning. The other student must catch the ruler between thumb and forefinger, and note the number of centimeters the ruler has dropped, by observing the position on the ruler of his/her thumb.

4. Calculate the average distance over ten ruler drops. Use the graph to convert this distance into response time in seconds.

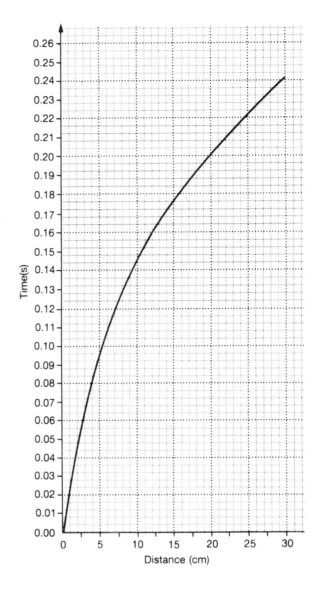

Questions

1. What is a reflex action, and what reflexes occur:

   a) when dust blows into the eyes (2 examples);

   b) when a bright light suddenly shines in the eyes (2 examples);

   c) when a person runs fast in hot weather (3 examples);

   d) when a person moves suddenly from a warm room to a very cold one (2 examples);

   e) when a hungry person smells cooking food (1 example);

   f) when food accidentally enters the wind-pipe?

2. Explain how each of the reflexes given in answer to 1 protects the body and, in certain cases, conserves energy and body materials.

3. List the receptor and effector organs concerned in each of the reflex actions given in answer to 1.

4. A piece of thread was tied tightly round an animal's pancreatic duct. The animal subsequently had difficulty in digesting food, but did not get diabetes. Explain.

5. Describe the main differences between a sensory neurone and a motor neurone.

6. The diagram overleaf shows a cross-section of the spinal cord.

   a) Name the parts labelled A to G.

   b) What is the name of the region indicated by label H (where two neurones meet)? Briefly describe what happens at this point.

   c) The arrows on the diagram represent nerve impulses. Briefly describe what a nerve impulse is.

   d) Name four places where nerve impulses shown on fibre A could have come from.

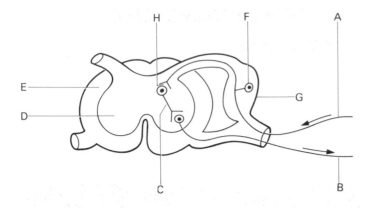

e) What is the destination of impulses moving along fibre B?

f) Describe the main difference between region D and region E.

7. *a)* What is a a stimulus? Give three examples.
   b) What is a response? Give three examples.
   c) What are receptors and effectors? Give an example of each.

8. If all nerve impulses are alike how can the brain distinguish between a strong stimulus and a weak stimulus?

9. *a)* What is a synapse?
   b) Why are synapses important?

10. List the parts of the nervous system, and the peripheral nervous system.

11. What part of the central nervous system:
   a) consists of two cerebral hemispheres;
   b) regulates body temperature;
   c) co-ordinates muscles;
   d) contains the cerebral cortex;
   e) has spinal nerves attached to it;
   f) allows us to remember things?

Summary and factual recall test

The central nervous system consists of (1–name its two parts). It is connected to all parts of the body by nerve (2) which are (3) from each other by a sheath of fatty material. A sensory neurone conducts impulses from receptors such as (4–name five) along a fibre called a (5) to the cell (6) of the neurone and from here along another fibre called an (7) into the (8). The (9) body of a motor neurone is embedded in the (10). Here, it collects impulses through fine fibres called (11) and passes them through a long fibre called an (12) into an effector organ such as (13–name two types).

Sense organs sent out impulses only when stimulated above their (14) level. These impulses either occur or they don't, which is called the (15) principle. The only feature of impulses which ever varies is (16). As impulses pass from one neurone to the next they cross a gap called a (17). A reflex action is (18). Examples are (19–name four).

The medulla forms the (20) part of the brain where it merges with the (21). This region controls unconscious processes such as (22–name four).

The cerebellum receives impulses from the (23–name two places). This information is used to achieve muscle (24) and (25) co-ordination, for example when (26–name two examples).

The cerebrum has a thin outer layer of (27) matter called the (28) which consists of (29). The cerebrum is concerned with (30) activities such as (31–name four).

The endocrine system consists of glands which have no (32), and which secrete chemicals called (33) into the (34). The pituitary gland is situated beneath the (35). It is sometimes called the 'master gland' because (36). Apart from this, its main functions are (37–list three).

The thyroid gland is situated in the (38). Its main functions are to control the rate of (39) consumption in cellular (40), and influences (41) and (42) development from birth to old age.

Adrenalin prepares the body for (43). In wild animals it is produced in situations such as (44), and in humans during (45–name three situations).

Insulin controls the level of (46) in the blood by (47).

Female sex hormones are known collectively as (48). Their main functions are (49–name three). The (50) produce male sex hormone called (51) which controls the development of (52).

14

Animal senses and behaviour

Sense organs are the **receptors** of the body. They receive information about conditions both inside the body and in the world around it. This information is known as the **stimulus** because it stimulates the sense organs. But no matter how the sense organs are stimulated, whether by the sound of the latest pop tune, the sight of a beautiful face, the smell of bacon and eggs, the smooth touch of a cat's fur, or a pain in the stomach, the stimulus is turned into only one thing: a pattern of nerve impulses which speed down sensory nerves to the brain. Here, the patterns of impulses are interpreted and transformed into **sensations** such as touch, sight, hearing, taste, and smell.

14.1 Types of sense organ

Many people if asked how many senses they possess would reply 'five: sight, hearing, touch, taste, and smell.' In fact, there are many more senses than this. Sense organs are grouped according to the type of stimulation they receive.

Light receptors (photoreceptors)
Eyes are obviously included in this group, but there are organisms without eyes which are still sensitive to light. The protozoan **Euglena**, for example, has a tiny eye spot which allows it to swim towards the light it needs to make food by photosynthesis. Many insects have two kinds of eye. There are **simple eyes** which have only one lens, and **compound eyes** with up to 30 000 lenses (see photograph).

Mechanical receptors
These are sensitive to 'mechanical' stimulation such as touch and sound. The human skin is rich in touch receptors (section 14.2). Ears are not the only sound receptors. Fish have a sense organ along each side of their bodies called a **lateral line**. This allows them to detect sound waves passing through water. Grasshoppers have 'ears' on their front legs. Moths

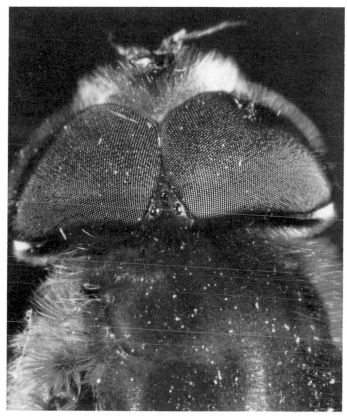

A dronefly showing two compound eyes and three simple eyes

have similar organs on their antennae which detect the ultrasonic 'squeaks' of insect-hunting bats.

Chemical receptors
These include taste buds and 'smelling' or **olfactory organs**. Taste buds detect chemicals which dissolve in liquid on the tongue. There are four types of taste receptor: sweet, sour, salt, and bitter. The countless different flavours of food and drink are detected according to the level of stimulation they produce in each type of receptor (Fig. 14.1).

Smell receptors are situated high in the nasal cavity inside the nose. In general, harmful substances such as decaying food have unpleasant smells, while pleasant smelling substances are harmless (Fig. 14.2).

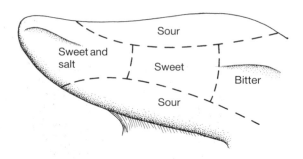

Fig. 14.1 A map of the tongue's taste areas. Taste buds which respond to salt, sweet, sour, and bitter tastes are concentrated into the areas shown in the diagram

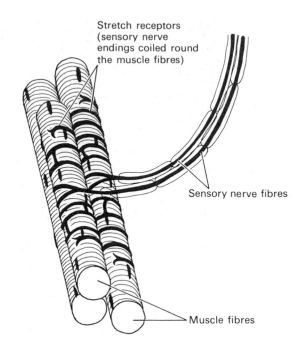

Fig. 14.3 Stretch receptors in muscle fibres

Proprioceptors

These include stretch receptors inside muscles and organs of balance. Stretch receptors provide the brain with information about the degree of tension in muscles, and the angle at which each joint is bent (Fig. 14.3). This information is essential for properly co-ordinated movement.

Organs of balance include the **semi-circular canals** embedded in bone behind each ear (Fig. 14.4). These detect changes in the *direction* of movement. The brain uses this information to help maintain balance.

Other types of sense organ

There are **temperature receptors** in human skin and insect antennae which detect changes in temperature. Certain arteries contain **baro-receptors** which respond to changes in blood pressure. Some fish have **osmoreceptors** sensitive to changes in the strength of surrounding sea water.

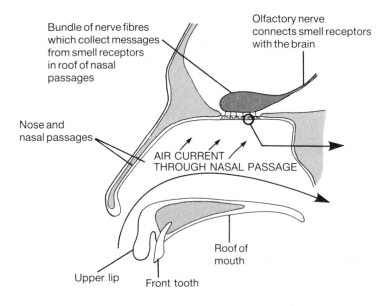

Fig. 14.2 Smell receptors are called olfactory organs. They are situated in the roof of the nasal passages and detect chemicals in the air. First the chemicals must dissolve in a film of moisture which covers the receptors

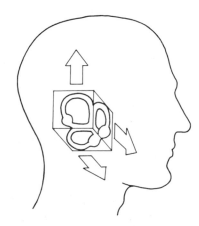

Fig. 14.4 The position of semi-circular canals in the head (canals are drawn much larger than life)

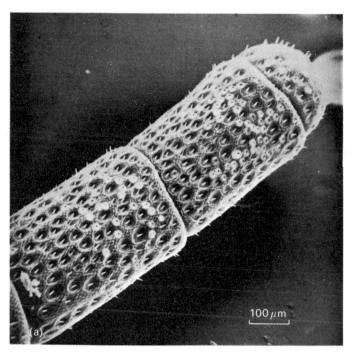

The antenna of a locust. The white pegs are nerve endings sensitive to chemicals. The white rings contain nerve endings sensitive to changes in humidity

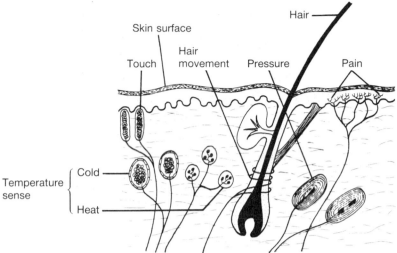

Fig. 14.5 Cross-section through mammal skin showing sense organs

14.2 Mammal skin senses

The skin contains millions of separate tiny sense organs with several different functions. Microscopic examination of the human skin shows that it contains at least five different types of sensory nerve ending. It is not yet certain how these receptors work or exactly what their individual functions are. Nevertheless, they have been named according to their supposed functions, which are: touch, pressure, temperature, pain, and hair movements.

For various reasons the functions of each type of skin receptor are difficult to establish. First, certain parts of the body, like the cornea of the eye (Fig. 14.7), are sensitive to touch, pressure, and pain and yet they have only one type of nerve ending, and these look like the pain receptors illustrated in Figure 14.5. Second, it is possible to find the exact location in the skin of, say, a pain receptor only to discover a few hours later that this exact spot has become sensitive to heat or cold and not pain.

Sense of touch

The sense of touch enables a person to distinguish between a variety of textures, from sandpaper roughness to glassy smoothness, and between hard, soft, and liquid substances. What is more, touch can give a vivid impression of three-dimensional shape, and it is sensitive to fine detail. For example, blind people can read with their finger-tips using a system of raised dots known as Braille. Touch receptors are not evenly distributed over the body surface. They appear to be especially close together in the tongue and at the fingertips (exercise A).

Sense of pressure

Pressure receptors are also concentrated in the skin of the tongue and finger-tips, where pressure differences as small as $2\,g/mm^2$ can be detected. Humans have the ability to 'project' pressure sense into objects held in the hand. A cook stirring a pudding can feel lumps deep inside it just as if there were live pressure receptors at the end of his spoon. This illusion occurs because pressure changes felt at the handle are projected into the spoon. Projection is important in the skilled use of tools such as files, spanners, and chisels.

Temperature sense

There are separate 'heat' and 'cold' receptors in the skin, but these cannot be used, like a thermometer, to tell the exact temperature of something. Temperature sense is limited to comparing temperature differences. If a person from the Arctic and a person from the tropics both arrive in London by air on a warm spring day, the man from the Arctic is likely to feel hot, while the man from the tropics feels cold. This happens because they are each comparing English spring temperatures with those of different climates. However, after a few days, when their heat and cold receptors have undergone adaptation, they both feel roughly the same temperature sensations.

Temperature receptors in the fingers can distinguish differences as small as 0.5°C. However, the tongue is surprisingly insensitive to temperature, which explains why drinks that scald the fingers can be sipped without discomfort.

Pain

Pain receptors are more evenly distributed through the skin, except that in certain areas they can be partly obscured by thick layers of epidermis, as under callouses which form on the palms and fingers after prolonged manual work.

Pain receptors are not restricted to the skin; they are located inside muscles, tendons, ligaments, in the walls of the digestive system, in fact everywhere except in the brain. Even though pain is unpleasant it is important because it acts as a warning that something is going wrong in the body. Moreover, the brain can usually work out the exact location, and sometimes the nature, of the trouble according to the position of affected pain receptors.

There is a rare abnormality which greatly reduces, or completely removes, a person's sensitivity to pain (in Britain there are fewer than 100 cases). These people are in constant danger because they are unaware of cuts, burns, and pressure which could break their limbs, and unaware of diseases which show no visible symptoms.

Hair movements

Most hair follicles (Fig. 14.5) have a sensory nerve ending attached to their bases. These receptors are stimulated when the hair is moved by objects close to the skin or by air movements. Some animals, such as cats and mice, have very long hairs, or whiskers, which extend outwards from the face to approximately the width of the body. These sensitive whiskers are extremely helpful in avoiding obstacles, especially when the animals move about at night.

14.3   Mammal eyes

Most living things, including green plants and even some unicellular organisms, are sensitive to light. But relatively few organisms are capable of vision; that is, few have the ability to form picture images of the outside world. This section describes the image-forming eyes of mammals (e.g. humans).

An eye resembles a camera in at least three ways.

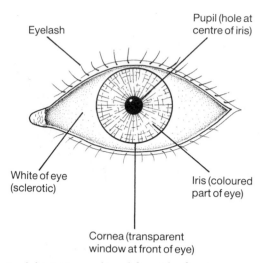

Fig. 14.6   A human eye viewed from the front

First, cameras and eyes both have a mechanism which focuses light. In the eye this consists of the transparent **cornea** and **lens**, which act like the glass lens of a camera in forming a clear, upside-down, full-colour image (Fig. 14.7). Second, in the eye this image falls on a layer of receptors called the **retina**, which, like the film in a camera, is sensitive to light. Third, eyes and some cameras have an apparatus called an **iris diaphragm**, which is an opaque disc with a hole at its centre. The size of the hole can be increased or decreased to control the amount of light reaching the light-sensitive surface.

Unlike a camera the eye does not make a permanent record of images which fall on its retina. The retina transforms light into a stream of nerve impulses which pass down the optic nerve to the brain. The frequency and pattern of these impulses vary according to the patches of colour, light, and shade which make up the retinal image. The visual area of the brain interprets these impulses to form a three-dimensional, full-colour, moving impression of the outside world.

Protection of the eyes   The eyes are protected in many ways. The skull has two deep cavities about 2.5 cm in diameter called the **orbits**, that enclose and protect all but the front of the eyes (Fig. 14.8). The exposed region is covered by a transparent, self-repairing skin called the **conjunctiva**. The conjunctiva is kept moist and clean by a slow continuous stream of liquid from the **tear glands**, and every few seconds it is wiped by the eyelids during their automatic (reflex) blink movements. When dust or chemicals reach the conjunctiva the rate of tear flow and blinking is automatically increased until the eye is clean. The blink reflex also protects the eye by

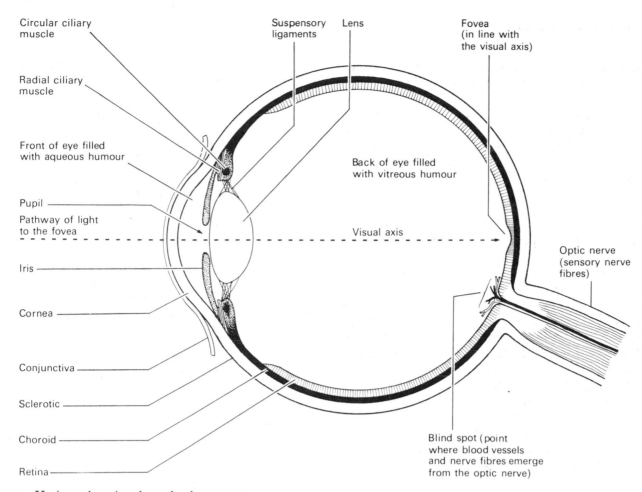

Circular ciliary muscle

Radial ciliary muscle

Front of eye filled with aqueous humour

Pupil

Pathway of light to the fovea

Iris

Cornea

Conjunctiva

Sclerotic

Choroid

Retina

Suspensory ligaments

Lens

Fovea (in line with the visual axis)

Back of eye filled with vitreous humour

Visual axis

Optic nerve (sensory nerve fibres)

Blind spot (point where blood vessels and nerve fibres emerge from the optic nerve)

Fig. 14.7   Horizontal section through a human eye

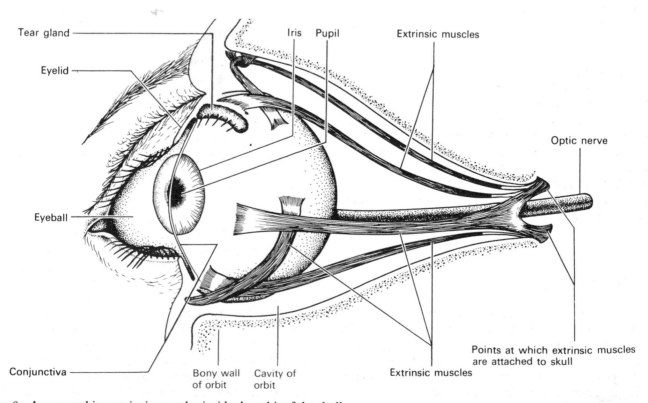

Tear gland

Eyelid

Iris   Pupil

Extrinsic muscles

Optic nerve

Eyeball

Conjunctiva

Bony wall of orbit

Cavity of orbit

Extrinsic muscles

Points at which extrinsic muscles are attached to skull

Fig. 14.8   An eye and its extrinsic muscles inside the orbit of the skull

closing them whenever an object moves quickly towards the face. Finally, the eyelashes form a net in front of the eyes, which traps large airborne particles.

Movements of the eyeballs Each eyeball is held in place, and moved within its orbit, by six **extrinsic muscles** attached to the outer surface of the eyeball (Fig. 14.8). The extrinsic muscles can rotate the eyes to follow moving objects, and direct the gaze to a chosen object. These movements are precisely co-ordinated so that both eyes work together, and at all times are directed at the same spot. This means that the eyes converge inwards to watch objects which move towards the face.

Nourishment and support of eye tissues Eyes are nourished and supplied with oxygen by blood vessels which enter through the optic nerve. These vessels spread out through the **choroid** layer, and over the surface of the retina. It is hardly surprising that the cornea and lens have no direct blood supply, since a network of capillaries would impair their ability to focus light. The cornea and lens obtain oxygen and food by diffusion from blood vessels through a liquid called the **aqueous humour**. This liquid is secreted into, and absorbed from, the front cavity of the eye. It is renewed about every four hours. (Incidentally, tiny particles in the aqueous humour are the cause of spots which sometimes appear to float before the eyes.) The aqueous humour, together with a jelly called the **vitreous humour** in the rear cavity of the eye, exert an outward pressure on the eyeball, and this maintains its rounded shape.

The iris The iris is the coloured part of the eye, and has a round hole at its centre called the **pupil**. The iris consists of radial muscles which contract to enlarge the size of the pupil, and circular muscles which make it smaller in size.

The iris regulates the amount of light which reaches the retina by opening the pupil in dim light and reducing it to pin hole size in bright light (these movements are reflex actions). As all photographers should know, reducing the size of the hole in a camera iris increases the **depth of focus** of the lens, which means that both near and distant objects are in focus at the same time. The same is true of the eye, and this explains a second function of the iris. When the eyes are focused on near objects the depth of focus of the lens is low. Therefore, vision would be poor were it not for an automatic reduction in pupil size (irrespective of light intensity). When the lens is focused on a distant object its depth of focus is far greater, and the iris opens again.

The lens The lens of an eye consists of layers of transparent material arranged like the skins of an onion, which are enclosed in an elastic outer membrane. The whole lens is held in place by **suspensory ligaments** attached to its outer rim, and these ligaments are attached to a ring of muscle fibres, the **ciliary muscles**, which run around the eyeball next to the iris (Figs 14.7 and 14.9).

Fig. 14.9 Diagram of the changes which take place during accommodation movements

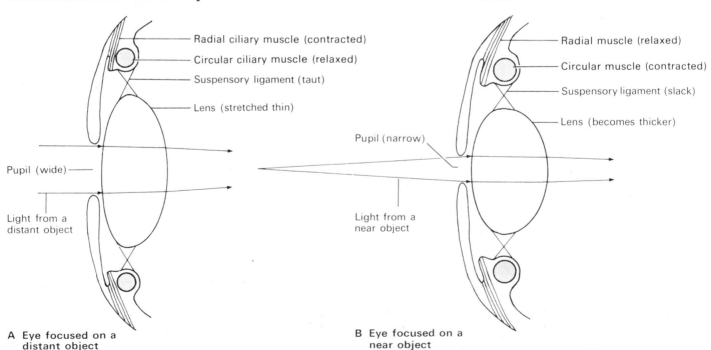

A Eye focused on a distant object

- Radial ciliary muscle (contracted)
- Circular ciliary muscle (relaxed)
- Suspensory ligament (taut)
- Lens (stretched thin)
- Pupil (wide)
- Light from a distant object

B Eye focused on a near object

- Radial muscle (relaxed)
- Circular muscle (contracted)
- Suspensory ligament (slack)
- Lens (becomes thicker)
- Pupil (narrow)
- Light from a near object

The cornea alone is sufficient to form an image on the retina of distant objects. However, the lens makes it possible for this image to be re-focused during shifts of vision from distant to near objects, and back again. The process of changing focus is called **accommodation**.

Accommodation   Unlike a camera lens, the lens of the eye is not moved in and out to change focus. In the eye, focusing is accomplished by changing the shape of the lens (Fig. 14.9). This is possible because the lens is made of an elastic substance, and it can be stretched into a slim (less convex) shape, but when this tension is released it reforms into a fatter (more convex) shape of its own accord. These changes of shape are brought about by the action of the ciliary muscles.

An eye is focused on a distant object by changing the lens to a flattened (less convex) shape. This reduces to a minimum the power of the lens to bend (refract) light. A flattened shape is needed to focus the almost parallel light rays from distant objects (Fig. 14.9A). The lens is flattened by contraction of the radial ciliary muscles. These pull against the suspensory ligaments, which pull against the lens stretching it into a flatter shape.

An eye is focused on a near object by changing the lens to a rounded (more convex) shape. This increases the lens's power to refract light, which is necessary in order to focus onto the retina the diverging light rays from a near object (Fig. 14.9B). The lens is made more rounded by contraction of the circular ciliary muscles and relaxation of the radial ones. Circular ciliary muscles contract to form a circle with a smaller diameter. This reduces tension on the suspensory ligaments and allows the lens to become rounded in shape. Several sight defects may upset this focusing mechanism. Three of them are listed below:

Presbyopia, or old sight   The lens continues to grow throughout life, but at a very slow rate after adolescence. By the age of about sixty the centre of the lens is so far removed from supplies of oxygen and food that its cells die. This process reduces the lens's elasticity and it can no longer change in shape. It then becomes more or less fixed into a shape suitable only for distant vision. Therefore, old people usually require 'reading glasses' which have converging lenses to give the eyes extra power for close work.

Hypermetropia, or long sight   Long sight occurs when the distance between the lens and the retina is shorter than normal (Fig. 14.10B). An image cannot

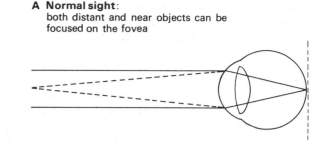

A Normal sight:
both distant and near objects can be focused on the fovea

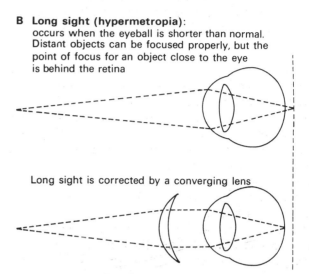

B Long sight (hypermetropia):
occurs when the eyeball is shorter than normal. Distant objects can be focused properly, but the point of focus for an object close to the eye is behind the retina

Long sight is corrected by a converging lens

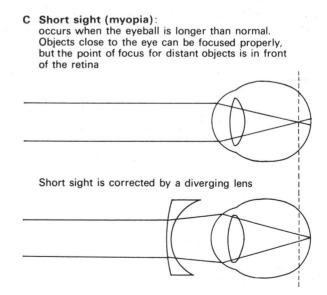

C Short sight (myopia):
occurs when the eyeball is longer than normal. Objects close to the eye can be focused properly, but the point of focus for distant objects is in front of the retina

Short sight is corrected by a diverging lens

Fig. 14.10   Some eye defects and their correction

be focused in so short a distance; in fact the point of clear focus is somewhere behind the retina. Long sight can be corrected by fitting spectacles with converging lenses, which add to the refractive power of the eye.

Myopia, or short sight One cause of short sight is an abnormally elongated eyeball (Fig. 14.10C). That is, one in which the distance between the lens and the back of the eye is so great that the point of clear focus is in front of the retina. This defect can be corrected by fitting spectacles with diverging lenses, which reduce the refractive power of the eye.

The retina The retina consists of 126 million light-sensitive receptors, which are of two types: six million are known as **cones**, and the rest are **rods** (because of their shape).

Surprisingly, rods and cones do not face the light. Like a film which has been put into a camera back to front, the light receptors of the eye are mostly buried under the nerve fibres which conduct impulses from the retina to the brain, and under a layer of capillaries. This results from the way in which the eye develops in the embryo. These obscuring layers are absent from an area directly opposite the lens. This area, the **fovea**, is where the clearest image is formed.

Rods and cones are not evenly distributed over the retina. The fovea consists exclusively of cones, packed tightly with 125 000/mm². In fact, the fovea contains almost all of the 6 million cones in the eye. Elsewhere, the retina consists mostly of rods, packed together with only 6000/mm². However, groups of up to 150 rods are connected to the brain by only one nerve fibre, whereas each cone has its own exclusive nerve fibre connection to the brain or shares it with very few others.

This means that the fovea gives a much clearer visual impression in the brain than the rest of the retina, since an image that falls on the fovea is minutely analysed by the tightly packed cones, which individually or in small groups send separate impulses to the brain. In comparison, an image which falls on rods is less closely analysed, since rods are not so tightly packed as cones, and large patches of them send just one set of impulses to the brain. Accordingly, when a person wishes to examine something carefully he moves his eyes, or the object, until its image falls on the fovea.

Rods at the outermost edge of the retina do not form images at all. They serve only to trigger reflexes which turn the eyes towards objects just beyond the limits of normal vision. This is what happens when something is seen 'out of the corner of the eye'.

There are several other important differences between rods and cones. Rods continue to work in dim light, but they are not sensitive to colour. Cones, on the other hand, work only in bright light, and they are sensitive to colour. These facts explain the differences between day and night vision. Daylight vision is in colour and has precise detail because it results mostly from images which fall on the cones of the fovea. However, the cones stop working as daylight fades, and vision relies more and more on the rods alone. Towards evening, a person gradually loses the ability to see colours, and his vision becomes less distinct.

The **blind spot** in the retina consists of blood vessels and nerve fibres leading to the optic nerve. This part of the eye, as its name implies, is entirely insensitive to light (exercise E).

14.4 Stereoscopic vision and distance judgement

Stereoscopic or three-dimensional vision is best developed in animals whose eyes face forwards (e.g. humans, apes, cats, owls, and to a certain extent fish such as pike). This type of vision depends on both eyes looking at the same object.

In humans, the eyes are about 6.3 cm apart, and when the eyes are focused on an object they each receive a slightly different view (Fig. 14.11). This can be demonstrated by closing one eye at a time and comparing the view from each. The brain puts these two views together and makes from them one three-dimensional impression.

Stereoscopic vision makes it possible to judge distances, but only up to about 50 metres; beyond this point objects produce an almost identical image in both eyes. Distances are also judged by using information from proprioceptors in the eyes. These detect tension changes in the ciliary muscles during accommodation, and in the extrinsic muscles as they swivel the eyes inwards or outwards to look at an object.

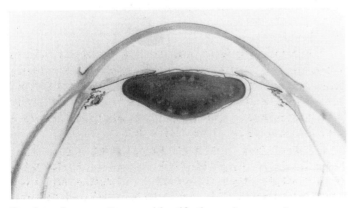

Section of an eye. Can you identify the various parts?

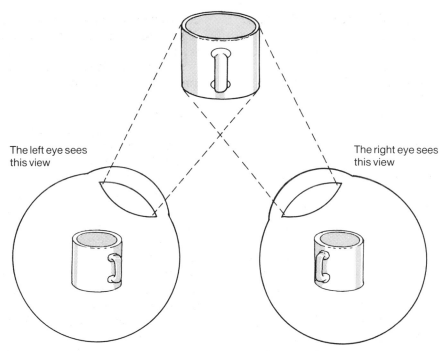

Fig. 14.11 Stereoscopic vision. Each eye sees a slightly different view. The brain puts these two views together to make a three-dimensional impression

The left eye sees this view

The right eye sees this view

14.5 Invertebrate response and behaviour

Invertebrates can respond to a stimulus with quick 'simple' behaviour similar to the reflex actions described in section 13.5. An earthworm, for example, can quickly withdraw into its burrow in response to vibrations in the soil caused by an approaching animal; houseflies can take to the air with incredible speed when they see an approaching fly swatter.

Some responses to stimuli are more complex than these examples. They involve behaviour which is more than a reflex action. **Orientation behaviour** is one example of such responses.

Orientation behaviour

This kind of behaviour allows an organism to orientate itself; that is, to move in a particular direction depending of the stimulus it is receiving. For instance, the organism can orientate itself to move towards a desirable stimulus, or away from an unpleasant stimulus.

Orientation behaviour includes plants growing towards light, animals moving towards food, or away from fire; parasites moving towards their host; and sperms swimming towards an ovum.

Kinesis

The simplest kind of orientation behaviour is known as kinesis. A kinesis occurs when an organism responds to a stimulus which is coming from no particular direction. If the stimulus is unpleasant the organism responds by moving more quickly and by turning more frequently to one side or the other until the unpleasant stimulus is less intense. Speed of movement and turning then decrease. If there is no unpleasant stimulus, or if conditions are entirely satisfactory, then movement stops.

Response of woodlice to dry conditions is an example of a kinesis. Woodlice are placed in a **choice chamber** (Fig. 14.12), half of which is dry and the other half humid. The animals move rapidly and turn frequently when in the dry half. But they slow down, and eventually stop when in the humid half. As a result they eventually gather in the humid half.

Taxis

A taxis is orientation behaviour which occurs when an organism responds to a stimulus coming from one direction. For example a plant's phototropic response to light is also known as positive phototaxis. The movement of sperms towards chemicals secreted by ova is called positive chemotaxis.

The response of blowfly larvae (maggots) to light is another example of phototaxis. The apparatus used to investigate this response is a transparent tube shaped like a letter 'T'. One arm of the 'T' is painted black. One fly larva is placed in the tube, and illu-

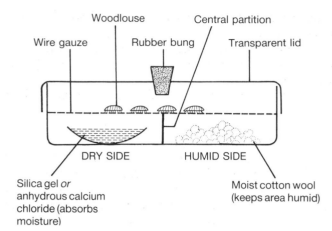

Fig. 14.12 A choice chamber. Used to investigate orientation behaviour (kinesis) in woodlice

minated as shown in Figure 14.13. The larva moves rapidly away from the light (i.e. it demonstrates negative phototaxis). When it reaches the T-junction it has a choice between entering the dark (black painted) arm, or the illuminated (transparent) arm. If it enters the dark arm it slows down and stops. If it enters the illuminated arm it continues to move up and down. This movement continues indefinitely, unless it finds the entrance to the dark arm.

Orientation behaviour helps organisms to find conditions which best suit their way of life.

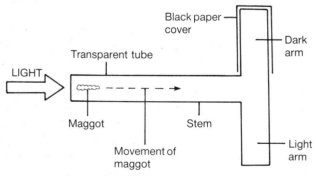

Fig. 14.13 A T-shaped maze. Used to investigate orientation behaviour (taxis) in blowfly larvae

Investigations

A   *An investigation of skin receptors*

1. *To discover the distribution of touch receptors in the skin.*
a) Pass two needles through a piece of bottle cork so that their points protrude the same distance

and are 1.5 cm apart. Prepare another cork in the same way but with the needle points only 0.5 cm apart.

b) Ask someone to sit with his eyes closed and one arm bare to the elbow.

c) Beginning with the needle points 1.5 cm apart, touch the skin of the person's forearm gently with either one or both points. Ask the subject if he feels one or two points. He must not guess. Repeat, using one or two points in random order until two points have been used five times on the forearm. Record the answers received when two points were used.

d) Repeat this procedure on the palm of one hand and then on the finger-tips. Repeat the whole experiment using needle points 0.5 cm apart.

e) Compare the number of correct answers obtained with each set of needles on the forearms, palms, and finger-tips. What do these results indicate about the density of touch receptors in these three regions?

f) How is it possible for a person to be touched with two needle points but feel only one?

2. *An investigation of heat and cold receptors*
a) Obtain several knitting needles (preferably size 12 or 13 – $2\frac{3}{4}$ or $2\frac{1}{4}$).

b) Place half the needles in hot (not boiling) water, and the rest in ice cold water for at least two minutes.

c) Take a needle from the hot water, dry it, then run its point gently but slowly over the skin on the back of the hand. Place an ink mark wherever the heat sensation is felt most strongly. Continue, changing the needles as they cool, until several 'hot spots' have been found.

d) Use dried cold needles to find out if 'hot spots' are sensitive to cold, or if there are separate 'cold spots'.

e) Devise an experiment to find out the smallest difference in temperature to which the fingers are sensitive.

3. *An investigation of sensory adaptation*
a) Obtain three 500 cm³ beakers. Fill one with hot (not boiling) water, a second with ice-cold water, and a third with lukewarm water.

b) Place the fingers of one hand in the hot water, and fingers of the other hand in the cold water. After about 1 minute place the fingers of both hands into the lukewarm water.

c) Does the lukewarm water feel different to each hand? If so, does this difference persist for long? How does this experiment illustrate sensory adaptation?

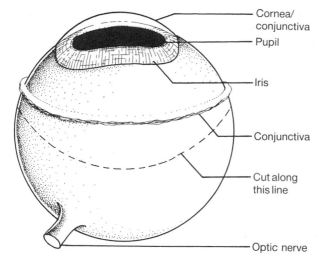

Fig. 14.14   A bull eye. See Practical B

B   *Dissection of an eye*
   1. Look for the following external features of a bull or sheep eye: pupil, iris, sclerotic, optic nerve, remains of the extrinsic muscles.
   2. Using sharp scissors cut around the eyeball (see dotted line in Fig. 14.14). This will separate the eye into front and rear portions.
   3. Lay the two portions in a dish of water so that the inside of the eyeball can be examined. In the front part of the eye look for: the lens and suspensory ligaments. Cut the lens away from the wall of the eyeball and observe its shape. Look for the iris diaphragm. In the rear part of the eye look for the vitreous humour; black pigment of the choroid; the retina, fovea, and blind spot.

C   *Changes of pupil size*
   1. Hold a mirror close to the face, close the eyes and cover them with the hands for about 15 seconds. Quickly remove the hands and open the eyes, and observe how the pupils alter in size.
   2. Observe changes in pupil size when a person shifts his gaze from a distant object to an object about 1 metre away.
   3. What changes in pupil size take place in each situation, and what is the significance of these changes?

D   *Observing accommodation movements in the human eye*
   1. In a darkened room it is possible to see three reflections of a point of light (such as a candle flame) in a person's eye. These reflections come from the front surface of the cornea, and the front and rear surfaces of the lens.
   2. Observe these reflections carefully while the subject shifts his gaze from a near to a distant object and back again.
   3. Which of the three reflections change in size? What causes these changes?

E   *Detecting the blind spot*
   1. Hold this book with Figure 14.15 at arm's length. Close the left eye and stare at the cross with the right eye. (Note that the black circle is still visible.) Bring the book slowly towards the face. At a certain point the circle will disappear. This happens when its image falls on the blind spot.
   2. Why don't the blind areas in each eye interfere with normal vision?

Fig. 14.15   Blind spot experiment (exercise E)

F   *Investigating invertebrate behaviour*
   1. Using a choice chamber.
   Obtain a long glass trough, or perspex box with a tight-fitting lid, which has a central hole closed by means of a rubber bung (Fig. 14.12).
   a) Leave the apparatus to stand for 10 minutes to establish a humidity gradient, then place 10 woodlice of the same species in the apparatus – drop them through the bung hole. Replace the bung.
   b) After 10 minutes, count the number of animals in each half of the chamber. Remove the animals and repeat the experiment at least five times (ten times if possible).
   If humidity has no effect on woodlice, roughly equal numbers should be found on either side. You can, therefore, compare your actual result with this possible result. Serious study of these results involves the use of a statistical quantity known as the **standard error**, details of which can be found in any statistical text book.

   2. *a*) Investigating the effect of humidity on behaviour, without a choice chamber.
   Prepare two perspex or glass containers with tight-fitting lids, so that one has a dry atmosphere and one has a humid atmosphere. Place woodlice or flour beetles in each container and note any difference in behaviour.
   b) Investigating the effects of light on behaviour.
   i) Illuminate a number of woodlice or fly larvae with a lamp positioned exactly above their container, and observe their behaviour.

ii) Set up two lamps at right angles to each other. Observe behaviour with both lamps switched on, and then with the lamps switched on alternately. Do the animals exhibit positive or negative phototaxis?

Questions

1. Explain how each of the following facts illustrates a difference between the fovea and the remainder of the retina.

a) When the eyes are fixed on a single letter in the middle of a word in the middle of a printed page, no more than two or three letters on either side of that particular letter are clearly visible.

b) If a very dim star is looked at directly it disappears from view, but it reappears when looked at indirectly 'out of the corner of the eye'.

c) When a person buys a coloured garment in a dimly lit shop he often finds that it appears to be a different colour when seen in daylight.

d) It is difficult to read small print in dim light, and those who persist in doing this often develop defective sight.

2. Match the technical terms listed below with the phrases which are examples of each.

Negative chemotaxis	Woodlice moving out of dry conditions
Positive phototaxis	Sperms swimming towards an ovum
Positive chemotaxis	Moths flying towards light
Kinesis	Earthworms moving away from an unpleasant chemical

3. The diagram below shows a section through the human eye

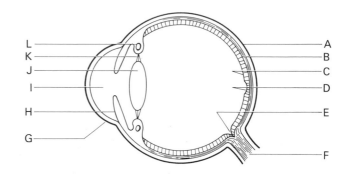

a) Name the parts labelled A to L.

b) What changes take place in the shape of J: when the eye is focused on a near object; when the eye is focused on a distant object?

c) What changes take place in L: to focus the eye on a distant object; to focus the eye on a near object?

d) What type of light-sensitive cell: is more numerous at D than C; is more numerous at C than D; is sensitive to colour; can work in dim light?

Summary and factual recall test

Photoreceptors are sensitive to (1). Mechanical receptors are sensitive to stimuli such as (2) and (3). Chemical receptors include (4)-buds and smelling or (5)-organs. Examples of proprioreceptors are (6– name two).

Touch receptors enable a person to distinguish between (7) and (8) substances and (9) and (10) textures. Temperature receptors cannot be used to tell the exact (11) of something; they can only (12). Pain receptors are useful because (13). Taste receptors are called (14). There are separate ones for (15–list four different tastes). Smell receptors are called (16). Unpleasant tastes and smells are useful because (17).

Human eyes are protected in the following ways (18–list four). They are held in place by six (19) muscles, which also control eye (20).

The iris has a (21) in its centre called the (22). This controls the amount of (23) entering the eyes, and the (24) of focus of vision in the following way (25).

To focus the eye on a near object the (26) ciliary muscles (27), which (28) the tension on the (29) ligaments, which allows the lens to become (30) in shape.

This (31) its power to (32) light. To focus on a distant object the (33) ciliary muscles contract, which makes the lens (34) in shape.

(35), or old sight is caused by reduced (36) of the lens, so that (37) objects cannot be seen clearly. (38), or long sight, occurs when (39), and it is corrected by (40) lenses. (41), or short sight, occurs when (42), and is corrected by (43) lenses.

An image is formed on the (44) of the eye. This layer consists of light-sensitive cells called (45–two names). These differ from each other in the following ways (46–describe their sensitivity and in what light they work best).

Good stereoscopic vision depends on having eyes which face (47), as in (48–name three animals) Stereoscopic vision helps judgement of (49).

Behaviour which allows invertebrates to direct their movement according to stimuli received is called (50) behaviour. Examples are (51–name two). Movement in response to stimuli from no particular direction is called (52). An example is (53). A taxis is (54). Examples are (55).

15

The reproductive process

Organisms have a limited life span. They all die eventually, either from accidents, diseases, or old age. And yet life continues, because organisms have the capacity to create new organisms by reproducing. Indeed, some organisms from each species must reproduce or that species will die out and disappear from the earth. Reproduction gives rise to new organisms with the same basic characteristics as their parents: sunflower seeds always grow into sunflowers and not cabbages; and cats always produce kittens and not puppies. This and the following two chapters describe many examples of reproduction among protists, animals, and plants. First, however, it is necessary to deal with the general features of the reproductive process.

15.1 The reproductive process

How is a new organism created? To answer this question it is necessary to refer to what was said in chapter 2 about the structure of a cell nucleus, and its division by mitosis.

Every cell in every organism contains a set of 'instructions', in chemical form, for building the whole organism. These instructions exist as a pattern of molecules that make up structures called chromosomes, which are situated in the nucleus of each cell. Since every cell contains these instructions it follows that a new organism can, theoretically, be created from any cell of a living organism. If this were true in practice, an organism could produce another one like itself by releasing any cell from its body under conditions which would allow that cell to divide by mitosis.

In fact, this is more or less what does happen, except that new organisms do not arise from just any cell from the parent. They usually arise from specialized **reproductive cells**, and these are often produced by specialized **reproductive organs**.

Thus, the answer to the question: 'How are new organisms created?' is that they arise from cells released from the bodies of parent organisms, and develop according to building instructions contained in the chromosomes of these cells. These building instructions are called the **hereditary information** of the species, and, by means of the reproductive process, they are passed from one generation to the next in the chromosomes of the reproductive cells. If a species becomes extinct it can never arise again because its hereditary information is lost forever.

There are two ways in which organisms give rise to new organisms: by **asexual reproduction**, and by **sexual reproduction**.

15.2 Asexual reproduction

Asexual reproduction involves only *one* parent. It occurs mainly in organisms whose bodies have a simple structure. *Amoeba*, for example, reproduces asexually by dividing into two, and mould fungi reproduce asexually by producing special reproductive cells called spores. Spores develop into new mould colonies.

New organisms produced by asexual reproduction are almost identical to their parent, since they have only one source from which they can inherit hereditary information.

In asexual reproduction, the parent organism produces by mitosis a number of cells with chromosomes that are precise copies of those in its own body cells (section 2.6). When these cells develop they form new organisms with the same hereditary information as their parent. When the young grow up and reproduce asexually the same hereditary information is copied by mitosis again, and passes unchanged to form the next generation and so on as long as the organisms continue to reproduce asexually.

15.3 Examples of asexual reproduction

Amoeba

Amoebae live in mud at the bottom of ponds and slow-moving streams. The largest are 0.5 mm in diameter, and are just visible to the naked eye as white blobs of jelly-like protoplasm. Their most noticeable feature is that their shape is constantly changing.

Amoebae and many other unicellular organisms reproduce asexually by the simplest of all methods: they divide into two parts. The technical name for this is **binary fission**. First, the nucleus divides by mitosis. Then the cytoplasm divides into two parts, one part around each nucleus, and eventually the organism separates into two daughter cells (Fig. 15.1). When this happens the original parent ceases to exist, and becomes instead two new identical individuals. There is no evidence of sexual reproduction in *Amoebae*.

Bread mould (mucor)

Like other moulds, *mucor* reproduces asexually by forming enormous numbers of **microscopic spores**. Spores are produced at the tips of upright hyphae. Cytoplasm and nuclei flow into these hyphae, making them swell so that each eventually forms a rounded spore case or **sporangium** (Fig. 15.2A and B). At this stage sporangia on mouldy food look like thousands of tiny black dots. Soon the cytoplasm in the sporangium separates into hundreds of tiny round spores, each containing several nuclei. A fully formed sporangium bursts open, exposing powdery spores which are blown away by air currents. If spores land on food in warm damp conditions they split open and produce hyphae which become new mycelia (Fig. 15.2C and D).

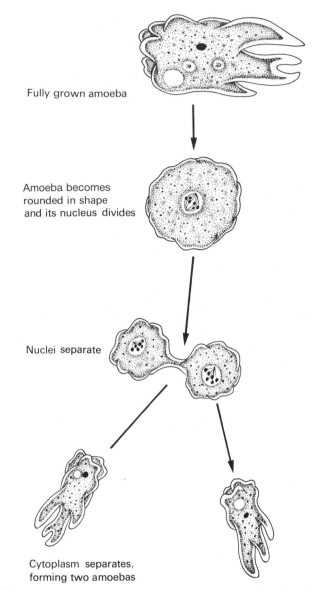

Fully grown amoeba

Amoeba becomes rounded in shape and its nucleus divides

Nuclei separate

Cytoplasm separates, forming two amoebas

Fig. 15.1 Asexual reproduction by binary fission in *Amoeba*

15.4 Advantages and disadvantages of asexual reproduction

Advantages

The main advantage of asexual reproduction is that it is very simple and very fast.

For example, provided it has food and warmth a single bacterium can produce 16 million bacteria in 8 hours simply by dividing in two every 20 minutes! A mould colony the size of a 50 pence coin produces tens of thousands of spores a day. In fact there are countless billions of fungal and bacterial spores floating in the atmosphere, as far as the boundaries of space.

Disadvantages

Young produced asexually are identical to their parents, and are referred to collectively as a **clone**. Consequently, asexual reproduction does not allow a species to change from one generation to the next, except in very minor ways. Young, therefore, share any defects of their parent. In addition, it is impossible for evolutionary change to take place so that new and more advanced species are produced, or so that a species can adapt to changes in its environment.

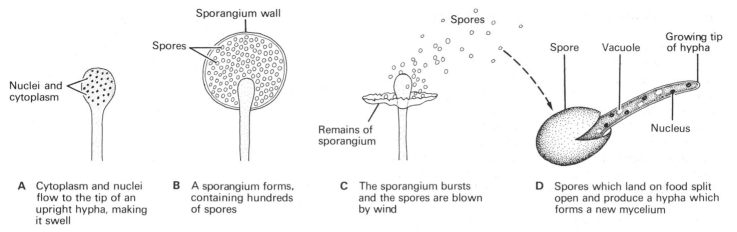

A Cytoplasm and nuclei flow to the tip of an upright hypha, making it swell

B A sporangium forms, containing hundreds of spores

C The sporangium bursts and the spores are blown by wind

D Spores which land on food split open and produce a hypha which forms a new mycelium

Fig. 15.2 Diagram of asexual reproduction in bread mould

15.5 Sexual reproduction

Unlike asexual reproduction, sexual reproduction almost always involves two parent organisms. The parents give rise to reproductive cells called **gametes** by a type of cell division called meiosis (section 20.3). A gamete contains a copy of part of the hereditary information of the organism which produced it.

During sexual reproduction a gamete from one parent fuses with a gamete from another parent. This process is called **fertilization** and results in a single cell called a **zygote** which contains two different sets of hereditary information: one set from each parent.

When the zygote develops into a new organism it does not use all this information; it selects some from each parent according to 'rules' which are studied in the science of genetics. Consequently, the new organism is not an exact copy of either parent. It shows some features of both.

The fact that sexually produced organisms differ in several ways from their parents is one of the most important factors in the process of evolution because these differences make it possible for a species as a whole to change, in time, and give rise to an entirely new type of organism. Figure 15.3 summarizes these points, and gives the main differences between

Fig. 15.3 Diagram of the differences between asexual and sexual reproduction

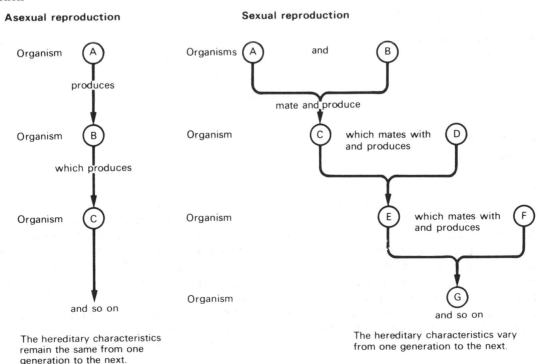

Asexual reproduction

Organism (A)

produces

Organism (B)

which produces

Organism (C)

and so on

The hereditary characteristics remain the same from one generation to the next.

Sexual reproduction

Organisms (A) and (B)

mate and produce

Organism (C) which mates with and produces (D)

Organism (E) which mates with and produces (F)

Organism (G)

and so on

The hereditary characteristics vary from one generation to the next.

asexual and sexual reproduction. Several aspects of sexual reproduction should be understood before actual examples are studied. The following are the most important.

Gametes and sex organs

In most species there is a clear difference between the male and female organisms, and between the gametes which they produce. Where this is so the female produces gametes which are larger than the male's. The female gamete is larger because it contains stored food in its cytoplasm. This food is used to nourish the embryo which develops from the zygote. Examples of female gametes are the large shelled eggs of birds, the smaller eggs of other animals, and the ovules of plants.

Male gametes move, or are moved, to the larger female gamete. Sperms of mosses, ferns, and animals are examples of male gametes which move. Sperms are equipped with tail-like structures with which they swim to the female gamete. Pollen grains of flowering plants are examples of male gametes that are moved, by wind and insects for instance, to the female.

A species with distinctly different male and female gametes usually has specialized reproductive organs, often called **sex organs**. Examples of sex organs are the testes (singular testis) which produce sperms in male animals, and the ovaries which produce eggs in female animals.

15.6 Types of fertilization

In some organisms fertilization occurs after gametes have been shed from the parents' bodies. This is called **external fertilization**, and is typical of animals such as fish and amphibia, which breed in water. Animals whose gametes fuse externally generally have patterns of mating behaviour which ensures that sperms are shed either directly on to the eggs, or nearby in the water. Furthermore, sperms do not swim at random through the water until they reach an egg by accident; they swim towards the nearest one, attracted by chemicals which it produces.

Internal fertilization takes place inside the female parent's body, usually in her reproductive organs. This is typical of land animals such as reptiles, birds, and mammals, but is also found in certain types of fish and amphibia. Internal fertilization entails a mating behaviour called **copulation**, during which the male passes sperms from his sex organs directly into the female's body. Once inside her body the sperms behave as in external fertilization: they swim to the eggs attracted by chemicals (Fig. 15.4C).

Advantages of internal fertilization

Internal fertilization has certain advantages over external fertilization. First, it increases the chance that fertilization will occur, since sperms are shed into a confined space – the female reproductive system – which they share with the eggs. Second, the increased chance of fertilization is probably related to the evolution of females which produce a small number of eggs. A frog, for example, produces an enormous number of eggs, whereas a human female produces only one egg a month, which has the advantage of saving body materials. Third, internal fertilization means that the zygote can be retained in the female's reproductive system where it develops in a protected environment.

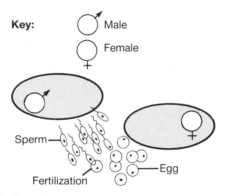

Fig. 15.4A Random external fertilization. Eggs and sperms are released at random in water. Large quantities of each are produced since many sperms will fail to find an egg

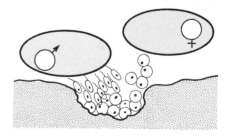

Fig. 15.4B External fertilization in a nest. Eggs and sperms are deposited into a specially prepared hollow or other nest. This greatly increases the chances of fertilization

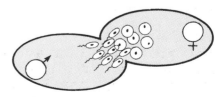

Fig 15.4C Internal fertilization. Sperms are passed into the female's body where fertilization occurs. Few eggs are produced since eggs and sperms are in a confined space

Life cycles

Reproduction is part of a sequence of events called a **life cycle** in which new organisms are created, develop to maturity, and repeat the process. In addition to details of asexual and sexual reproduction, this chapter and the following two include information about the life cycles of various organisms.

15.7 Birth rates and death rates

There is a limit to the number of organisms which can live in the world, because there is a limit to the amount of land and water they can occupy. Each individual organism breeds as fast as it can in a race that decides whose babies will fill this available space.

Living things try to win this race in two different ways. They can produce huge numbers of small young which are scattered uncared for into the environment. Alternatively, they can produce a few large young which are cared for until they can look after themselves.

The first of these methods involves a high birth rate, while the second makes use of a low birth rate.

High birth rates

Section 15.4 has mentioned the uncountable numbers of spores produced asexually by moulds and other fungi. Many organisms achieve a huge output of young using sexual methods. Houseflies, gnats, midges, sycamore trees, and weeds such as dandelions and groundsel all follow the high birth rate strategy. It has obvious advantages.

Flood the environment with young, and it is likely that some will find whatever living space is available – like freshly dug soil which is quickly covered with weeds, and open wounds which are infected with germs.

The cost of a high birth rate strategy is grim. If huge numbers of tiny young are shed straight into the environment without any parental care, most will die. Many are eaten by predators, and many fail to reach a suitable living space.

A high death rate is an almost inevitable consequence of a high birth rate.

Low birth rates

Organisms which employ the low birth rate strategy put much time and effort into producing a low number of big, strong, young. Each baby has a good chance of survival. It is big to start with, and is fed and protected by its parents until old enough to make its own living.

This breeding strategy is employed to a certain extent by fish such as sticklebacks, and by reptiles such as crocodiles. Birds and mammals have the most highly developed parental care system.

Frogs produce large numbers of small young which are released, uncared for, into the environment. What are the advantages and disadvantages of this strategy?

Birds have small numbers of large young which are fed and protected until old enough to look after themselves. What is the main problem with this breeding strategy?

The success of this method depends on one condition. Producing the right number of young. If too many are produced, there may not be enough food to go round and some will die. If too few young are produced, neighbours with bigger families will take over more of the available living space.

Questions

1. What is the main difference between asexual and sexual reproduction?

2. Match the words in the left-hand column with the appropriate descriptions in the right-hand column:

testis	male sex cell
sperm	male sex organ
ovary	the fusion of a male and female gamete
fertilization	female sex organ
zygote	female sex cell
ovum	a fertilized egg cell

3. a) Name two animals which reproduce by external fertilization and two which reproduce by internal fertilization.

b) Which group of animals reproduces by both internal fertilization and internal development?

c) Why does external fertilization require the production of many more gametes than internal fertilization?

d) What are the advantages of internal development over developments outside the female's body?

Summary and factual recall test

In (1) reproduction there is one parent. This parent organism produces cells, by a type of cell division called (2), containing (3) information which is a precise (4) of that contained in the parent's cells. This type of reproduction does not allow a species to (5) from one (6) to the next.

In (7) reproduction there are usually two parents. These produce cells called (8) by a type of division called (9). The fusion of two of these cells is called (10), and results in a single cell called a (11). This cell develops into an offspring which differs from both parents because (12).

Most organisms are unisexual which means (13), whereas some are hermaphrodite which means (14).

Most fish carry out (15) fertilization. Mammals carry out (16) fertilization which takes place within the (17) of the female by means of mating behaviour called (18). The main advantage of this type of fertilization is (19).

Amoeba reproduces asexually by (20)-fission. Mucor reproduces asexually by forming huge numbers of (21) in the following way (22).

Other organisms with a high birth rate are (23–name four). A high birth rate ensures that some young will find (24) space, but most will (25). Examples of animals with low birth rates are (26–name two). Most have highly developed (27)-care. The advantage of this is that (28).

16

Reproduction in plants

16.1 Flower structure

Flowers are the most advanced and complex reproductive structures in the plant kingdom. Their function is sexual reproduction, as a result of which they form fruits and seeds that give rise to the next generation. There is an immense variety of shape and structure among flowers, but all of them have certain features in common. These features, with their scientific names, are described below, and they are illustrated in Figure 16.1.

Pedicel
Flowers grow on a specialized reproductive shoot called the flower stalk or pedicel. In some plants, such as the tulip, there is only one pedicel bearing a single flower. In others, such as lilac, the pedicel is branched in various ways and bears many flowers.

Receptacle
The receptacle is the region of a flower stalk to which the parts of the flower are attached. Receptacles can be flat, dome-shaped, or concave depending on the species. The parts of a flower are attached to the receptacle in rings, or **whorls**. Starting with the innermost whorl the floral parts are arranged as follows:

Carpels
Carpels are situated at the centre of the receptacle, and are known collectively as the **gynoecium** or **pistil**. Carpels are the female reproductive organs. Each carpel consists of an expanded hollow base called the **ovary**, above which is a narrow region called the **style** which ends in a pointed, flattened, or sculptured region called the **stigma**. The stigma receives pollen grains from the same or another flower during pollination.

Within the ovary are varying numbers of **ovules**. At the centre of each ovule is a large cell called the **embryo sac**, and this contains several nuclei, one of which is the female gamete, or **egg nucleus**. It is this egg nucleus which is fertilized by a 'male' nucleus from a pollen grain, after which the whole ovule develops into a seed, and the ovary wall becomes the **pericarp**.

Some plants have pistils made up of separate carpels (e.g. buttercups), but in most the pistil is a number of carpels fused together (e.g. poppy).

Stamens
Surrounding the gynoecium is a whorl of stamens. Stamens are known collectively as the **androecium**. These are the male reproductive organs. Each consists of a stalk, or **filament**, bearing an **anther** which is made up of four **pollen sacs** in which **pollen grains** are formed. Pollen grains contain the male gametes.

Fig. 16.1 Diagram of a flower (cut in half)

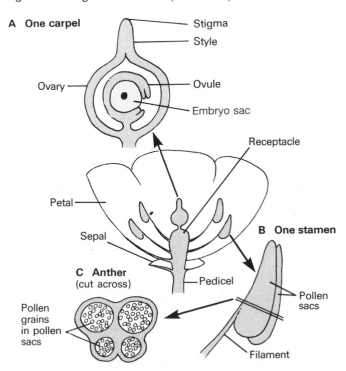

A Whole flower

Anther

Petal

Sepal

B Reproductive organs

Stigma

Anther

Style

Filament

Ovary

Position of sixth stamen
(removed to expose ovary)

Pedicel (flower stalk)

Flower bud

Stigma

Style

C Carpel cut in half

Ovary wall

Ovules

Pedicel (flower stalk)

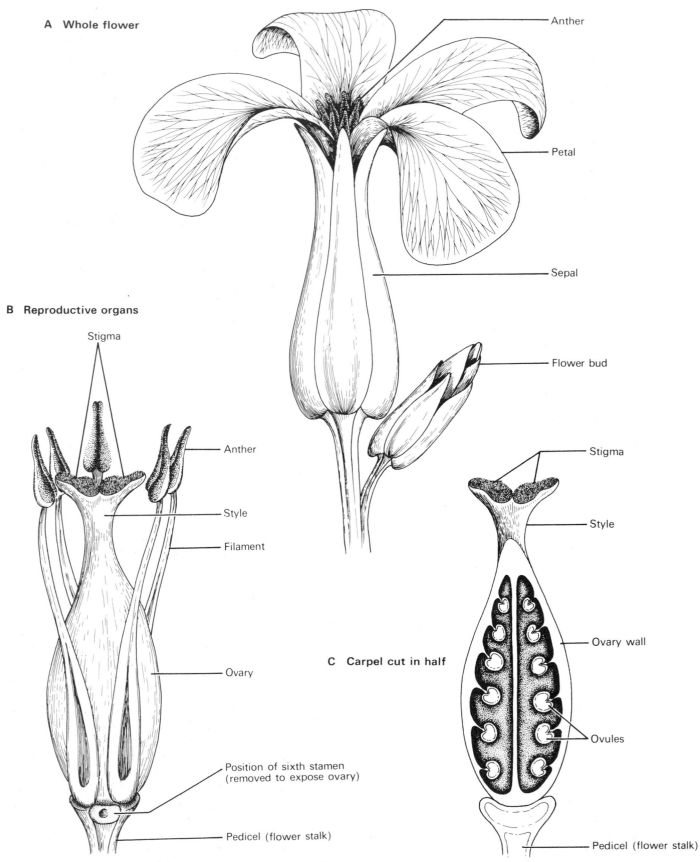

Fig. 16.2 Structure of a wallflower

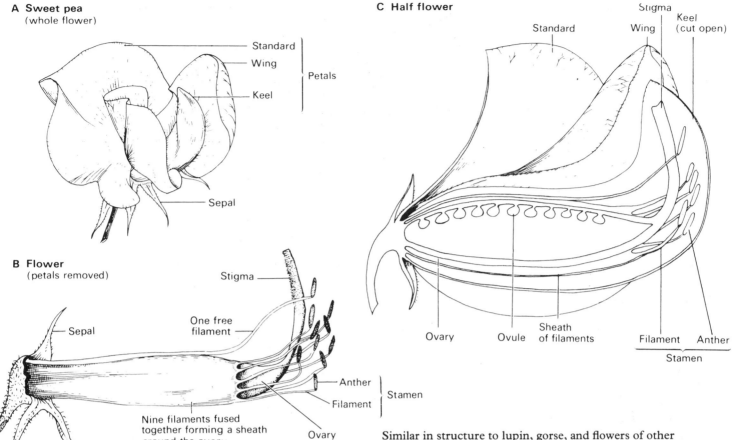

A Sweet pea
(whole flower)

Standard
Wing
Keel

Petals

Sepal

B Flower
(petals removed)

Stigma

Sepal

One free
filament

Nine filaments fused
together forming a sheath
around the ovary

Anther
Filament

Stamen

Ovary

Fig. 16.3 Structure of a sweet pea flower

C Half flower

Standard

Stigma
Wing
Keel
(cut open)

Ovary

Ovule

Sheath
of filaments

Filament Anther

Stamen

The structures described so far are characteristic of all flowers, but this is not true of the following parts. One or all of them may be absent, according to the species.

Petals

In the majority of flowers the reproductive organs are surrounded by a whorl of petals. Petals are known collectively as the **corolla** of the flower. Some flowers have coloured and scented petals with a **nectary** at the base which produces sugary nectar. Petals of this type attract insects which come to collect the nectar and by doing so transfer pollen from one flower to another.

Sepals

In many flowers there is an outermost whorl of sepals, known collectively as the **calyx**. Sepals are usually green and look like small leaves. They enclose and protect the central region of the flower when it is in the bud stage of development. The calyx and corolla together are referred to as the **perianth**.

Similar in structure to lupin, gorse, and flowers of other leguminous plants. Pollination is mostly by bees, which land on the 'wings' and 'keel'. A bee inserts its tongue into the keel in search of nectar which is produced at the base of the nine fused filaments, and gathers in the trough formed by this structure. The insect's weight depresses the petals so that the anthers and stigma protrude from a hole at the end of the keel, and touch the insect's underside. The flowers are protandrous, and so fertilization occurs only when bees visit young flowers and then older flowers

16.2 Flower shape and symmetry

Compare a wallflower (Fig. 16.2) with a sweet pea (Fig. 16.3) and a white dead nettle (Fig. 16.4). Wallflower petals are about the same size and shape as each other, and the flower as a whole can be cut in two along many vertical planes to produce identical halves. Such flowers are said to be **regularly** shaped, and **radially symmetrical**. Other regular flowers are the rose, tulip, poppy, and daffodil.

In contrast the petals of sweet pea and white dead nettle flowers are of several different shapes and sizes, and some are fused together. Such flowers are said to be **irregularly** shaped, or **zygomorphic**. They are also **bilaterally symmetrical**, which means they can be cut into identical halves along only *one* vertical plane. Other irregular flowers are the foxglove, and antirhinum (snapdragon).

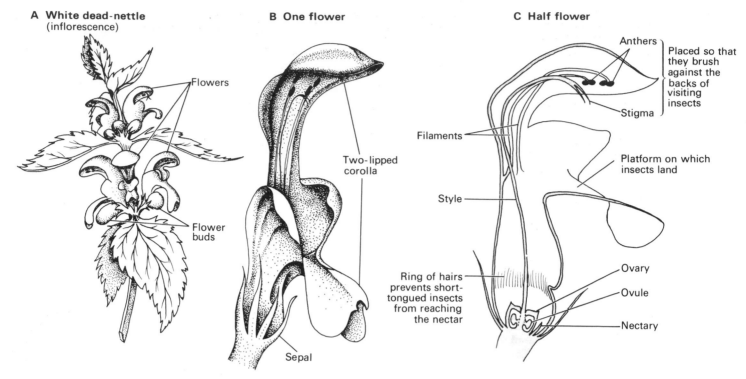

A White dead-nettle
(inflorescence)

Flowers

Flower
buds

B One flower

Two-lipped
corolla

Sepal

C Half flower

Anthers

Placed so that
they brush
against the
backs of
visiting
insects

Stigma

Filaments

Platform on which
insects land

Style

Ring of hairs
prevents short-
tongued insects
from reaching
the nectar

Ovary

Ovule

Nectary

Fig. 16.4   Structure of white dead nettle flower
Pollinated mostly by bees, which land on the lower lip and search
for nectar at the bottom of the tubular corolla. In this position
the anthers and stigma touch the insect's back. The flowers are
protandrous, and the forked stigma remains closed until the
pollen has been shed

Another very distinctive flower shape is found in
the largest group of flowering plants: the Compositae,
or **composite** flowers, e.g. daisy and dandelion (Fig.
16.5). They are called composite flowers because they
consist of many tiny flowers, the **florets**, packed
together on a large flattened receptacle. Dandelion
florets, for instance, have five petals but these are
fused together making a corolla which looks like one
long petal. This arrangement is called a **ray floret**.
In other composite flowers, e.g. thistle, the florets
are simple tubes called **disc florets**. Daisies and
sunflowers have ray florets around the edge, and disc
florets in the centre.

16.3   Pollination

Pollination is the transfer of pollen grains from
anthers to stigmas, and eventually leads to fertilization. **Self-pollination** is the transfer of pollen
from anthers to stigmas in the same flower, or
between flowers on the same plant. **Cross-
pollination** is the transfer of pollen from one plant
to the stigmas on another plant of the same species.

Fertilization does not result when pollination occurs
between different species.

In the course of evolution, plants have developed
characteristics which help to prevent self-pollination
and favour cross-pollination. This ensures that, more
often than not, fertilization involves the inter-mixing
of hereditary information from two plants, thereby
increasing the chance of further evolutionary change.
Self-pollination is obviously impossible in unisexual
species: that is, species in which some plants have
flowers containing stamens, while other plants have
flowers containing carpels. Bisexual plants (that is,
plants with flowers containing both stamens and carpels) have more elaborate methods of preventing self-
pollination. In the dandelion, for example, stamens
ripen and release their pollen before the carpels in
the same flower are fully developed. This is called
protandry. In the plantain, however, the carpels
mature before the stamens. This is called **protogyny**.

There are two main ways in which pollen is transferred from anthers to stigmas: on wind currents, and
on the bodies of insects. Most flowers are so adapted
that they can be pollinated by one or the other of
these methods.

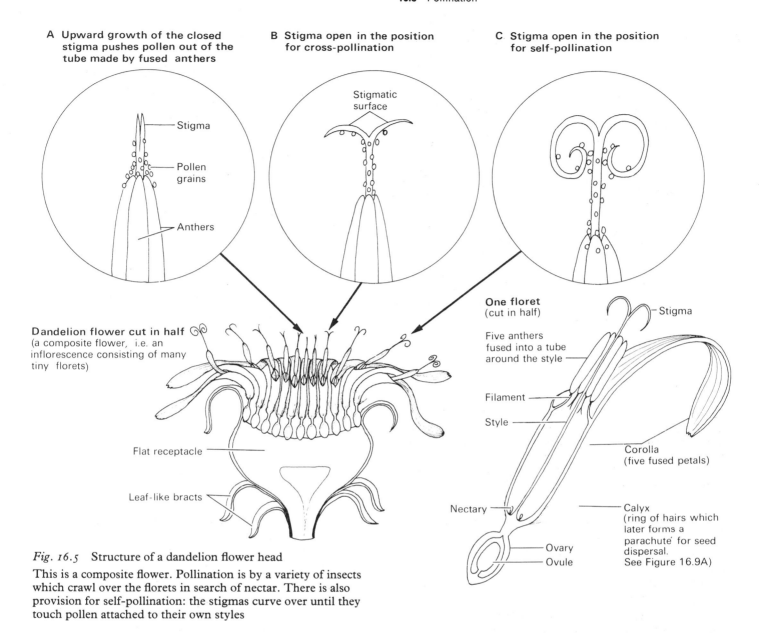

A **Upward growth of the closed stigma pushes pollen out of the tube made by fused anthers**

B **Stigma open in the position for cross-pollination**

C **Stigma open in the position for self-pollination**

Stigma

Pollen grains

Anthers

Stigmatic surface

Dandelion flower cut in half
(a composite flower, i.e. an inflorescence consisting of many tiny florets)

Flat receptacle

Leaf-like bracts

One floret
(cut in half)

Five anthers fused into a tube around the style

Filament

Style

Nectary

Ovary

Ovule

Stigma

Corolla (five fused petals)

Calyx (ring of hairs which later forms a 'parachute' for seed dispersal. See Figure 16.9A)

Fig. 16.5 Structure of a dandelion flower head

This is a composite flower. Pollination is by a variety of insects which crawl over the florets in search of nectar. There is also provision for self-pollination: the stigmas curve over until they touch pollen attached to their own styles

Wind pollination

Grass (Fig. 16.6), stinging nettle, hazel, and willow are examples of plants pollinated by wind. Wind-pollinated flowers usually have no petals, but some have small inconspicuous petals which are often white or green. Nectaries are absent altogether. Most wind-pollinated flowers have all or some of the following characteristics which increase the chances that pollen will reach their stigmas:

a) Abundant pollen production. This compensates for the fact that each grain has an extremely small chance of reaching a stigma;

b) Small, light pollen grains, which float on the lightest breeze;

c) Large anthers which often have long filaments so that they hang well outside the flower, and are attached so that they sway and shake out pollen in the lightest breeze;

d) Spreading, 'feathery' stigmas which act like a net, catching pollen as it floats through the air;

e) Flowers which are either on long stalks well above the leaves, or which develop from flower buds that open before the leaf buds. Both these features increase the flowers' exposure to air currents.

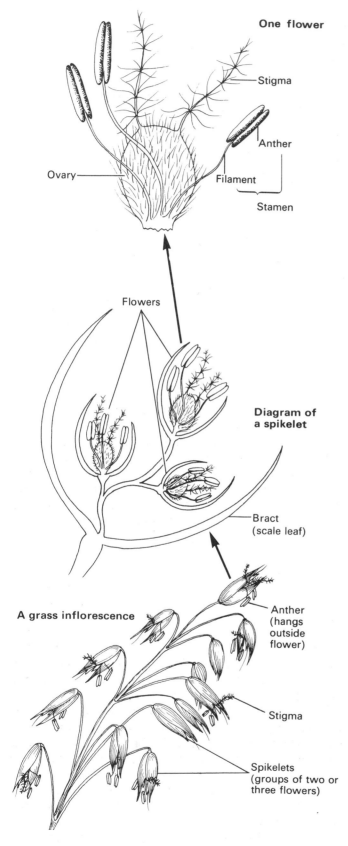

One flower

Stigma

Anther

Filament

Ovary

Stamen

Flowers

Diagram of a spikelet

Bract (scale leaf)

A grass inflorescence

Anther (hangs outside flower)

Stigma

Spikelets (groups of two or three flowers)

Fig. 16.6 Structure of a typical grass flower
Note the features of wind pollination

Insect-pollinated flowers

Buttercup, sweet pea, dead nettle, and dandelion are some examples of flowers pollinated by insects. These flowers have some or all of the following characteristics which attract insects, and ensure pollination:

a) Large, often brightly coloured and scented petals with nectaries. Some flowers have petals with grooves or dark lines leading from the petal border to the nectaries. These are **honey guides**, and are thought to 'guide' insects to the source of nectar, e.g. pansy and foxglove.

b) The production of a few large pollen grains. In some plants the grains are smooth and sticky, in others they are covered with spiky hairs. Both of these features help to make the grains cling to the insects' bodies;

c) Anthers tend to be small, and situated inside the flower where insects are likely to brush against them;

d) Stigmas are also situated so that visiting insects brush against them. Stigmatic surfaces often produce a sticky, sugary fluid to which pollen grains become attached.

16.4   Fertilization

Of the thousands of pollen grains released from the anthers of a flower only a tiny fraction reach the stigmas of another flower of the same species. But even when a grain has made this long journey and become firmly attached to a stigma, the male gamete within it is still separated from the egg nucleus (female gamete) by seemingly impassable barriers.

First, there is the hard wall of the pollen grain, then there is the tissue of the stigma and style, and finally the egg nucleus itself is embedded in the embryo sac of the ovule, both of which are suspended in the empty space of the ovary. Even if the male gamete were able to swim like a sperm it could not overcome these obstacles. Its pathway to the egg is cleared in a most remarkable fashion (Fig. 16.7).

The stigma produces a sticky fluid which nourishes the pollen grains and stimulates each one to burst open and develop a long, hollow, tubular outgrowth called a **pollen tube**. This tube pushes its way between cells of the style from which it absorbs more nourishment. It grows towards the ovule, probably guided by chemicals, then enters it by a tiny hole, the **micropyle**, through which it reaches the embryo sac. Here, the tip of the pollen tube bursts open forming a clear pathway through which the male

Anther with pollen sacs split open
(cut horizontally to show structure)

Pistil with germinating pollen

Pollination
(transfer of pollen
to stigma)

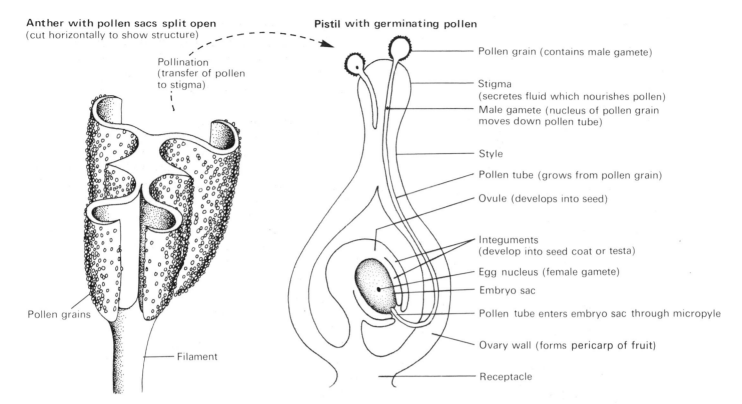

Pollen grain (contains male gamete)

Stigma
(secretes fluid which nourishes pollen)

Male gamete (nucleus of pollen grain
moves down pollen tube)

Style

Pollen tube (grows from pollen grain)

Ovule (develops into seed)

Integuments
(develop into seed coat or testa)

Egg nucleus (female gamete)

Embryo sac

Pollen tube enters embryo sac through micropyle

Ovary wall (forms pericarp of fruit)

Receptacle

Pollen grains

Filament

Fig. 16.7 Pollination and fertilization

gamete reaches the female. In some species the whole process takes only a few minutes.

The male gamete is a nucleus from one of the cells inside the pollen grain. This nucleus passes down the pollen tube into the embryo sac and brings about fertilization by fusing with the egg nucleus. After fertilization the stamens, petals and eventually the sepals shrivel up and drop off. At the same time, the ovule develops into a seed, enclosed within the ovary wall which becomes the fruit.

16.5 Seed and fruit formation

The fertilized ovule divides by mitosis, forming a tiny embryo plant. This embryo consists of a young root, called the **radicle**; a shoot, or **plumule**; and one or two 'seed-leaves', or **cotyledons**. The embryo also develops a supply of stored food. Depending on the species, this food is deposited either inside the cotyledons, or in a mass of cells called the **endosperm** which surrounds the embryo. These two methods of storing food are illustrated in Figure 16.8.

While this is happening the embryo sac and the membranes around it, called **integuments**, expand as the embryo grows inside them. The integuments become much thicker and form a tough protective 'seed coat', or **testa**, around the embryo. A seed, then, is made up of an embryo plant and stored food, enclosed within a protective testa.

Lastly, most of the water is withdrawn from the seed making it hard and extremely resistant to cold and other adverse conditions. In this state a seed is said to be **dormant**. Dormant seeds can survive for months or years. There are unconfirmed reports that corn seeds germinated after being stored for thousands of years in Egyptian tombs.

The fruit

The fruit develops from the ovary after fertilization; therefore a fruit consists of the ovary wall and the seeds it contains.

In **true fruits** the ovary wall changes in various ways to form a protective layer called the **pericarp** which surrounds the seeds. For example, a bean pod; the skin, flesh, and stone of a plum; and the hard shell of a nut are all types of pericarp.

In **false fruits** the ovary wall may grow a little but otherwise remains unchanged. For example, in strawberries the fruits are the tiny seed-like objects attached to a large fleshy structure which develops from the receptacle of the flower. In apples and pears the receptacle becomes fleshy and completely encloses the whole ovary, which remains as the core (Fig. 16.9E and F).

In both false and true fruits, the development of the ovary wall or receptacle is usually associated with a mechanism which disperses the seeds.

Vertical section of broad
bean seed showing embryo
(testa removed)

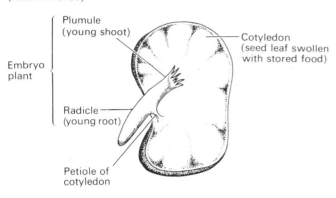

Maize grain
(cut in half and stained to
show food reserves)

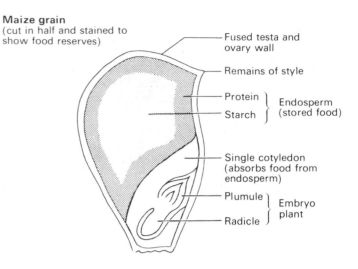

Fig. 16.8 Seed structure

16.6 **Dispersal of fruits and seeds**

It is important that seeds are carried away from the parent plant. This avoids overcrowding. It can also lead to the spread of plants into new and different environments. There are three main types of dispersal mechanism (Fig. 16.9).

Wind dispersal (Fig. 16.9)

In some plants the surface area of the fruit or seed is increased in various ways, as in the 'parachutes' of a dandelion, or the wings of the sycamore and ash. In a poppy the seeds are shaken through pores in the ovary wall as the plant sways back and forth in the wind.

Animal dispersal (Fig. 16.9)

Hooked fruits and seeds, e.g. burdock and goosegrass, can be carried long distances attached to animals' fur. Succulent fruits and nuts attract animals as a source of food. Small hard-coated seeds, e.g. blackberry, strawberry, and rose hip, can pass through an animal's digestive system unharmed and can therefore be carried some distance before being deposited on the ground – along with a convenient supply of fertilizer. In other plants, such as apple, plum, and cherry, the seeds in their hard coats are discarded after the soft fruit has been eaten by animals.

Self-dispersal (Fig. 16.9)

Several plants have mechanisms which throw seeds some distance from the parent. Most of these depend on tension caused by the drying of the fruit wall. The ripe pods of sweet pea, gorse, broom, and lupin, suddenly split open and the two halves curl outwards, scattering the seeds. In the geranium, the styles curl up and out, throwing seeds from the cup-shaped ovaries. A similar mechanism occurs in the wallflower fruit, which splits open from the base upwards.

16.7 **Germination**

A seed at the time of planting is dry, hard, and appears to be dead. But a few days later the seed suddenly 'comes to life' sprouting a seedling root and shoot. The development of a seedling from a seed is called **germination**.

In order to germinate, seeds require water, warmth, and air. (This can be confirmed by carrying out investigation A at the end of this chapter.)

The first thing which happens to a newly planted seed is that it absorbs water in large quantities. The inner part of the seed swells, which smoothes out any wrinkles in the seed coat (testa). Swelling continues until the seed coat bursts open.

Food stored in the cotyledons or endosperm soaks up water and soluble substances dissolved in it. The life processes which lay dormant all winter can now begin again. Enzymes digest the stored food, and respiration begins.

The food thus supplies energy and raw materials for cell division and growth. A radicle appears and later a plumule (Fig. 16.10).

At first, production of new tissues proceeds at the expense of stored food, so that the **dry weight** of the seedling falls; its **fresh weight** increases however due to absorption of water.

Wind dispersal

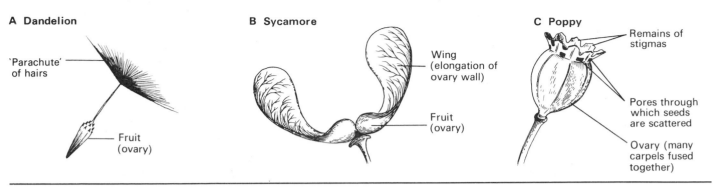

A Dandelion

'Parachute' of hairs

Fruit (ovary)

B Sycamore

Wing (elongation of ovary wall)

Fruit (ovary)

C Poppy

Remains of stigmas

Pores through which seeds are scattered

Ovary (many carpels fused together)

Animal dispersal

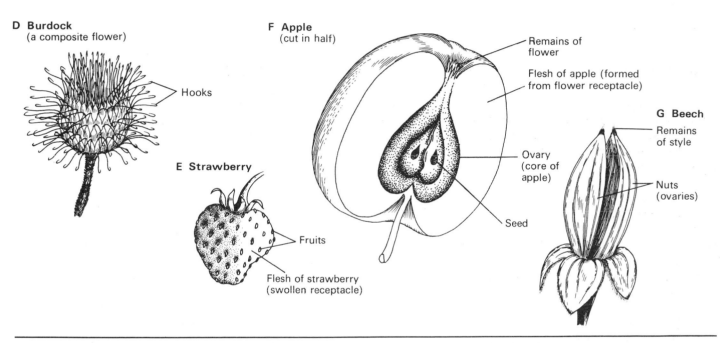

D Burdock (a composite flower)

Hooks

F Apple (cut in half)

Remains of flower

Flesh of apple (formed from flower receptacle)

Ovary (core of apple)

Seed

E Strawberry

Fruits

Flesh of strawberry (swollen receptacle)

G Beech

Remains of style

Nuts (ovaries)

Self dispersal

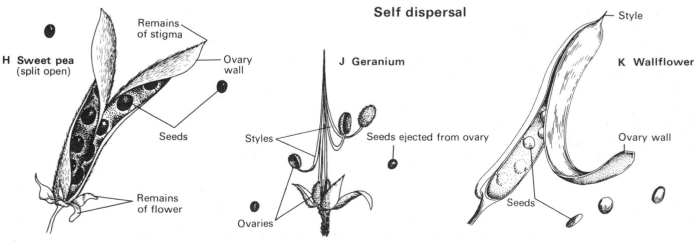

H Sweet pea (split open)

Remains of stigma

Ovary wall

Seeds

Remains of flower

J Geranium

Styles

Seeds ejected from ovary

Ovaries

K Wallflower

Style

Ovary wall

Seeds

Fig. 16.9 Dispersal mechanisms of fruits and seeds

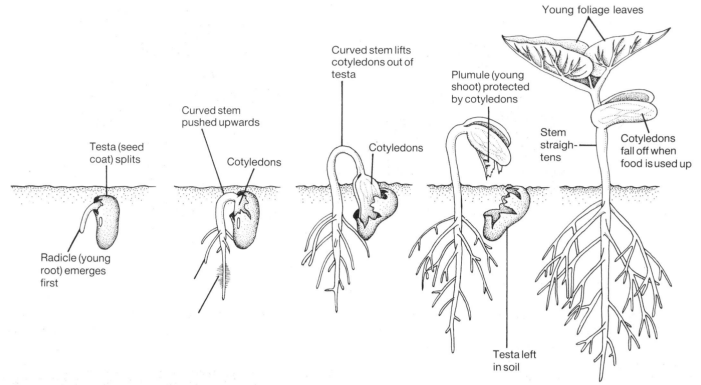

Fig. 16.10 Stages of germination of the french bean

Dry weight continues to decrease until enough leaves have matured to produce food by photosynthesis faster than it is used for growth. By this time, food stored in the seed is mostly used up. Figure 16.10 illustrates the stages of germination in a French bean.

16.8 Vegetative reproduction

Many flowering plants are capable of some type of asexual reproduction. The parent plants either separate into parts which continue growing independently, or they produce outgrowths which eventually separate and grow into new daughter plants. This type of growth is called **vegetative reproduction**.

Stem tubers (e.g. potato, Fig 16.11)
Stem tubers are formed by outgrowths from the lowest axillary buds which turn downwards into the soil. Eventually the tip of the underground stem fills with food (mainly starch) and swells rapidly to form a tuber. Stem tubers are distinguished from root tubers (e.g. dahlia) by their origin, and the presence on their surfaces of scale leaves, and axillary buds, which form the 'eyes'.

Tubers lie dormant until spring, when shoots arise from one or more of the eyes, using the stored food to establish a new plant. Stem tubers are formed on potatoes and artichokes.

Fig. 16.11 Stem tubers of potato

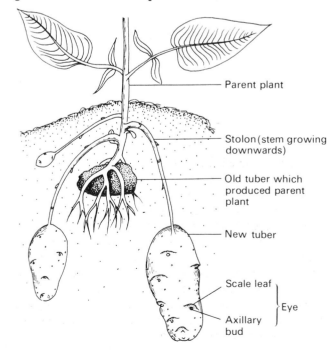

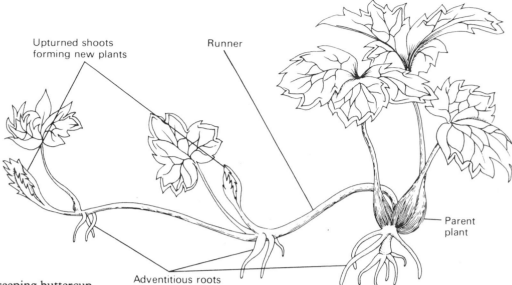

Upturned shoots
forming new plants

Runner

Parent
plant

Adventitious roots

Fig. 16.12 Runners of creeping buttercup

Runners (e.g. creeping buttercup, Fig. 16.12)

Runners grow out horizontally from axillary buds, and 'run' over the surface of the soil forming several new plants. The terminal bud of a runner turns upwards and roots form behind it to produce a daughter plant some distance from the parent. The daughter plant, in turn, produces runners and more daughters, until a long chain of plants is formed. When the daughter plants are well established their connections with the parent plant die away.

Runners are formed on creeping buttercups, strawberries, and houseleek (which forms a very short runner somtimes called an **offset**).

16.9 Artificial propagation

There are several ways in which new plants can be produced, or propagated, artificially from parent plants.

Cuttings

A cutting is any portion of a root or shoot which, after being severed from a parent, can be induced to grow into a new individual. Propagation by cuttings is an easy and inexpensive method of obtaining many plants with the same characteristics as a parent.

Chrysanthemums, coleus, and geraniums can be propagated by **stem cuttings**. These are taken from non-flowering side shoots. A sharp knife is used to cut off a shoot just below a leaf. Leaves are trimmed off the part of the stem to be inserted into moist rooting compost. Root growth can be encouraged if, before planting, the end of the cutting is dipped into powder containing rooting hormone.

Phlox and primulas can be propagated from **root cuttings**, and begonia rex plants can be produced from a single leaf.

This *Tradescantia* cutting grew roots in two weeks

Budding and grafting

It is possible to cut part of the stem from one plant and join it to a cut surface on another plant of the same genus in such a way that the two parts unite and become one plant.

The portion with roots is called the **stock,** and the portion transferred to the stock is either a single bud with a small piece of bark attached (Fig. 16.13), or a **scion,** which is a whole shoot (Fig. 16.14).

Care is taken to have the cambium (Fig. 7.2) of the bud or scion in contact with the cambium of the stock.

The type of bud or scion used decides what the flowers and fruit will be like. The type of stock used decides the ultimate size of the plant and the time it takes to mature.

Budding is a popular way of producing large num-bers of rose bushes from a parent with desired charac-teristics. Grafting is used to propagate apple, pear, and plum, and ornamental trees like laburnum and rhododendron, some of which cannot be grown from seed.

Advantages of artificial propagation

Seeds from a red rose will not necessarily grow into plants with red roses; and seeds from Cox's Orange Pippin apple are likely to grow into trees with entirely different apples. This happens because seeds nearly always result from cross-pollination between plants with different characteristics. Consequently, seeds grow into plants with characteristics from two parents. By using artificial propagation, however, gardeners can produce any number of plants with exactly the same characteristics.

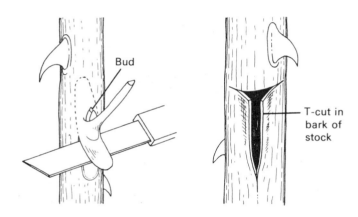

A Bud is removed   **B T-cut is made in stock**

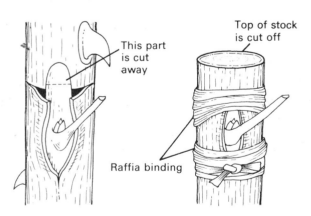

C Bud is fitted into cut   D Bud is bound in place

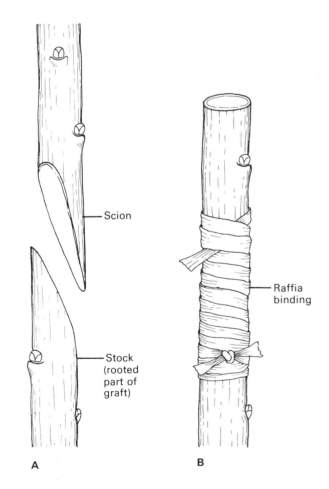

Fig. 16.13 Budding of roses. **A** A well-developed bud is cut from one parent. **B** A T-cut is made in the stock (root and stump of other parent). **C** and **D** The bud is fitted into the T-cut and bound in place with moist raffia. Further growth from the stock is cut away. Only growth from the grafted bud is allowed to remain

Fig. 16.14 Splice grafting (roses, clematis, and broom). Stock and scion should be equal in diameter. **A** Slanting cuts are made in scion and stock. **B** The two cut surfaces are fitted together and bound with raffia

Investigations

A *An investigation of the conditions necessary for the germination of seeds*

1. Obtain a quantity of mustard seeds, and five test-tubes numbered 1–5.

a) Tube 1. Put a few seeds on dry cotton wool in the bottom of the tube. Leave the tube in the light, in a warm place.

b) Tube 2. Put a few seeds on wet cotton wool in the tube. Leave the tube in the dark in a warm place.

c) Tube 3. Put a few seeds on wet cotton wool in the tube. Leave the tube in a refrigerator.

d) Tube 4. Put a few seeds in the tube and cover with boiled and cooled water, then cover with a layer of olive oil. Leave it in the light in a warm place.

e) Tube 5. Put a few seeds on wet cotton wool in the tube. Leave it in the light, in a warm place.

2. This experiment is designed to investigate the influence of five different factors on germination: light, dark, warmth, cold, and oxygen.

a) Which tube is designed to investigate which factor?

b) Note the changes in each tube after 2 or 3 days. Which of the five factors must be present for germination to occur?

B *An investigation of germination*

1. Obtain seeds of broad beans and french bean, and some maize grains. Obtain three jam-jars. Roll short lengths of blotting paper into cylinders and drop one into each jam-jar. Fill each jar with dry sawdust or clean sand by pouring it into the blotting paper cylinder. Place three seeds between the glass and blotting paper of each jar. Use a separate jar for each type of seed. Lastly, pour enough water into each jar to dampen the paper thoroughly.

2. Make drawings each day to record the stages of germination up to the establishment of root and shoot.

3. *a*) Germination in which the cotyledons remain below ground is called **hypogeal**, and where they rise above the ground it is called **epigeal**. Which of the seeds has hypogeal, and which has epigeal germination?

b) Of what advantage to the plants are the hook-shaped plumules which emerge from broad beans and french beans, and the coleoptile sheath in maize?

C *An investigation of seed formation*

1. Plant a few runner beans in well-manured soil in the second half of May. Support them with canes as they grow.

2. Dissect a mature flower, and compare it with Figure 16.3.

3. Pick a flower every day or two from the moment they begin to shrivel (i.e. after fertilization). Cut out the ovaries and open them to observe the seeds.

4. Draw the changes which take place between fertilization and the formation of a large bean pod.

Questions

1. The diagram below illustrates the structure of an insect-pollinated flower.

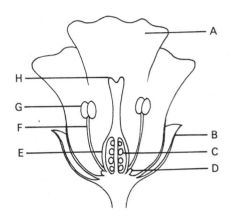

a) Identify the parts marked **A** to **H**.

b) Which part produces pollen grains?

c) Which part receives pollen during pollination?

d) Which two parts show that this flower is insect pollinated?

e) Name two things which insects collect from flowers.

f) What changes will take place in parts **A** and **C** after fertilization has occurred?

2. Two dishes were filled with agar jelly containing soluble starch. In dish **A** a germinating maize grain was cut across and placed on the agar. In dish **B** a boiled maize grain was cut across and placed on the agar. After 24 hours both dishes were flooded with iodine solution, with the results illustrated below.

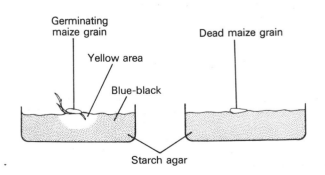

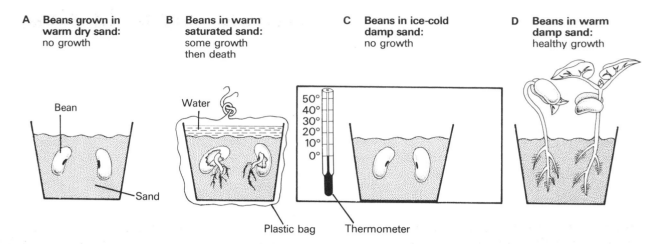

A Beans grown in warm dry sand: no growth

B Beans in warm saturated sand: some growth then death

C Beans in ice-cold damp sand: no growth

D Beans in warm damp sand: healthy growth

Bean

Water

Sand

Plastic bag

Thermometer

a) What must be absent from the agar surrounding the germinating maize grain in dish **A**?

b) Suggest a reason for this result (i.e. what process occurring in the maize grain could have caused this result?)

c) Why was dish **B** a necessary part of this experiment?

3. Study the results of the experiment illustrated above.

a) This experiment shows that bean seeds require three things before they will germinate. What are these three things?

b) Would the beans in **D** have germinated if they had been placed in the dark?

Summary and factual recall test

The stalk or (1) of a flower ends in a region called the (2) to which the flower parts are attached. These parts are arranged in rings called (3). Carpels are the (4) reproductive organs, and are known collectively as the (5). Each carpel has a hollow base called the (6). This contains one or more (7) each of which has a cell called the (8) at its centre with several nuclei, one of which is the (9) nucleus, or female (10). After fertilization, the (11) becomes a seed while the (12) becomes the fruit surrounding the seeds. Stamens are the (13) reproductive organs, and are known collectively as the (14). Each consists of a stalk called the (15) bearing an (16) with four (17). These produce (18) grains which contain the male (19). Petals make up the (20) of a flower, and can attract insects in three ways (21–list three). As a result the insects (22) the flower. Sepals make up the (23) of a flower. Their function is (24).

A buttercup flower is said to be regularly shaped because (25), and (26) symmetrical because (27). The white dead nettle is said to be irregularly shaped because (28), and (29) symmetrical because (30).

Self-pollination is the (31–define the term); whereas cross-pollination is the (32). Many flowers have characteristics which favour (33)-pollination, which ensures that (34), and thereby increases the chance of (35) change. Three examples of these characteristics are (36).

Wind-pollinated flowers usually have no (37), and

never have (38). Most of them have some or all of the following features (39–list five). These features (40) the chances of (41) reaching their (42). Three examples of wind-pollinated flowers are (43). Insect-pollinated flowers usually have the following characteristics (44–list at least three). These features (45) insects and ensure that (46). Three examples of insect-pollinated flowers are (47).

After pollination, pollen grains produce a (48) outgrowth called a (49). This grows towards the (50) which it enters through a hole called the (51). When it reaches the (52) sac its tip (53), releasing the (54) gamete which fuses with the (55) gamete.

A seed consists of a small root or (56), a shoot or (57), and one or two (58)-leaves called (59). There is also a store of food either in the (60) or in a mass of cells called the (61). The whole seed has a tough outer coat called a (62).

Seed dispersal is important for at least two reasons (63). The three main ways in which seeds are dispersed are (64).

Germination is (65). To germinate, a seed needs (66–name three things). At first a seedling's (67)-weight falls, though its (68)-weight increases.

Vegetative reproduction is a form of (69) reproduction. Examples are (70–name two). Examples of artificial propagation are (71–name two). Artificial propagation is useful because (72–give two reasons).

17

Human reproduction

Humans, and all other mammals except egg-laying monotremes, have the most advanced reproductive systems in the animal kingdom. Not only do they have internal fertilization, they have internal development as well.

This means that the embryo develops inside the female's body where it lives as if it were a parasite, absorbing food and oxygen from its mother's blood. Unlike birds' eggs, which are exposed to cooling and other dangers whenever the female leaves the nest, human eggs have the advantage of developing in the continuous warmth of their mother's body where they are fed and protected from injury and predators.

Internal development has an added advantage; since the embryo develops inside the female she does not have to remain on a nest incubating the eggs. She is free to lead a reasonably normal life until a few hours before the birth of her young.

For some time after birth all young mammals remain entirely dependent on milk produced by their mother's mammary glands. Milk is a complete and balanced diet, and is used to nourish the young until they are old enough to take solid foods (section 17.6).

17.1 Female reproductive system

The reproductive system in female mammals consists of ovaries, oviducts, a uterus, and a vagina (Fig. 17.1).

Ovaries
The ovaries are oval-shaped structures, about 3 cm long in humans, which are attached to the back wall of the abdomen below the kidneys. The ova, or female gametes, begin to develop inside the ovaries while the female is still a developing embryo, and a newly born baby girl has several hundred thousand partly developed ova in her ovaries. After birth no new ova are produced; in fact the majority disintegrate. The remaining ova gradually complete their development, and between the ages of about eleven and fifteen years, when a girl reaches sexual maturity or **puberty**, the ova are released from the ovaries one at a time by a process called **ovulation**. Between puberty and about fifty years of age when ovulation ceases, a woman releases about five hundred ova from her ovaries. At ovulation an ovum is released from an ovary into a tube called an oviduct.

Oviducts
The uppermost part of each oviduct is funnel-shaped and is lined with cilia which create a current that draws the released ovum inside. A narrow tube leads from the funnel to a wider, thick-walled tube about 7.5 cm long called the **uterus** or **womb**. It is here in the uterus that a fertilized ovum undergoes its embryonic development. A ring of muscle, the **cervix**, closes the lower end of the uterus where it joins another tube, the **vagina**. The vagina extends for about 10 cm before reaching the exterior at an opening called the **vulva**.

Between puberty and about fifty years of age the female reproductive system passes through a regular monthly sequence of events called the **menstrual cycle**. These events are controlled by hormones produced by the ovaries and the pituitary gland.

17.2 Menstrual cycle

During one menstrual cycle an ovum is released from an ovary, and the uterus is prepared to receive this ovum should it be fertilized and begin to develop into a baby. The ovaries of a sexually mature female contain ova at various stages of development (Fig. 17.2). The final stage is a structure called a **Graafian follicle**, which consists of an ovum, and a mass of follicle cells which enclose a large bubble of liquid. A fully developed Graafian follicle can reach 1 cm in diameter and often bulges from the surface of the ovary.

During the last days of its formation the cells of a Graafian follicle produce a hormone (oestrogen) which causes a layer of cells lining the uterus to grow rapidly and develop a dense network of blood vessels. This is the first stage of preparations for the reception of a fertilized ovum.

Ovulation now takes place. Fluid pressure in the Graafian follicle increases to a level which bursts the follicle, shooting the now ripened ovum onto the surface of the ovary. From here it is drawn into an oviduct, and begins its journey to the uterus. At this stage, the remains of the follicle in the ovary collapse and its cells become yellow in colour forming a solid structure called a **corpus luteum**. The corpus luteum produces hormones which stimulate cells lining the uterus to complete the preparations described above ready for a fertilized ovum.

If the ovum is not fertilized within thirty-six hours after ovulation it dies. This is followed by a slow disintegration of the thickened lining of the uterus,

and about twelve to fourteen days after ovulation the dead ovum together with the uterus lining and a quantity of blood are passed out of the body through the vagina. This process is called **menstruation**.

Counting the onset of menstruation as day one, ovulation usually takes place on about day fourteen, but it may occur on the thirteenth or the fifteenth day. If the ovum is not fertilized menstruation begins again on about the twenty-eighth day. The menstrual cycle is summarized in Figure 17.3.

Menopause

Men can continue producing sperm throughout life but women lose their ability to have children between forty-two and fifty-five. **Menopause** is the technical name for this loss of fertility, but it is commonly known as 'change of life'. At this time ovulation and menstruation stop and the reproductive organs decrease in size.

Fig. 17.1 Reproductive system of a human female

Female reproductive system (front view)

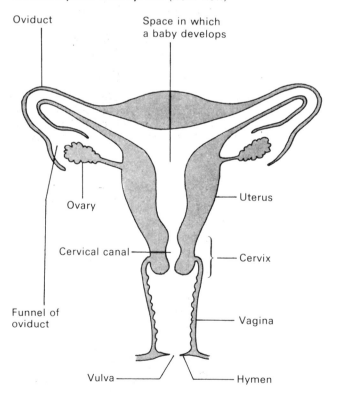

Female reproductive system (side view)

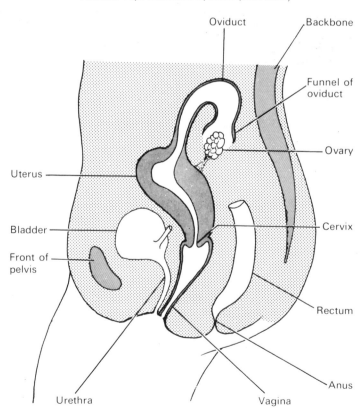

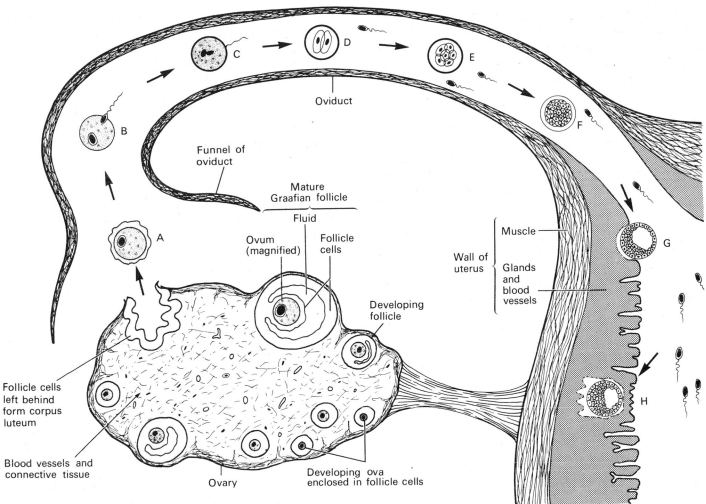

Fig. 17.2 Ovulation, fertilization, and first stages of development

Fig. 17.3 Menstrual cycle in the human female

A Ovulation. **B** Sperm penetrates ovum. **C** Sperm nucleus fuses with ovum nucleus (fertilization). **D, E,** and **F** Cell division of zygote produces a ball of cells (embryo). **G** and **H** The embryo digests its way into the uterus wall and becomes completely embedded

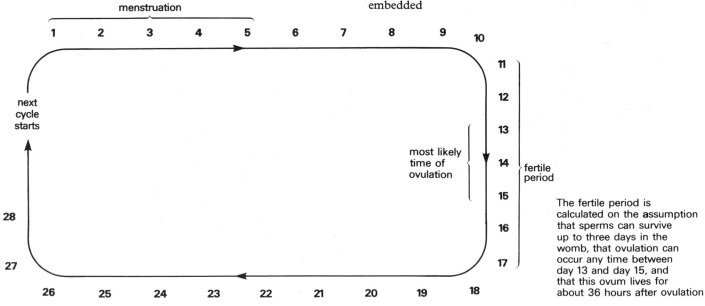

The fertile period is calculated on the assumption that sperms can survive up to three days in the womb, that ovulation can occur any time between day 13 and day 15, and that this ovum lives for about 36 hours after ovulation

17.3 **Male reproductive system**

The reproductive system of male mammals consists of testes and sperm ducts (Figs. 17.4 and 17.6).

Testes

The testes are oval in shape, about 5 cm long, and are located in a sac called the **scrotum**. Here, they are at a slightly lower temperature than the rest of the body, which appears to be necessary for proper formation of sperms. Unlike ovulation, sperm production is not a monthly event; it proceeds continuously from about twelve to seventy years of age.

The inside of a testis is divided into about three hundred compartments each containing three coiled and twisted tubules about 50 cm long. These coiled tubules are lined with rapidly dividing cells which produce male sex hormone (testosterone). The hundreds of sperm-producing tubules join together forming a smaller number of collecting ducts which convey sperms out of the testis into a single coiled tube, called the **epididymis**. This tube is 6 m long and forms a temporary storage area for sperms, which at this stage are completely dormant and immobile.

Sperm ducts

Each epididymis leads into a sperm duct, which has thick muscular walls. The sperm ducts, one from each testis, rise up the body and are joined by a duct from a **seminal vesicle**, and finally join a single tube called the **urethra** near the base of the bladder. At this point the urethra and sperm ducts are surrounded by the tissues of the **prostate gland**. The urethra leads to the outside of the body through an organ called the **penis**. During copulation the urethra carries sperms, and when the bladder is emptying during urination it carries urine.

Sperm (highly magnified)

Head   Nucleus   Tail

Fig. 17.5 Sperms are sex cells. The head of a sperm contains the cell nucleus and this is propelled by the tail

Fig. 17.4 Reproductive system of a human male

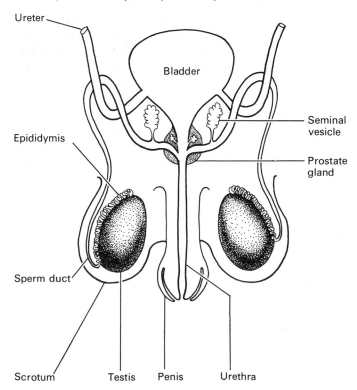

Male reproductive system (front view)

Ureter

Bladder

Epididymis

Seminal vesicle

Prostate gland

Sperm duct

Scrotum   Testis   Penis   Urethra

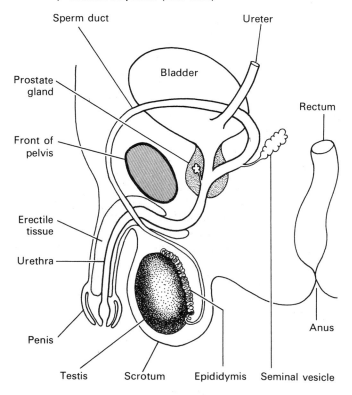

Male reproductive system (side view)

Sperm duct   Ureter

Prostate gland

Bladder

Rectum

Front of pelvis

Erectile tissue

Urethra

Penis

Anus

Testis   Scrotum   Epididymis   Seminal vesicle

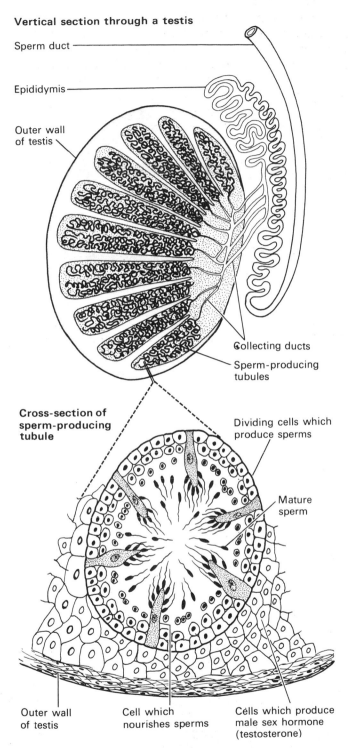

Vertical section through a testis

Sperm duct

Epididymis

Outer wall
of testis

Collecting ducts

Sperm-producing
tubules

**Cross-section of
sperm-producing
tubule**

Dividing cells which
produce sperms

Mature
sperm

Outer wall
of testis

Cell which
nourishes sperms

Cells which produce
male sex hormone
(testosterone)

Fig. 17.6   Structure of a testis

17.4   Copulation and fertilization

The thick walls of the penis contain sponge-like spaces which fill with blood whenever a male is sexually stimulated. This makes the penis erect and firm. During copulation the erect penis is inserted into the vagina of the female and moved back and forth. These movements stimulate sense organs in the penis and eventually cause an ejaculation in which about 5 cm^3 of a liquid called **semen** is passed from the epididymis and sperm ducts into the female reproductive system. Semen is forced out of the penis during an ejaculation by rhythmic contractions of the sperm ducts and other muscles. The reflex action of ejaculation and the physical excitement associated with it are known as an **orgasm** (Fig. 17.7).

The semen ejaculated into the female contains up to 100 million sperms from the epididymis, together with liquid produced by the seminal vesicles and prostate gland. This liquid contains chemicals which stimulate swimming movements of the sperm tails, and other chemicals necessary for the nourishment and survival of sperms inside the female's body.

Copulation usually causes the female to experience an orgasm, during which various muscular contractions draw a little of the semen into the uterus. From here a few thousand sperms may manage to swim up into the oviducts, and if a ripe ovum is present at the same time fertilization may occur (Fig. 17.2).

As they swim through the female system, sperms undergo changes which prepare them to penetrate an ovum. One change is to develop enzymes which will dissolve the outer membrane of the egg. Another change is brought about by chemicals from the ovum. These dissolve away part of the sperm head so these enzymes can be released.

Out of the millions of sperms which enter the female only one brings about fertilization. As this sperm penetrates the ovum it instantly triggers off the formation of an extra membrane around this cell so that no other sperms can enter. The sperm tail is left behind and the head, which contains the nucleus, moves towards and then fuses with the nucleus of the ovum.

Sperms can live for two or three days after entering the female, and therefore copulation two days before ovulation can still result in fertilization.

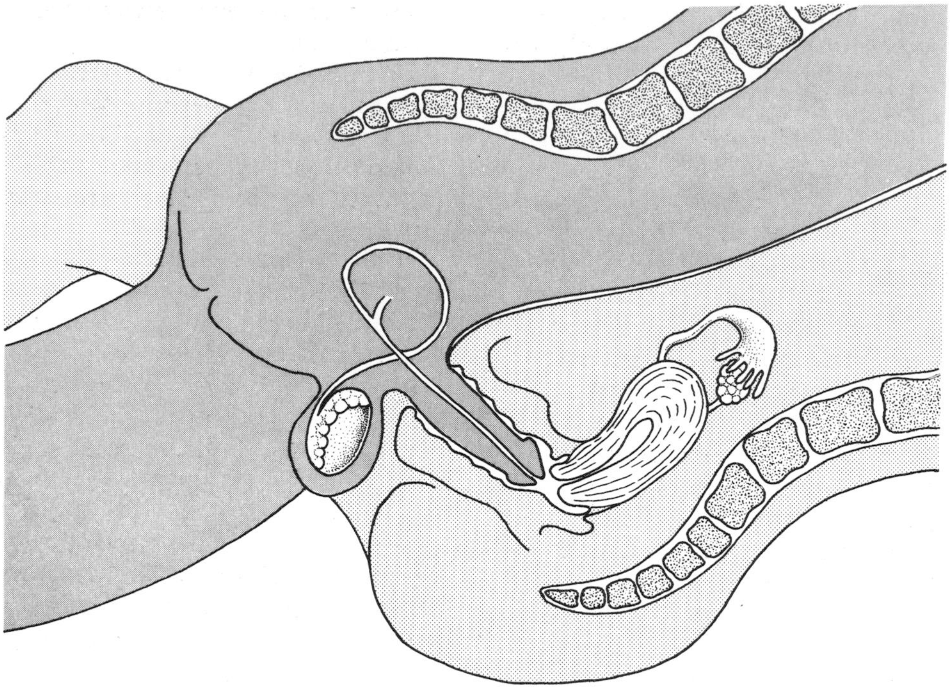

Fig. 17.7 Copulation. This diagram shows how the erect penis fits into the vagina

17.5 Pregnancy

The period of development between fertilization and birth is called pregnancy. In humans pregnancy lasts 280 days, plus or minus about 7 days (i.e. approximately 9 months). One of the shortest periods of pregnancy is found in hamsters, which give birth only two weeks after fertilization.

The fertilized ovum begins rapid cell division immediately after fertilization, and during its seven-day journey down an oviduct to the uterus it develops into a hollow ball containing hundreds of cells. It is now called an **embryo**. Meanwhile, the uterus wall has developed its thick inner lining of cells and blood vessels, and in the next four days the embryo produces enzymes and digests its way into this layer, using broken-down cells and substances produced by the uterus wall as food. In this way the embryo becomes firmly embedded in the uterus wall.

This process is called **implantation**. When it is finished, hormones appear in the female's blood which prevent the menstrual cycle from taking place until well after the baby has been born. In a sense, the mother's body is 'aware' of the presence of the embryo, and is now preparing to care for it.

Clearly, the first requirement of the embryo is oxygen and food since, unlike the eggs of birds, mammalian eggs are not exposed to the air and contain very little yolk. From the beginning the embryo absorbs food and oxygen by diffusion from nearby capillaries in the uterus wall, and the efficiency of this system is greatly increased by the development of a **placenta**.

The placenta

The placenta develops partly from the embryo's tissues and partly from the uterus wall. The whole structure takes the form of a large disc-shaped mass of tissue which spreads over and deep into the lining of the uterus as the embryo grows (Fig. 17.7). The embryo is attached to this disc by a tube called the **umbilical cord**, which carries an artery and a vein from the embryo's developing circulatory system. These blood vessels lead to an immense network of capillaries which extend throughout the disc of the placenta, and into millions of finger-like villi which grow into the uterus wall. The capillaries in these villi carry blood from the embryo to within a fraction of a millimetre of the mother's blood supply; in fact the two blood systems are separated by only the capillary walls and the membrane covering each villus. Food and oxygen diffuse across these membranes from the mother's blood into the embryo's blood, and carbon dioxide and nitrogenous wastes diffuse out of the embryo's blood into the mother's blood supply. It is important to realize that the mother's blood does not flow into the embryo, but it does flow close enough to it for these vital exchanges to take place (Fig. 17.10).

To a limited extent, then, the placenta has functions which will be taken over by the embryo's lungs, digestive system, and kidneys after it has been born. At the same time the placenta is an endocrine gland; that is, it produces hormones. These hormones stimulate further growth of the uterus, and stimulate milk-producing glands in the breasts (mammary glands), ready for the infant's birth.

Development

At fertilization the ovum is the size of a full stop on this page (and the sperm is 20 000 times smaller). After five weeks the embryo is almost 10 mm long, its brain has begun to develop, and its heart is pumping blood through the placenta and around its body. After two months the embryo is recognizably human. It is nearly 6 cm long, has limbs with fingers and toes, and a well formed face. After six months the embryo is almost 30 cm long, has hair, finger- and toe-nails, and milk teeth are developing in its jaws. At birth a baby weighs, on average, between 3 and 3.5 kg, and is about 50 cm long.

From the first few weeks of development the embryo is enclosed in a water-filled sac, called the **amnion**. The fluid in the amnion acts as a shock-absorber and helps protect the embryo from damage should anything hit or press against the mother's abdomen.

Twins

A woman's ovaries usually release one ovum a month. But sometimes two or more are released or, alternatively, an ovum can divide into two or more completely separate cells. This is how twins, triplets, quadruplets etc., are produced.

Identical twins It is possible for a fertilized ovum to produce more than one baby. This happens if, after fertilization, it divides into not one, but two or more balls of cells. The babies produced in this way are identical to one another, because they are all from the same ovum and sperm. Identical twins almost always share the same placenta (Fig. 17.9A).

Non-identical twins If a woman's ovaries release two or more ova at the same time and all are fertilized, they will develop into babies which are born at the same time, but are different from one another. This happens because each baby has developed from a different ovum, and each ovum was fertilized by a different sperm (Fig. 17.9B).

Fig. 17.8 An embryo and its placenta

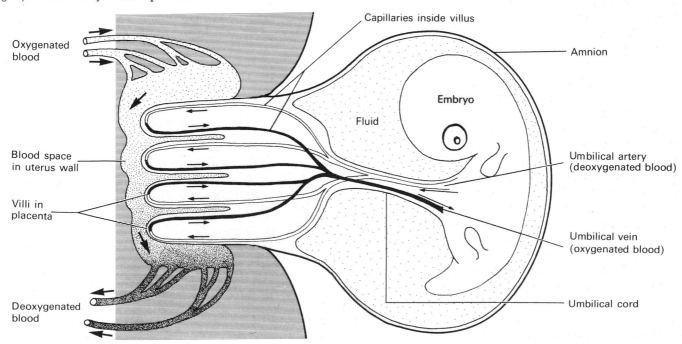

Oxygenated blood

Capillaries inside villus

Amnion

Embryo

Fluid

Blood space in uterus wall

Villi in placenta

Umbilical artery (deoxygenated blood)

Umbilical vein (oxygenated blood)

Deoxygenated blood

Umbilical cord

A Identical twins are produced when the ball of cells which develops from a fertilized egg cell splits in two

B Non-identical twins are produced when two egg cells are released at the same time and both are fertilized

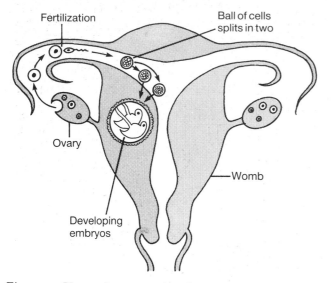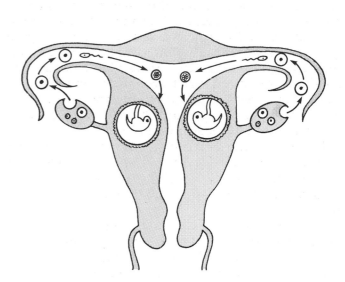

Fig. 17.9   How twins are produced

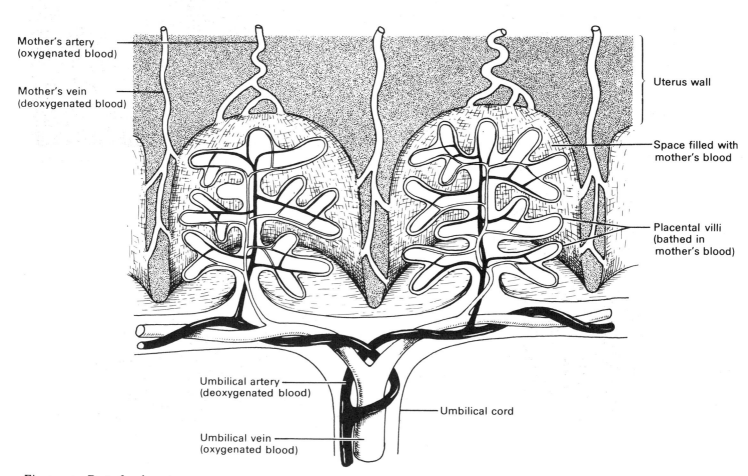

Fig. 17.10   Part of a placenta

17.6 **Birth**

Birth is a dangerous time for any baby mammal. It is the moment when it leaves the warm protection of its mother's womb, emerges into the cold air, and is suddenly cut off from its supply of food and oxygen through the placenta. Within seconds its lungs must inflate with air for the first time, and soon afterwards its digestive system must absorb the first meal of milk. Thus, birth is the moment when a mammal ceases to be a parasite, and begins the process of becoming an independent organism.

During the months before birth the uterus walls develop the muscle fibres which will be used to expel the baby from the mother's body. A few weeks before birth the baby turns within the uterus until its head lies towards the cervix (Fig. 17.11).

Exactly what causes birth to begin is not fully understood, but it is almost certainly controlled by changes in the mother's hormone output. The uterus walls begin rhythmic muscular contractions, which are intermittent at first, but become increasingly more powerful and frequent. The cervix opens and the baby's head passes into the vagina. This bursts the amnion and its fluid escapes. Soon, contractions of the uterus, aided by voluntary contractions of the abdominal muscles, propel the baby out of the mother's body.

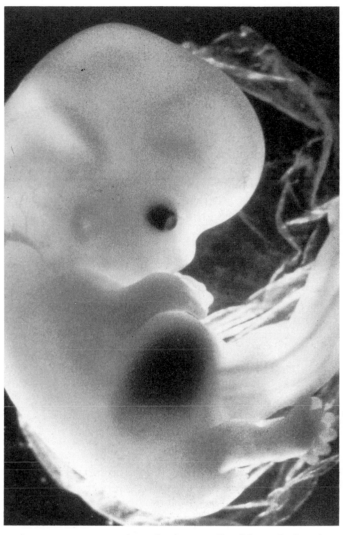

Human embryo approximately seven weeks old attached to the placenta

Fig. 17.11   Birth

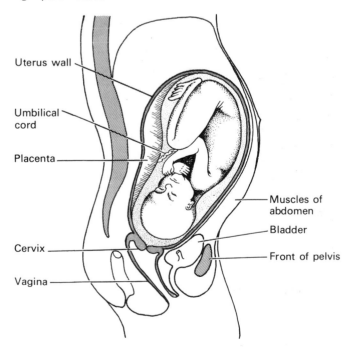

Position of baby immediately before birth

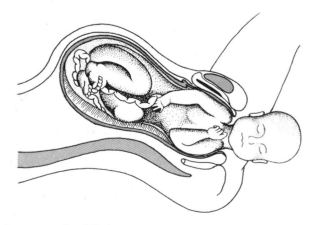

Baby emerges head first

In humans, the umbilical cord is cut and tied to prevent excessive bleeding and infection, but in wild mammals the mother simply bites through the cord. In any case, bleeding stops within a few seconds. Shortly after the baby's appearance further contractions of the uterus expel the placenta from the mother's body. This is called the **after-birth**.

On leaving the mother the baby experiences a sudden drop in temperature, and this stimulates the reflex action of its first breath. Human babies are ready to take their first food after about 24 hours. Normally, they are breast-fed on the mother's milk which contains an ideal, balanced diet. Gradually, the baby is weaned on to more solid foods.

Breech birth

As mentioned above, the majority of babies turn in the uterus so that they are born head first. Approximately 4% of babies fail to turn so that the baby's bottom is facing the cervix as labour pains commence.

It is sometimes possible to turn the baby round under the influence of anaesthetic, but if this cannot be done the baby is born bottom first. This is called a **breech birth**.

Caesarian section

A caesarian section is the surgical removal of a baby from the womb. This is necessary whenever a normal birth proves especially difficult, or impossible. For example, a caesarian section is necessary when a baby is so large that it cannot pass through the ring of bones formed by the mother's pelvis, or when the mother's pelvis is so small that it cannot accommodate a normal-sized baby. A caesarian section can also be used to replace a difficult breech birth, and when the placenta has developed so that it covers the cervix.

Miscarriage (spontaneous abortion)

A miscarriage is the loss of a developing embryo before the 28th week of pregnancy. Development stops and the contents of the uterus are passed out of the body. If a miscarriage occurs at a very early stage of development the mother may not be aware that it has happened. As many as one in four pregnancies can end in miscarriage. The main reasons include failure of the embryo to implant properly in the womb, failure of the placenta to develop to sufficient size and, in certain cases, the development of a deformed embryo. One of the first signs that a miscarriage may occur is bleeding from the vagina. If bleeding is accompanied by pain due to uterine contractions then a miscarriage is inevitable.

17.7 Feeding and care of babies

Babies can be fed on either breast milk or bottle milk. Both can produce normal healthy babies, although breast feeding has certain advantages over bottle feeding.

Breast feeding

If a baby is to be breast-fed it should be put to the breast as soon as possible after birth. When a baby is first put to the breast it does not receive milk. It receives a thick liquid called **colostrum**, which is rich in proteins, vitamins, and antibodies which help the baby fight early infections before its own immune system develops. The sucking of a baby at the breast stimulates the mother's pituitary gland to produce hormones which cause milk production in her breasts. Milk appears on about the third day after birth. Milk production is called **lactation**.

The main advantages of breast feeding can be summarized as follows:

1. Both colostrum and breast milk contain antibodies and living cells which destroy germs. This helps babies to fight early infections such as those which cause diarrhoea, nappy rash, and bronchitis.

2. Breast milk is pure and fresh and its contents are constantly changing to exactly meet the needs of a growing baby. Moreover, it is available the instant it is needed.

3. Breast milk is digested more quickly and easily than bottle milk, which is why breast-fed babies rarely suffer from constipation.

4. There is evidence that chemicals in breast milk aid development of the central nervous system.

Breast milk is perfect for human babies

Some mothers prefer to bottle-feed

Bottle feeding

Some mothers cannot breast-feed their babies and have to rely on commercial bottle-feeding milks. Many bottle-milks are based on cow's milk. But untreated cow's milk is different from human milk. It cannot be fully absorbed by a baby's body, and often causes digestive upsets.

The main advantages of bottle feeding are that the baby's food intake can be measured and that people other than the mother can help with feeding.

The main disadvantages of bottle feeding are that bottle milk is expensive and, unless the bottle is carefully cleaned and sterilized, and pure water used to dissolve milk powder, germs can be transmitted to the baby.

Parental care

Section 15.7 explains how parental care in wild animals make it possible for them to have a small number of large young, rather than large numbers of small young which are released uncared for into the environment.

Humans have the most highly developed system of parental care in the animal kingdom. Young are often fed, clothed, educated, and housed until they are old enough not only to earn a living, but to have families of their own. Even then they need not be entirely independent since, while their parents live, they are often available for advice and other types of support.

17.8 Diseases associated with reproduction

Venereal diseases (V.D.)

Venereal diseases are those which are passed from person to person during sexual intercourse. Each year 250 000 men and women in Britain receive treatment in V.D. clinics. It is very important, therefore, that V.D. is discussed openly so that no one is ignorant about the nature and symptoms of these diseases, or the precautions needed to avoid risk of infection.

There are two major diseases which are caught *only* by sexual contact. These are syphilis and gonorrhoea.

Syphilis   This is quite a rare disease although its incidence is increasing, especially among homosexual males. The first symptoms usually appear between fourteen and twenty-eight days after contact with an infected person. Symptoms include sores around the sex organs, anus, or mouth. The sores look like craters which vary in size between a pinhead and a pea, and have a red centre. They are painless and soon disappear without treatment.

The second stage of the disease appears weeks or months later in the form of a rash. It too is painless, and soon disappears.

It is the fact that the early stages of syphilis are inconspicuous and often go ignored and untreated that makes this disease so dangerous. This is because the third stage, which can appear years later, results in blindness, heart disease, and insanity.

A pregnant woman with syphilis can pass the infection to her unborn baby. Often the baby dies in the womb, or dies soon after birth. If the baby does survive it can develop deafness and defective vision later in life.

It is essential, therefore, that anyone who develops the symptoms described above after a sexual relationship, should seek immediate medical attention. The early stages are easily treated with antibiotics.

Gonorrhoea   This is more common than syphilis, and the number of cases is also increasing. It affects both males and females. Symptoms appear about ten days after contact with an infected person. Symptoms include a burning sensation during urination followed by a greenish-yellow discharge from the tip of the penis in males, and from the vagina in females. Unfortunately, two out of three women infected with gonorrhoea have no symptoms at all, *except* an infected partner.

Later stages of the disease include inflammation of the lining of the heart, the synovial membranes of the joints, and the iris of the eye. A baby can pick up gonorrhoea from an infected mother as it passes through the vagina during birth. All patients with gonorrhoea can be cured with antibiotics.

Precautions against V.D. The risk of infection is greater among people who have many different sexual partners. In other words, one-night stands, sexual relations with strangers, and partner swapping increase the risk of infection.

People who take these risks can also try other precautions.

1. Men should use a condom during intercourse (section 18.2). These protect against gonorrhoea but not syphilis.

2. Women should use a spermicidal pessary (section 18.8). The spermicide kills some V.D. germs.

3. Urinate and wash the genitals after sexual intercourse.

4. If there is any sign of infection visit a doctor or V.D. clinic. V.D. clinics have specialist staff and facilities for dealing with sexual infections. Staff are trained to give sympathetic treatment, and not to give lectures on morals.

German measles (rubella)

Rubella is not a sexually transmitted disease. But it has important associations with reproduction because the rubella virus can cross the placenta and affect a developing baby.

If a mother is infected with rubella during the first twelve weeks of pregnancy (counting from the first day of the last period) the virus can damage the foetus causing deafness, blindness, and heart disease. It is estimated that a woman with rubella during early pregnancy has five times a greater chance of having an abnormal baby than an uninfected mother.

Once a girl has had rubella she cannot be infected again. So it is very important that skin rashes and spots in young girls are seen by a doctor so that a firm diagnosis can be made. Innoculation against rubella is also advisable for girls and young (non-pregnant) women.

Precautions against rubella are important during the first twelve weeks of pregnancy.

1. Avoid all contact with people infected with rubella.

2. If there is a rubella infection going round a school, children of pregnant mothers should be kept at home.

3. If a pregnant woman has contact with rubella she can get an injection which prevents the disease developing.

17.9 New medical techniques and reproduction

New medical techniques are being developed in this field which could bring many couples great benefits, but also cause certain moral and ethical problems.

Artificial insemination

Human semen can be rapidly frozen using liquid nitrogen and stored in **sperm banks** for several years without losing its fertile condition. It is then thawed, and introduced into a woman by means of a syringe at a time when ovulation is taking place. This is called **artificial insemination**. In the United States 60 000 pregnancies a year are started in this way.

This technique makes it possible for couples in which the husband is infertile to have a baby using semen donated by another man. It allows men to store sperm so that their wives can conceive when they are away from home for long periods, or even after the husband has died.

In theory it could offer a woman the chance to choose the father of her child as she can select semen from sperm banks throughout the world. What moral and ethical problems would this cause?

Fertility drugs

Some women become infertile (sterile) because their ovaries fail to develop the Graafian follicles needed to release ripe ova into their reproductive systems (Fig. 17.2). This usually happens because the pituitary gland is not producing a substance called **follicle stimulating hormone**, or FSH.

It is now possible to artificially stimulate follicle production by injecting sterile women with a so-called **fertility drug** containing FSH obtained from animals. But if the woman's ovaries happen to be unduly sensitive to FSH, more than one follicle may develop, or follicles may develop in both ovaries so that six or more ova are released at the same time. If all these ova are fertilized a **multiple birth** will result.

Test-tube babies

Some women are unable to have babies because their oviducts (fallopian tubes) are blocked, preventing ova from being fertilized. This problem can be overcome by what is known as the **test-tube baby** technique, even though the baby does not develop in a test tube.

One or more ripe ova are sucked from a woman's ovaries using a special syringe inserted into her abdomen. The ova are placed in a dish containing sperms from her partner and kept warm for a few hours.

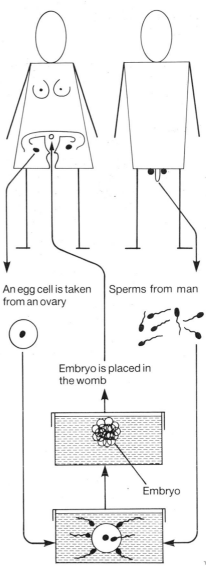

An egg cell is taken from an ovary

Sperms from man

Embryo is placed in the womb

Embryo

A sperm fertilizes the egg

Fig. 17.12 Test tube babies are produced by taking an egg from a woman and placing it in a container with sperms from a man. If the egg is fertilized the resulting embryo is placed in the woman's womb to develop normally

Louise Brown – the world's first test-tube baby

During this time, sperms fertilize the ova which divide forming embryos. One embryo is then inserted into the woman's womb where there is a chance it will implant and develop into a baby (Fig. 17.12).

Embryo transplants

It is possible to flush developing embryos from an animal before they have implanted into the wall of the womb. These embryos can be kept for a few days and then placed in the womb of a different animal where it can implant and develop normally. This technique is valuable in agriculture where it is used to take embryos from prize animals and transfer them to females, perhaps thousands of miles away. This enables a farmer to improve his stock without buying expensive prize animals. It also allows the prize animal to 'have babies' without becoming pregnant.

What moral and ethical problems would embryo transplants cause if used with human subjects?

Questions

1. What are the main differences between the eggs of amphibia, birds, and mammals, and what is the significance of the differing amounts of yolk which they possess?

2. 'A placenta performs functions which are taken over by the lungs, digestive system, and kidneys after the baby is born.' How far is this true?

3. Identical twins are produced when a zygote separates into two cells which then develop independently, whereas fraternal (non-identical) twins are produced when two ova are released from the ovaries at the same time. Explain why fraternal twins have fewer features in common than identical twins.

4. *a)* What are labour pains and their cause?
   b) What is the after-birth?

5. List the advantages and disadvantages of breast-feeding and bottle feeding.

6. What is the amnion and its functions?

7. Study the graph below.
   a) What has happened to the death rate of infants and newborns between 1961 and 1981?
   b) Try to account for this trend.

8. The drawing below shows a developing human embryo inside the uterus.

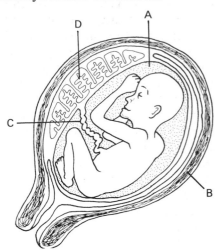

a) Name the parts marked **A** to **D**.
b) Name four substances which pass from the mother to the embryo.
c) Name one substance which passes from the embryo to the mother.
d) What are the functions of the parts labelled **A** and **D**?

Infant and neo-natal mortality rate 1961–81

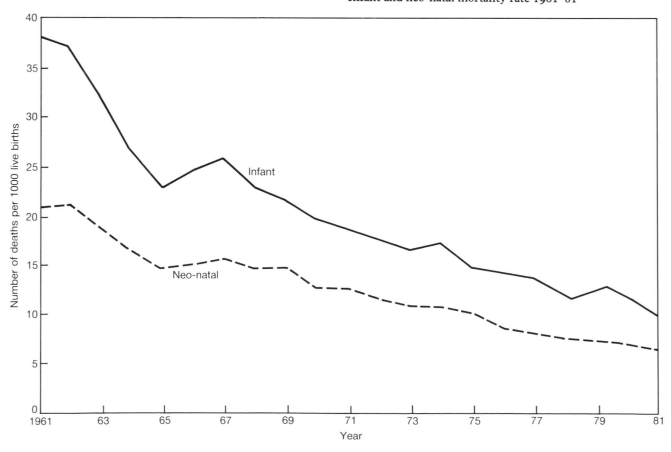

Summary and factual recall test

In mammals, internal development of the young occurs which means (1). Compared with bird reproduction this has at least four advantages (2). For a time, a newborn mammal is fed on (3) from the (4) glands of the female. At sexual maturity or (5) a girl's ovaries release ova by a process called (6) about once every (7). An ovum is released from a structure called a (8) follicle which afterwards turns yellow and is called a (9). This structure produces hormones which (10). The released ovum is drawn into a tube called the (11) and from there passes to the (12) or womb. If fertilization does not occur within (13) hours after the ovum is released it (14) and a process called (15) occurs about (16) days later.

Sperms are produced in organs called (17), and then pass into a temporary storage tube called the (18). During an ejaculation, sperms – together with chemicals from the (19) and the (20) gland – are propelled by contractions of the (21) through an organ called the (22) into the female. The sperms and chemicals together form a liquid called (23). The chemicals are necessary to (24–give two functions).

The period from fertilization to birth is called (25).

In humans it lasts about (26) months. The fertilized ovum divides to form an (27) which becomes embedded in the (28) wall in the following way (29). A large disc-shaped organ called the (30) now develops, through which the embryo absorbs (31) and (32) from the mother's (33), while (34) and (35) pass in the opposite direction. Birth is a dangerous moment for a baby because (36). Contractions of the (37) and (38) muscles push the baby, (39)-first, from the mother's body. The (40) cord is then cut and tied to prevent (41).

A (42) birth is one in which the baby is born bottom first. The surgical removal of a baby is called a (43).

Milk production in mammals is called (44). At first colostrum is produced which contains (45) that help fight early infant infections. The main advantages of breast-feeding are (46–list three).

Two diseases spread by sexual contact are (47) and (48). Rubella, or (49) virus can cross the (50) and infect a developing baby causing (51–name three conditions). Sterile women can be injected with a fertility drug called (52). This can cause multiple births if (53).

18

Population growth and control

18.1 Population growth

A group of organisms of the same kind which are limited to a particular environment are called a population. There are examples everywhere: the populations of frogs and sticklebacks in a pond; the populations of starfish on a rocky sea shore; populations of sea birds on a cliff face.

Human populations are rather different from wild populations because humans can live in most of the environments which this planet can offer. Also, humans have learned how to protect themselves from predators, diseases, bad weather, and other factors which tend to limit the size of wild populations. Consequently it is best to study wild and human populations separately.

Growth of wild populations

If a population of animals finds itself in an environment where food is plentiful and there is lots of space, shelter, and fertile mates, its numbers will rapidly increase. This happens because extra food makes the population more fertile and so birth rate exceeds death rate.

A high growth rate cannot be maintained for long because available space is filled up and competition for food begins.

At a certain stage competition for a limited food supply, and in some cases limited nesting places, uses energy which at first was available for reproduction. Fewer eggs or babies are produced when females are hungry, so birth rate falls. Attacks from predators will occur, and parasites and diseases can spread rapidly in overcrowded conditions – death rate will increase. These, and other factors which limit population growth, add up to what is called **environmental resistance**.

As environmental resistance builds up, growth rate slows down because birth rate decreases, death rate increases, and animals may migrate to other areas. These things happen as a population nears the maximum number which its environment can support. This maximum number is called the **carrying capacity** of the environment for that particular species. At this point the population growth stops, and numbers remain fairly constant.

A population of barnacles

Growth curves If this type of population growth is traced on a graph, the result is a growth curve which looks like a flattened letter 's'. This shape is called a **sigmoid curve**. Figure 18.1 is an example obtained by following the population growth of the protozoan *Paramecium* in tubes containing the bacteria on which it feeds.

Similar results are obtained by breeding many different organisms in captivity. Mice, flour beetles, and fruit flies are the favourite laboratory animals for such experiments.

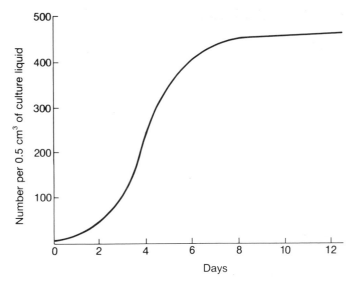

Fig. 18.1 A sigmoid (s-shaped) growth curve for a population of *Paramecia* in a tube of the bacteria on which they feed

game reserve. In order to preserve the deer herds, farmers were ordered to move out their livestock, and predators such as cougars, wolves, and coyotes were hunted and killed.

With few predators, and few competitors for food, the deer herds increased from 4000 to 100 000 by the summer of 1924. By then over-grazing had damaged the vegetation – the carrying capacity of the area had been passed.

The winters of 1924 and 1925 were hard and early. Snow covered the remaining deer food and in those two winters 60 000 deer starved to death.

This disaster shows that it is possible for a population to temporarily outgrow its food supply. But the important point to note is that deer could not have done this when their numbers were limited by predators. Under natural conditions predators, competition for food, and disease keep populations in balance with food supplies. These natural checks and balances to wild populations have so far had little effect on the overall growth of human populations.

Long-term population changes When a population reaches the carrying capacity of its environment it does not always remain constant from then onwards. Changes in the factors which make up environmental resistance may occur and cause fluctuations.

The snowshoe hare and the lynx live in the forests of Antarctic Canada. Figure 18.2 shows how their populations changed between 1856 and 1940. The snowshoe hare is one of the lynx's most important sources of food, but the hare population is not stable. It undergoes a series of population *explosions* and *crashes* which are probably due to the build up of disease when hares are plentiful.

Compare the growth curves of the lynx and hare in Figure 18.2 and note that the hare's high and low points usually occur before the high and low points of the lynx's population. The most likely explanation is that these curves illustrate a close predator–prey relationship. A population explosion amongst the hares provides more food for the lynx which, in turn, increases its numbers. When the hare population declines, the lynx has less food so *its* population declines. In this way the two populations achieve a natural balance so that changes in one are followed by changes in the other.

Upsetting predator–prey relationships The Rocky Mountain mule deer lives on the Kaibab plateau of North Arizona U.S.A. In 1906 the United States President proclaimed the Kaibab region a

A snowshoe hare

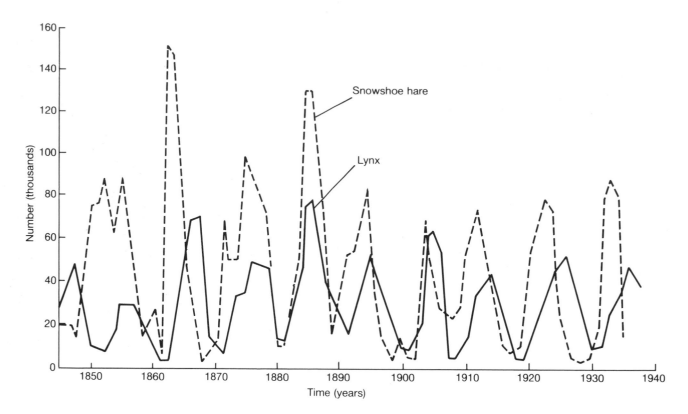

Fig. 18.2 Changes in the populations of the snowshoe hare and lynx. Explain why the lynx population usually peaks after the hare population?

Growth of human populations   No-one knows exactly how many people are alive today, but there are certainly more than 4 500 million, and the number is increasing by at least 200 000 a day!

Have we reached, or even passed, our planet's carrying capacity for people? The answer is – not yet. But certain parts of the world have more people than the land can support. This is why 1000 million people are struggling to stay alive.

The rate at which the world's population is growing is starting to slow down. Unless it slows much faster the problems of over crowding, pollution, and the scramble for diminishing raw materials could overwhelm us. When this happens the earth's resources – forests, grasslands, croplands, fisheries – will no longer be sufficient to meet our needs.

Human population growth curve   Research into historical records allows us to estimate how the

world's population has grown (Fig. 18.3).

Until the late eighteenth century population grew very slowly because the death rate and birth rate were almost equal. Many children failed to survive and adults did not live long, mainly because of disease and poor nutrition.

From the beginning of the nineteenth century the population began to grow with increasing speed. This happened because improved agricultural techniques led to better food supplies. Houses were built with pure water supplies and efficient sewage disposal, and advances in medicine greatly reduced death from diseases like cholera and diphtheria. Furthermore, there was ample space in America, Australia, and New Zealand so people were able to migrate when conditions became overcrowded.

From about 1940 onwards two different population trends began to appear. They occurred separately in the industrialized (developed) nations, and in the less industrialized (under-developed) nations.

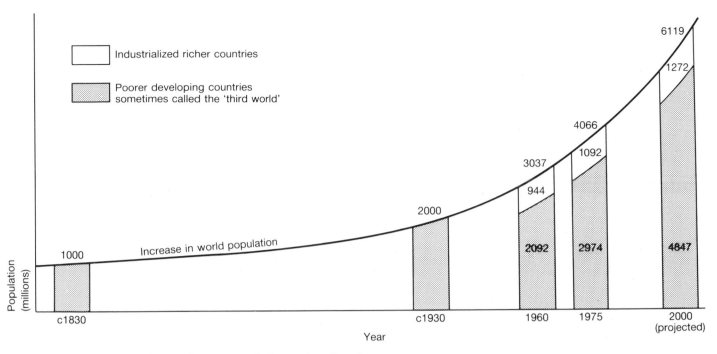

Fig. 18.3 Growth curve of world human population projected to the year 2000

Population trends in developed nations The increase in population in developed nations is slowing down. The main reason is that with increased mechanization in agriculture and industry fewer people are needed to produce food and other goods. So large families are no longer essential for survival.

Because of this many couples choose to avoid the cost of bringing up children by using birth control techniques to have smaller families.

In Britain this trend has reduced the birth rate until the population is almost stable. It may even begin to decline in the near future. A similar trend is appearing in most European countries.

Age structure in developed nations Reduced birth rate combined with reduced death rate due to better health care, results in a population made up of fewer young people and more old people.

Fears are often expressed that an ageing population will cause economic and social strains on society. These fears are based on the assumption that old people are a burden on the working population. In fact, old people have made a considerable contribution to the nation's wealth by the time they retire, and many are still able to contribute usefully to society for many years after retirement. Also, a reduced birth rate means there are fewer children and mothers in need of health services, and fewer children to educate.

A rough measure of those dependent on society is found by calculating the ratio of pensioners and children under 15, to those of working age. Table 1 gives these ratios for the United Kingdom. This table shows that the ratio of dependent to working people will not change to an alarming degree even if birth rate continues to decline.

Table 1 The ratio of children and old people to the working age population (United Kingdom)

Numbers per 100 people of working age	1981	1991 (estimate)	2001 (estimate)
Children aged 1–15	36	35	39
Old age pensioners	29	29	28
TOTAL	65	64	67
TOTAL POPULATION	55 697 000	56 712 000	57 535 000

(Source: Populations Projections 1976–2016 HMSO)

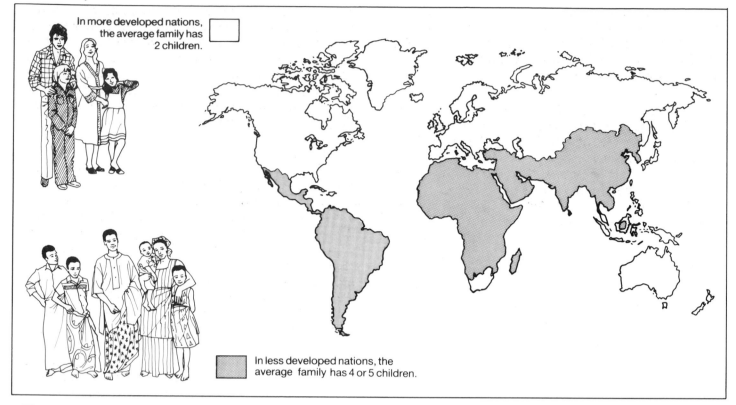

In more developed nations, the average family has 2 children.

In less developed nations, the average family has 4 or 5 children.

Fig. 18.4 Less developed nations have a higher birth rate than developed nations. Why is this so, and what are the consequences for both societies?

Population trends in less-developed nations In many less developed nations there has been a dramatic decline in death rates due to imported drugs, and public health measures against diseases such as malaria and yellow fever. But these countries still have a very high birth rate (Fig. 18.4). So there is a rapid population growth. Populations are expected to double in the next few years.

This will put an immense strain on countries already hard-pressed to support their present populations. Resources will be stretched to the limit and living conditions will get worse.

The only long-term solution to these problems is population control: to slow down and eventually halt population growth. This means the introduction of birth control (family planning) methods. It is widely agreed, however, that these measures can fail unless they go hand-in-hand with improved education.

18.2 Birth control

Birth control allows people to limit the size of their families by either preventing fertilization or implantation, or by limiting sexual intercourse to times when fertilization is unlikely to occur.

Preventing conception

Withdrawal The man withdraws his penis just before ejaculation occurs. The idea is to avoid shedding sperms into the woman but this method is extremely unreliable because many sperms are shed without the man's knowledge some time before ejaculation.

The condom A condom is a sheath of thin strong latex rubber which is unrolled onto the erect penis before intercourse. Sperms are trapped inside the condom and so do not enter the woman. This method is quite reliable, especially if used in conjunction with spermicides.

Spermicides These consist of a jelly, cream, or foam containing chemicals which kill sperms. Spermicides are inserted into the vagina before intercourse but, since they are unreliable on their own, they should be used in conjunction with a condom or diaphragm.

Diaphragm This is a dome-shaped disc of thin rubber which is kept in shape by a spring around the rim. A diaphragm is very reliable provided the woman obtains one which exactly fits her cervix. It must be smeared with a spermicide before fitting and it should be fitted half an hour before intercourse. The diaphragm should not be removed until six to eight hours after intercourse.

Intra-uterine device (IUD) An IUD is a piece of flexible plastic in the form of a loop or coil which is inserted into the womb. It must be fitted by a doctor, and regular medical checks are necessary to ensure that the device remains in place and does not irritate the lining of the womb. This method is very reliable. It appears to work by preventing a fertilized ovum from becoming implanted in the womb.

Sterilization This involves a surgical operation. In men the operation is called **vasectomy**, and involves cutting, tying, or blocking the tubes which carry sperms from the testes to the penis. In women the operation involves cutting, tying, or blocking the oviducts (fallopian tubes). The operation is effective immediately in women, but men can still produce live sperms for up to six months afterwards.

Contraceptive pills These contain one or more female hormones. They prevent ovulation, stop the mechanism which transports ova along the fallopian tubes, and cause mucus in the entrance to the womb to become sticky so that sperms are unlikely to swim through. Pills are almost 100% reliable when taken according to instructions. But they must not be taken by women with liver diseases or diabetes, and in rare instances they have been known to cause blood clots. There is evidence that women who smoke and take contraceptive pills are more likely than non-smokers to develop blood clots, migraine headaches, and allergies to certain foods.

Avoiding conception
The rhythm, or mucothermic, method This method involves avoiding intercourse during the time each month when the woman is fertile. This time can be discovered in two ways. As soon as her menstrual bleeding stops the woman must make daily observations to discover when a thin clear mucus is discharged from the entrance to her vagina. In addition she must record her temperature very carefully every day and note when it goes up by 0.1°C to 0.5°C and stays up for several days (Fig. 18.5). Ovulation occurs soon after the mucus appears and it is followed, within two or three days, by a slight rise in temperature. Consequently intercourse should be avoided after the mucus appears and should not start again until at least three days after the temperature rise occurs.

The temperature rise at ovulation is very small indeed. Consequently this method is only reliable when an accurate clinical thermometer is used correctly, and very careful records are kept.

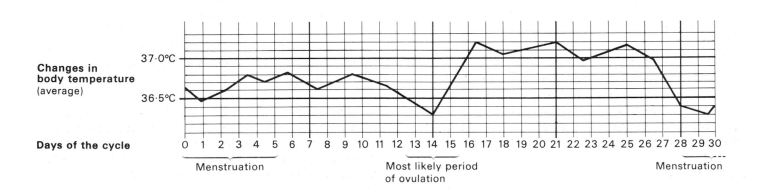

Fig. 18.5 Changes in a woman's temperature during one menstrual cycle. (Note these are average figures only, and may vary widely between one woman and another)

Questions

1. What is a population?

2. Under what conditions is birth rate likely to exceed death rate in a population?

3. *a)* List four factors which can control the rate at which a wild population grows.

   b) At what point in the growth of a population are these factors likely to have most effect.

4. What is the carrying capacity of an environment?

5. At the present rate of growth, Kenya's population will double within a few years, while Britain's population may even decline. Discuss some of the reasons why these two population trends are different.

6. In what ways does the age structure of a developed nation's population change as its growth slow down. Discuss the importance of these changes.

7. Why is the withdrawal method of contraception extremely unreliable?

8. Which contraceptives are best used in conjunction with a spermicide?

9. How do contraceptive pills prevent conception? Which women should avoid using these pills?

Summary and factual recall test

A population is (1). Birth rate will exceed death rate when (2). Factors such as competition for (3) and attacks from (4), which limit population growth, are called (5) resistance. The carrying capacity of an environment is (6). A graph of population growth to this point follows a (7) curve, which is shaped like a (8).

Human population grew slowly until the late eighteenth century because of (9–give two reasons). Then it grew rapidly mainly because of improved (10) techniques, improved (11) supplies, pure (12) and reduced deaths from (13).

In developed nations population growth is (14) because (15). But in less developed nations a decreased (16) rate and an increased (17) rate has lead to (18) population growth. The only long-term solution to this problem is (19).

A (20) is a sheath of latex rubber which is unrolled over the (21) before intercourse. Spermicides are chemicals inserted into the (22) to kill (23). A diaphragm is fitted over the (24) before intercourse, and should not be removed until (25) hours later. Male sterilization is called a (26), and involves (27), while in women the (28) are treated in the same way.

19

Evolution and natural selection

One of the most obvious characteristics of living things is their enormous variety. About two million different species are known and more are being discovered all the time. In addition, several million more species existed in the past but are now extinct.

But this variety is not a haphazard jumble of totally different organisms. Living things can be sorted into groups according to shared features. Members of the largest groups (kingdoms) have few features in common, but these groups can be divided into smaller groups (phyla, classes, etc.) whose members have more common features until, at species level, there are numerous similarities.

The fact that organisms can be sorted into groups according to shared features suggests that the members of each group are related in some way, and that they are more closely related to each other than to members of other groups. Mammals and insects illustrate this point. Despite differences of size and shape elephants and humans must be related because they both have hairy skins, and young which are born alive and suckled on milk. Butterflies and ants must also be related because they have three pairs of legs, antennae, a body divided into three parts, etc. However, the fundamental differences between insects and mammals suggest only a remote relationship between these two groups.

Several questions arise from these facts. Have living things always existed in their present variety or have they become varied with time? Furthermore, is it possible to explain the fact that this variety is not haphazard but displays many inter-relationships? This chapter describes two scientific theories which *suggest* answers to these questions. These are the theories of evolution and natural selection. Opposed to these theories is the theory of special creation, which in its earliest form proposed that all organisms on earth were created simultaneously at some time in the distant past and have remained the same ever since.

19.1 The theory of evolution

The theory of evolution proposes that species change with time. To be more precise, the theory states that the first living things were quite simple in structure and much less varied than at present. It is argued that these simple creatures gave rise to successive generations, some of which were slightly different and sometimes slightly more complex than their ancestors. Over hundreds of millions of years this process is thought to have produced a gradual sequence of changes leading from a simple state to more and more variety and complexity. Put simply, the idea of evolution implies a slow development over long periods, rather than a once-and-for-all creation of every living species.

The theory of evolution also explains the presence of groups with shared features. If evolution occurred it must have involved a number of stages. Thus, organism A produced organism B and this produced C and so on. It is unlikely, however, that when organism B appeared all the type-A organisms died off. It is more likely that they continued reproducing, and perhaps evolving, up to the present day, giving rise to a group with the basic features of their ancestor (Fig. 19.1A).

If this is so, it should be possible to arrange modern organisms into a sequence from relatively simple types (modern representatives of the earliest living things) to highly complex types (modern representatives of more recently evolved organisms). Moreover, it should be possible to construct an evolutionary tree showing how present-day groups can be traced back to ancestors whose features they share. This has been done in Figure 19.1B.

The theory of evolution also offers an explanation for the existence of **fossils** which represent the remains of extinct organisms. Indeed, fossils form one of the most important pieces of evidence in support of the theory.

19.2 How fossils support the theory of evolution

Normally, when an organism dies its remains decay and quickly disappear. However, under certain circumstances its remains are preserved as fossils, usually in the following way.

If the body of an animal or plant is washed into a river or comes to rest at the bottom of a shallow sea, it may become covered with sediment such as sand and other minerals which settle on top of it. The soft parts of the organism will probably decay, but the hard parts, such as animal bones or the cellulose and lignin of plant tissues, may survive long enough to absorb minerals from the water. These minerals gradually replace some or all of the organic materials in the body's remains, literally turning them to stone. When the body has been completely changed into stone, a fossil has been formed.

In time the fossils become covered by additional layers of sediment containing more trapped fossilizing remains. This causes a build-up of weight that presses on the deep layers, hardening them into **sedimentary rock**. Millions of years later these rocks and the fossils in them may be pushed upwards during movements of the earth's crust. Later still, the fossils may be exposed by cracking or faulting of the rocks, and by the action of the water which carves out gorges and valleys through them.

Since layers of sedimentary rock are laid down by the slow accumulation of material, the lowest layers are usually the oldest and contain the oldest fossils, while the topmost layers are the youngest and contain the most recently formed fossils. If life has evolved gradually through time it is reasonable to assume that fossils of the simplest organisms will be found only in the lowest and oldest layers, and that progressively higher layers will contain fossils of more complex organisms. This shows both the order in which organisms developed and the structural changes which they underwent.

This assumption has been confirmed by the discovery that only the most primitive organisms are found in the lower older layers of rock, and a sequence of progressively more varied and advanced animals can be found in successively higher, younger layers of rock.

Recently developed techniques make it possible to determine the age of rocks and fossils with great accuracy, so that they can be placed in a precise order according to age. These techniques have been used to establish with even greater certainty that the most primitive fossilized organisms are indeed the oldest.

A The theory of evolution Life began with simple creatures (**A**), which produced more complex ones (**B**), which produced (**C**) etc thus giving rise to an evolutionary sequence from simple to complex types. In addition, each organism in the sequence continued reproducing and evolving. This increased the variety of life by producing many groups of organisms each of which share the features of their ancestor.

B An evolutionary tree Present day organisms can be placed in a sequence from simple types (e.g. bacteria) to more complex types (e.g. mammals). This suggests that they are the products of an evolutionary sequence. Moreover, organisms can be arranged in groups which share certain features, and this suggests that each group is evolved from a common ancestor.

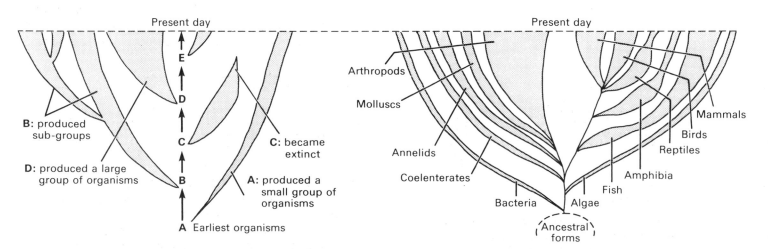

Fig. 19.1 Diagrams of evolutionary change

Sometimes these 'fossil records' as they are called, yield a sequence of fossils showing a possible line of descent to a present-day organism from an ancestor which lived millions of years ago. Such fossil records have been found for horses (Fig. 19.2), elephants, giraffes, and camels. Each fossil in the sequence appears to represent a stage in the series of changes right up to the modern animal.

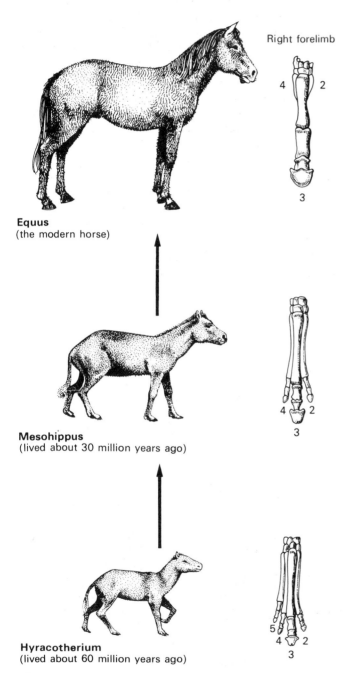

Equus
(the modern horse)

Right forelimb

Mesohippus
(lived about 30 million years ago)

Hyracotherium
(lived about 60 million years ago)

Fig. 19.2 Two of the many fossils which are thought to represent stages in the evolution of modern horses. Note the gradual loss of toes, and the development of toe 3, which forms the hoof of modern horses

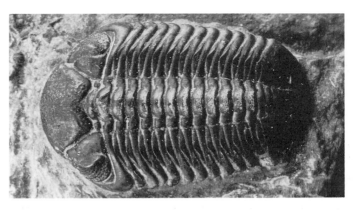

A trilobite fossil

19.3 The search for a mechanism of evolution

Many people have been involved in the search for an explanation of how species change, but among the most famous are Lamarck, Wallace, and Darwin.

The French biologist Jean Baptiste Lamarck suggested that evolution may have come about by the inheritance of **acquired characteristics**. By this he meant that the young of a species may inherit certain physical characteristics which their parents acquire in the course of their daily lives. In time this would produce organisms different in structure from their ancestors.

According to Lamarck, giraffes evolved their long necks in this way from short-necked ancestors. He believed that, in constantly striving to reach the leaves of trees, the ancestors must have stretched their necks. This acquired characteristic was then inherited by their young who through the same activity stretched their own necks still more and had young with even longer necks. Lamarck believed that this sequence, repeated from generation to generation, would have resulted in the very long necks of modern giraffes.

Lamarck's theory assumed that the body of an organism is 'plastic' in the sense that it will change in shape and form if some part, such as the neck, is put under strain or is required for more constant use than other parts. It also assumes that these changes will be inherited by the young. Men who have developed their muscles with prolonged, strenuous exercise have massive bodies. But the second assumption is not true. 'Muscle men' do not necessarily produce stronger or bigger children.

To be inherited an acquired characteristic would have to become incorporated into the hereditary information in an organism's chromosomes. At

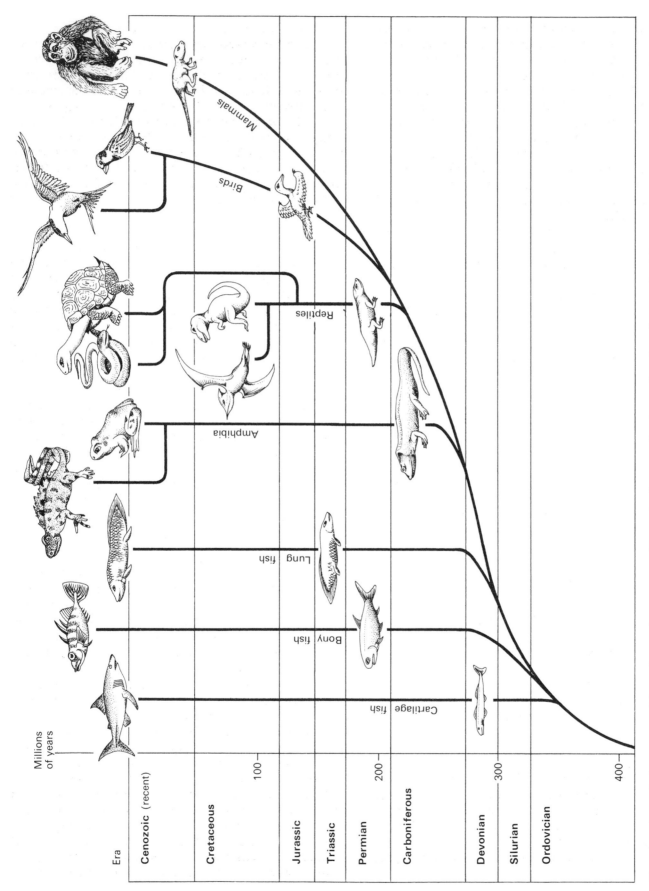

Fig. 19.3 An evolutionary tree of the main vertebrate groups, illustrating how vertebrates are thought to have evolved from common ancestors. All except animals along the top line are extinct

present there is no clear evidence how this could occur (this subject is discussed in section 19.6). Consequently Lamarck's theory of evolution cannot be accepted without modification.

In 1858 the British biologists Alfred Russel Wallace and Charles Darwin published an essay in which they stated an hypothesis which attempted to explain the mechanism of evolution. Unfortunately, their work aroused very little interest. In 1859 Darwin repeated this hypothesis in his book *The Origin of Species*, and almost immediately raised a sensational response which included both the highest praise and the severest criticism.

But Darwin's many critics served a useful function. By pointing out the weaknesses and loopholes in his arguments they stimulated many people to test his work thoroughly and examine all its implications. This research continues to the present day. The original hypothesis, now called **the theory of natural selection**, has been modified and extended into a form which is accepted by the majority of biologists as the most likely explanation of how species change with time and give rise to new species.

19.4 The theory of natural selection

Darwin based his theory of natural selection on a number of observations made in many parts of the world. These observations and the reasoned arguments which he derived from them are summarized below.

First observation

Each generation of a species has more offspring than parents. In fact most species reproduce at a rate which could at least double their numbers at each generation. An amoeba, for instance, divides into two, these two divide to produce four, and then eight, sixteen, thirty-two, sixty-four, and so on. The same is true of reproduction in higher organisms. Each set of parents has only to produce four offspring for the total population of the species to double at each generation.

Second observation

Despite their high rate of reproduction the total population in most species remains about the same once they are established in a particular environment. Of course there are seasonal variations – good years and bad years – but there are no instances of wild species which regularly double their population year after year.

First argument

Darwin argued that since populations do not increase as would be expected from their rate of reproduction, something must be controlling their numbers. He assumed that there must be a **struggle for survival**, especially among the young, so that many die before reaching reproductive age. This may happen through competition for food, water, light, warmth, and other factors affecting growth; through failure to escape from predators; death from disease or accidental injury; and all manner of hazards which wild organisms face.

Third and fourth observations

Darwin observed the variation which exists within a species, especially among those which reproduce sexually. Individual organisms in a species differ from each other in many small ways (this is easily verified by looking at any group of people); offspring differ slightly from their parents; and successive offspring of the same parents are different from each other. In fact, no two members of the same species are exactly alike.

At the same time, Darwin noted that although offspring differ from their parents in certain ways, they still inherit many parental characteristics. Darwin did not understand the mechanism of heredity. Indeed, like Lamarck, he believed in the inheritance of acquired characteristics, but this does not invalidate his arguments. Briefly, the third and fourth observations are: variation, and the inheritance of variations.

Second argument

Darwin reasoned that certain variations help an organism to survive in the struggle for existence while other variations do not. For example, in each batch of offspring those with favourable variations, such as strength and stamina, are more likely to win the competition for food, escape from predators, and withstand disease, than their weaker fellows. Consequently, organisms with favourable variations are likely to survive longer and reproduce more often than those with unfavourable variations.

Darwin called this **the survival of the fittest**, meaning that in the struggle for existence the fittest, i.e., those with favourable variations, will survive while the others die or are limited in numbers.

Darwin also argued that the survival of the fittest is a selection process. That is, 'nature' (i.e. the hazards of life in the wild) 'selects' those organisms best fitted for survival. Hence the phrase **natural selection**. Furthermore, after natural selection has taken place all the favourable variations are not lost when their owners eventually die; some at least are inherited by the next generation. In other words, nature selects the favourable variations and inheritance preserves them by transmitting them to the young.

But inherited variations are not the only ones which play a part in evolution. Darwin noted that organisms sometimes display unique variations possessed by neither parent. These new variations are now called **mutations**, and their evolutionary significance is discussed more fully in Chapter 20. Here it is enough to say that they add more characteristics to those upon which the forces of selection can operate.

Darwin now used his arguments to show how a new species can originate. He pointed out that variation, natural selection, and inheritance, operating on generation after generation of a species for millions of years, could limit unfavourable variations and lead to the accumulation of more and more favourable variations within a species. Moreover, the accumulation of variations with survival value could lead to a process of change, and probably of improvement. Ultimately change and improvement could give rise to a new species which is, in a sense, an advancement on its ancestors, because it possesses characteristics with survival value that its ancestors lacked. This represents one step in the long sequence of evolutionary changes from simple to complex organisms.

Summary of natural selection
The main points of Darwin's arguments can be summarized as follows:

1. Organisms reproduce at a rate which could potentially more than double their numbers at each generation.

2. Despite this, populations of wild organisms normally remain fairly constant in the numbers of their individuals.

3. Therefore something must be controlling population numbers. Darwin suggested that life involves a struggle for survival in which many of the young die before reaching reproductive age.

4. There is variation within every species and new variations are appearing all the time. No two members are exactly alike.

5. Offspring tend to inherit some of their parents' characteristics.

6. Some variations help an organism to survive in the struggle for existence; that is, these variations have survival value. Organisms with these variations will tend to grow and reproduce while their less favoured fellows will die off or be limited in numbers. This process is called natural selection.

7. Inheritance ensures that features with survival value are passed on to the next generation which will also have its own unique set of variations.

8. Over millions of years variation, natural selection, and inheritance may lead to the accumulation within a species of many features with survival value. Thus, a species may slowly change for the better, and may eventually produce an entirely new species.

It is easy enough to argue that one species may have evolved from another by a number of small changes, but it is another matter altogether trying to *prove* that this is so. In fact, concrete proof is not, and may never be, available. This is why it is necessary to speak of the *theories* of evolution and natural selection. Nevertheless, there is an immense amount of evidence available: Darwin collected it for twenty years before writing *The Origin of Species*, and since his death a vast amount of additional evidence has been discovered. Here is a small portion of it.

19.5 Evidence in support of natural selection

One of Darwin's most difficult tasks was to convince people that something as simple as selection can produce change within a species. He supported his argument by pointing out that man himself employs **artificial selection** to change animals and plants for his own use.

Evidence from artificial selection
Man produces completely new varieties of animals and plants by deliberately selecting and breeding those which possess the characteristics he desires.

Starting with wolves and hyenas which visited his camp sites at the dawn of civilization, man has produced hundreds of varieties of dog. Similarly, by breeding the largest and strongest wild horses, man has produced carthorses which can pull heavy loads; and by breeding only the fastest and sleekest animals he has produced racing and hunting horses. In the same way, by breeding sheep with the longest coats he has increased their yields of wool, and by breeding cows with the highest milk yield he has greatly improved the average yield per cow. Again, by selecting and breeding plants with the largest and tastiest fruit man has produced plums, apples, grapes, etc. of far better quality than those found on wild plants.

Man changes animals and plants by selecting certain specific qualities which fit his requirements, thereby giving them an *artificial* survival value. In this way man accelerates evolution.

Artificial selection allows us to *see* evolution taking place. But an objection to this as evidence supporting natural selection is that no one has yet produced a new species artificially. This is not strictly true.

Artificial production of new species

There are hundreds of examples of common plants such as species of iris, tulip, and crocus, as well as crop plants, which have been created artificially. Each is a true species in the sense than it can only breed with its own kind, and no other.

The method used can be illustrated by describing how the common flowerpot primrose *Primula kewensis* was produced from two different species (Fig. 19.4).

Two primrose species *P. verticillata* and *P. floribunda*, which do not normally interbreed, were induced to do so artificially. The seeds they produced grew into sterile plants. Their sterility was caused by the fact that their cells contained half the normal number of chromosomes.

Chemical treatment with colchicine caused a doubling of the chromosome number, and produced fertile plants which were bred in the normal way. There is evidence that new plant species could occur naturally in a similar manner.

Evidence of selection in nature

If selection operates in nature it should be possible to find organisms with variations which are favoured by selection in their environment, and so have become commoner than their less favoured fellows. A now famous example of such an organism was investigated by Dr H. B. D. Kettlewell of Oxford University in 1960.

At that time it was well known that a certain insect, the Peppered Moth, had a pale variety (white wings with black spots) and a dark variety (pure black wings). It was also known that over the past hundred years there had been a marked increase in the number of dark moths in the industrial areas of Britain and Europe. Dr Kettlewell tested the hypothesis that the black variety was favoured by natural selection in smoke-blackened industrial areas because its colouration made it practically invisible to insect-eating birds. He collected large numbers of pale and dark-coloured Peppered Moths and put some of each on trees in a clean country area, and others on blackened trees near factories. In clean woodland the dark moths were quickly taken by birds and the pale ones were overlooked, while in the woods near the factories the pale moths were quickly picked off and the dark ones were protected by their camouflage.

Fig. 19.4 The common flowerpot primrose, *Primula kewensis* is one of hundreds of new plant species created artificially

A Bulldog and a Toy Terrier. Note the many differences between the two breeds remembering that man has produced them from a common ancestor by artificial selection. This is clear evidence that selection can change a species

Dark and light varieties of Peppered Moth on a lichen-covered tree in unpolluted woodland. Note that the light variety merges with its background while the black is conspicuous and easily seen by insect-eating birds

Dark and light varieties of Peppered Moth on the smoke-stained bark of a tree near an industrial area. Note that the black variety is now camouflaged and the white one is now easily visible to bird-predators

It is now known that over eighty species of moths have produced dark varieties which are common in the industrial areas of Britain. In all these insects the principle of survival of the fittest seems to involve the natural selection of those varieties with the best camouflage. These favoured varieties become common simply because they are overlooked by birds, and so produce more offspring.

In this example a change in the moth population resulted from a change in their environment. When any environment alters there is a corresponding change in the processes of natural selection within that environment. Under changing circumstances a different set of variations within local organisms may suddenly prove to be an advantage. As a result the characteristics of the local organisms will probably alter. Environmental changes may therefore accelerate evolution.

There are many examples of environmental changes in the history of the earth which have probably influenced evolution: ice ages, mountain building, contintental drift, and the spread of deserts are a few of them. But there are other ways in which

organisms may be confronted with a new environment. The constant spread of plants and animals by migration and chance dispersal into new areas such as deserts, mountain-sides, and arctic regions may have resulted in the evolution of the specially adapted species found there today. Probably the most difficult migration of all took place when organisms spread from water on to land, and met a totally new environment with completely different characteristics, such as lack of water, lack of support against the force of gravity, and large variations in temperature. Over millions of years evolution gradually produced new organisms able to tolerate these new conditions, and then life on land began.

In the case of Peppered Moths a relatively simple environmental change favoured one variation and changed the population. During the millions of years since life began, large-scale geological and climatic changes, together with migration and dispersal, must have influenced countless variations in countless organisms, resulting in widespread evolutionary changes and the production of new species.

Darwin discovered an example of migration which

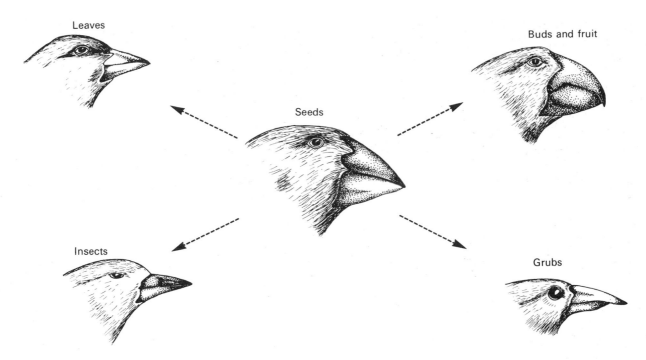

Fig. 19.5 Five of the sixteen types of Galapagos finch. The seed-eating species (centre) is thought to be closest to the seed-eating ancestors from which other types may have been derived by divergent evolution

appears to have produced new species. On a voyage around the world he visited the isolated Galapagos Islands located off the coast of South America. Here, he found several species of finch-like birds, obviously closely related but with differently shaped beaks and different feeding habits (Fig. 19.5). These particular birds appeared to be unique to the Galapagos Islands. Darwin suggested that long ago finches from the mainland, where only seed-eating species are to be found, accidentally reached these islands and found a number of different foods. Then, owing to the selective influence of competition for seeds, variations evolved with modified beak shapes suitable for eating fruit, leaves, grubs, and insects. These varieties eventually became separate species.

The Galapagos finches are an example of what may happen when organisms invade an isolated area. The invaders are physically separated from their main group by barriers such as water or mountains, and cannot cross-breed with its members. In their new environment the invaders are influenced by many selective forces so that not one but many variations are favoured. The organisms evolve, and become adapted in different ways which permit them to take advantage of all the new opportunities their isolated

environment affords. In this way a single invading species may produce many types of organism all adapted in different ways so that eventually they cannot cross-breed with each other. They have now become separate species. This is how divergent evolution probably occurs.

To summarize: **the inheritance of variations provides the raw material of evolution, because it produces new candidates to face the hazards of natural selection.** It is the death or survival of these candidates which determines the future course of evolution.

The weakest point in Darwin's account of this process was that he could not show how variations originate. This was unfortunate because variation is the source of evolutionary change. Darwin was also unable to understand the way in which variations are inherited, and was unaware that only six years after the publication of *The Origin of Species* an unknown Moravian monk called Gregor Mendel had published an account of the first principles of inheritance. Mendel's laws of inheritance are introduced in the next chapter.

19.6 Scientific critics of Darwinism

Natural selection is not a fact. It is a working hypothesis, which means that while it solves many problems there are others which it does not.

One unsolved problem is the presence of large gaps in the fossil record of some major events in evolution like the emergence of birds and mammals from reptile ancestors. Some gaps will occur in any fossil sequence but if, as Darwin proposed, evolution involves many small steps, it seems odd that we lack so many intermediate stages.

The jump theory of evolution

A modern viewpoint , called the **jump theory**, states that these intermediate stages have not been found because they do not exist. The jump theory proposes that, rather than evolution by many small steps, species change very little for millions of years. Then they abruptly give rise to something quite different and yet clearly related, in one large jump, with no intermediate steps (Fig. 19.6).

S. J. Gould, the originator of this theory, has suggested several ways in which these jumps may have occurred.

1. Jumps could have resulted from rapid evolution in small isolated groups, where new variations, or **mutations** as they are now called, would be less diluted than in a large population. Once established the new species would make use of its advantages to rapidly grow in numbers.

The polar bear, which seems to have evolved from the brown bear in the last 50 000 years, could have evolved in this way. Moreover, evolution from small isolated groups would leave few fossil traces, which would account for the gaps in the fossil record.

2. Another possibility arises from the fact that very strange things often happen to chromosomes, the carriers of hereditary information. Pieces of chromosome can turn back-to-front, jump from one part of a chromosome to another or even to another chromosome. Such rearrangements could lead to sudden, major evolutionary events.

Chromosomes of the eight living species of horse are a good example. The horses are similar in appearance but have very different chromosome arrangements.

'Jumpers' badly need proven mechanisms rather than theories. Until this happens there is little hope for more than intense argument.

Fig. 19.6 Did the horse evolve by continuous gradual change (Darwin's view) or by short bursts of rapid change (S. J. Gould's view)?

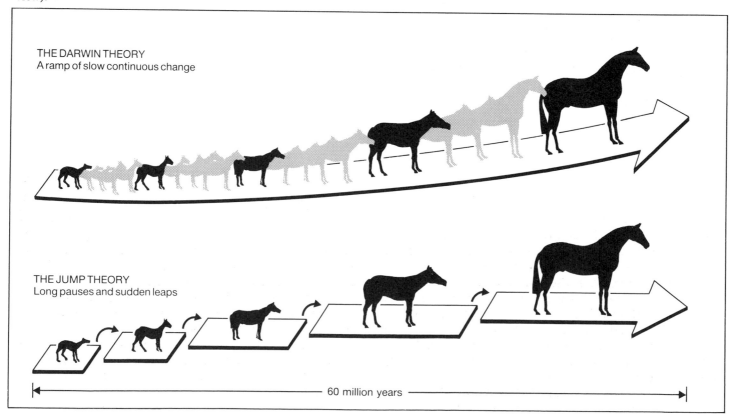

THE DARWIN THEORY
A ramp of slow continuous change

THE JUMP THEORY
Long pauses and sudden leaps

60 million years

Useful mutations are very rare

The essence of Darwin's theory is that evolution occurs by natural selection of chance mutations. Some people think that chance alone is not enough to produce the enormous variety and complexity of life. Is it possible, for instance, that a structure as complex as the human eye could have evolved by chance?

The eye consists of many inter-related parts, most of which are useless without the others. Consequently, for any one of these parts to have survival value they must have appeared at the same time as other related parts. In other words, several useful mutations must have occurred simultaneously. In fact, useful mutations are so rare that such an event is almost inconceivable. Perhaps other factors are involved in evolutionary change.

Could Lamarck be right after all?

Lamarck proposed that useful attributes acquired in an animal's lifetime could be inherited by its children. Recent investigations suggest that there may some truth in this idea. One experiment has shown that changes in height and weight of flax plants, produced by different growing conditions, can be passed on to subsequent generations.

In more recent experiments mice of one strain were given treatment which stopped them rejecting skin grafts from another strain of mice. The treated males were then mated with untreated females of the same strain and about half their young were able to accept the skin grafts without treatment.

Confirmation of these and similar results may make it necessary to rethink the role of chance in evolution, and allow for changes brought about by inheritance of acquired characteristics.

Questions

1. Lamarck believed that ducks developed their webbed feet through using them as paddles over a long period. Describe the reasoning on which this belief is based. How would Darwin have explained the evolution of webbed feet?

2. A dairy farmer found that his stables were infested with flies so he sprayed the area with D.D.T. (an insecticide). Nearly all the flies were killed. A few weeks later the number of flies was again large so the stables were sprayed again with the same chemical. Most of the flies were killed. Again the fly population increased and again the same spray was used, the sequence being repeated over several months. Eventually it was clear that D.D.T. was becoming less and less effective in killing the flies.

a) Construct as many different theories as possible to account for these facts.

b) Choose one theory which you consider to be correct and devise a controlled experiment which could be used to test it.

3. In man the muscles which in other mammals move the ears are present but poorly developed and useless. Man has more than 180 such structures, which are useless to him but are functional in other vertebrates. Explain how these structures suggest that man and other vertebrates share a common ancestor.

Summary and factual recall test

The theory of evolution states that species (1) with time. Evidence in support of evolution is that the most primitive fossilized organisms are found only in the (2) rocks, and progressively (3) layers of rock contain more and more complex forms.

Lamarck believed in the inheritance of (4) characteristics. By this he meant that (5). Darwin proposed the theory of (6). This is based on the following observations and arguments. Organisms reproduce at a rate which could (7) their numbers at each (8). Nevertheless, populations normally remain fairly (9) in numbers of individuals. According to Darwin this is because of a (10) for survival in which the young die before reaching (11) age. There is variation in every species, and some variations are said to have (12) value, because (13). (14) ensures that some of these variations are passed on to the next generation. Over millions of years (15), (16), and (17) may lead to the accumulation of (18) within a species, and as a result a new (19) may be produced.

Environmental changes can accelerate evolution because (20). Examples of such changes in the history of the world are (21–name four).

The jump theory states that evolution did not occur in (22) steps. Species may remain (23) for millions of years, then suddenly produce new species with no (24) steps.

20

Variation, heredity, and genetics

People, and all other living things, vary in many ways. In fact no two living things are exactly alike: even 'identical' twins differ in certain ways. Humans, for example, all have the same general shape and body organs. However, characteristics such as height, weight, shape of the face, knowledge, skills, body scars, etc., differ from one person to the next. These characteristics are examples of **variation**.

During sexual reproduction parents pass certain characteristics to their children. But not all characteristics can be inherited. A child inherits the shape of its face, nose, and ears from one or other of its parents, but there is no chance whatsoever of a child inheriting a knowledge of mathematics from its mother for example, or a scar which its father received in an accident.

Characteristics which can be inherited from parents are called **hereditary characteristics**. They include colouring of the hair, eyes, and skin; shape of the face, ears, nose, and mouth; and all the other characteristics which develop as a child grows from a fertilized egg. Characteristics such as knowledge, skills, scars, etc., are called **acquired characteristics**, because people acquire them during their lives.

Some hereditary characteristics show what is known as **continuous variation**. This means that there are many intermediate forms of the characteristic so that they can be graded from one extreme to the other. People, for example, occur in so many different sizes that it is possible to arrange even a small group, such as a class of students, into a continuous line from the smallest to the tallest. A block graph of a characteristic showing continuous variation has the shape shown on the left below.

Other characteristics have few or no intermediate forms, and so cannot be arranged in a continuous sequence. These characteristics are described as showing **discontinuous variation**. For example, humans are either male or female. Intermediate forms are very rare. Another example is the ability some people have to roll their tongues into a U-shape. Either people can do it or they cannot; there are no halfway stages.

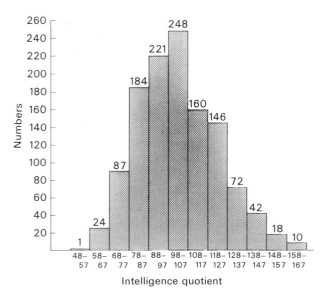

Intelligence shows continuous variation. Note that the majority of children given this intelligence test scored around the average mark of 100, while only a few scored very low or very high marks. A distribution of this type is typical of characteristics showing continuous variation.

Tongue-rolling is an example of discontinuous variation. Either you can do it or you can't

One of the most important facts about hereditary characteristics is that they are inherited according to certain rules or natural 'laws'. If an African man married a European woman with blond hair and blue eyes, their children would almost certainly have black hair and brown eyes (their father's features).

Why are brown eyes more likely to be inherited than blue eyes? Has the chance of inheriting blue eyes disappeared altogether or could it reappear in future generations? But most important of all, how is it possible to discover the laws which govern this, and the other ways in which offspring inherit parental features?

20.1 Mendel's experiments

Mendel was looking for the laws which govern the way in which hereditary characteristics pass from parents to their offspring. The quest for natural laws is the primary aim of all science, because a knowledge of them makes it possible to predict future events. In this case, a knowledge of the laws of inheritance would make it possible to predict the outcome of a particular mating, thereby helping man to breed particular types of animals and plants. Mendel was successful in this quest where others had failed because of the experimental methods he used.

First, whereas others studied the inheritance of several different characteristics simultaneously, Mendel studied only one characteristic, or character, at a time. For example, using pea plants he chose characteristics such as the length of the stem, i.e. tallness; the shape or colour of the pods; or the shape or colour of the peas (Fig. 20.1).

Second, and most important of all, the characteristics which he chose showed what is now called discontinuous variation. This means that they possessed variations which were distinctly different, and with no confusing intermediate forms. In the case of stem length for example, he used tall varieties and dwarf varieties which, when bred together, produced either tall or dwarf offspring with no intermediate forms that were difficult to classify. Similarly, he used plants with green pods and others with yellow pods which when bred together produced offspring with either green or yellow pods and no confusing intermediate colours. In other words, he used characteristics with *contrasting* variations. This was extremely important because the characteristics were easy to recognize during the course of the experiments. Had he used characteristics with many intermediate variations, i.e. with continuous variation, his results would have been extremely difficult to interpret.

Third, Mendel's use of pea plants in his experiments was a good choice for several reasons. Peas grow quickly and mature in one season; their pollination mechanism is easily controlled by hand so that they can be cross-pollinated – **crossed** – or self-pollinated – **selfed** – easily; and they are inter-fertile, which means different varieties produce fertile seeds after crossing.

Fourth, Mendel always began his experiments with **pure lines**. These are plants which, when self

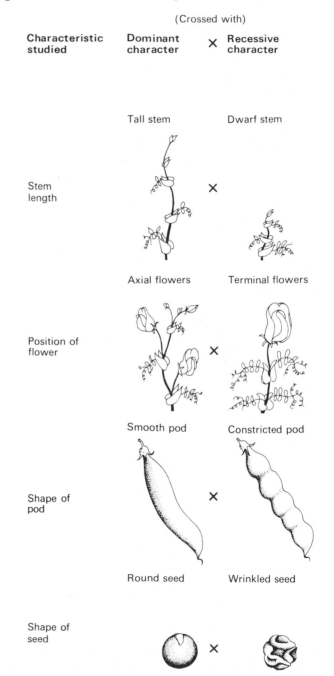

	(Crossed with)	
Characteristic studied	Dominant character	× Recessive character

Fig. 20.1 Some of the hereditary characteristics investigated by Mendel in his early experiments

pollinated generation after generation, always produce plants with the same characteristic, such as a tall stem. In other words, they breed true for a certain characteristic.

Mendel's basic method involved **hybridization**, which involves crossing two organisms which differ in some way and results in offspring called **hybrids**. In his earliest experiments Mendel produced hybrids by crossing plants which differed in only *one* way, such as stem length, or colour of pod, or shape of seed. These are called **monohybrid crosses**. The experiment described below was designed to investigate the inheritance of stem length, and involved a monohybrid cross between tall and dwarf plants.

Results of a monohybrid cross

In order to investigate the inheritance of stem length Mendel obtained pure lines of tall and dwarf plants and crossed them in the following manner, and with the following results.

1. In this, and all other experiments involving cross-pollination, Mendel had first to prevent self-pollination, and then ensure that cross-pollination took place between only the chosen opposite varieties. He prevented self-pollination by removing the anthers from a number of tall and dwarf plants before they had produced mature pollen grains. Later, when their carpels were mature, he transferred pollen to the stigmas of the same flowers from the anthers of the *opposite* variety, i.e. the stigmas of flowers on tall plants were dusted with pollen from dwarf plants, and the stigmas of flowers on dwarf plants were dusted with pollen from tall plants. These pollinated plants, which form the starting point of the experiment, are called the **first parental generation**, or **P₁**.

2. Mendel collected all the seeds from the P₁ plants, set them in soil and awaited the results. The new plants which grew from these seeds were called the **first filial generation**, or F₁. Without exception all the F₁ plants were tall, no matter whether the seeds which produced them had come from tall or dwarf parents (Figs. 20.2 and 20.3). There were no plants of intermediate size, and no dwarf plants. Clearly, the ability to produce the dwarf characteristic had either disappeared, or it had been suppressed in some way by the tall characteristic. Mendel said that since tallness had 'dominated' dwarfness in this way tallness should be called the **dominant characteristic** (Fig. 20.1).

3. Next, Mendel self-pollinated all flowers of the F₁ plants, and then covered the flowers to prevent any possibility of cross-pollination. Subsequently, he collected seeds from the F₁ plants, set them in soil

and waited for the F₂ generation to grow from them. In the F₂ generation, out of a total of 1064 plants, 787 were tall and 277 were dwarf. That is, roughly three-quarters were tall and one quarter were dwarf, giving a ratio of approximately 3:1 (Fig. 20.2). The ability to produce the dwarf characteristic had reappeared again, to a limited extent, and so Mendel

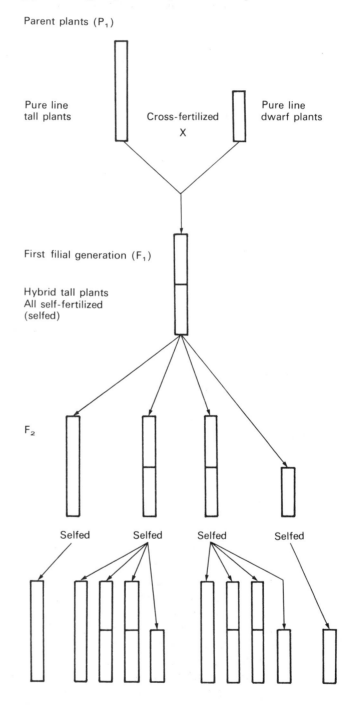

Fig. 20.2 Diagram of a monohybrid cross between pure line tall, and pure line dwarf pea plants

Table 1   Results of Mendel's monohybrid crosses

P_1 crosses			F_1	F_2		Ratios
tall	X	dwarf stems	all tall	787	tall	2.84:1
				277	dwarf	
				1 064	total	
round	X	wrinkled seeds	all round	5 474	round	2.96:1
				1 850	wrinkled	
				7 324	total	
yellow	X	green cotyledons	all yellow	6 022	yellow	3.01:1
				2 001	green	
				8 023	total	
coloured	X	white seed coats	all coloured	705	coloured	3.15:1
				224	white	
				929	total	
smooth	X	constricted pods	all smooth	882	smooth	2.95:1
				299	constricted	
				1 181	total	
green	X	yellow pods	all green	428	green	2.82:1
				152	yellow	
				580	total	
axial	X	terminal flowers	all axial	651	axial	3.14:1
				207	terminal	
				858	total	

called dwarfness a **recessive characteristic** because it had 'receded' in the F_1 generation only to reappear in the F_2 (Fig. 20.1).

4. Mendel now self-pollinated the F_2 plants and produced F_3 plants from their seeds. All the dwarf F_2 plants produced only dwarf F_3 plants. This meant that they were pure lines, producing nothing but dwarfs henceforth. However, the F_2 tall plants were found to be of two types: one third of them were pure lines giving rise to only tall F_3 plants, while two thirds proved to be hybrids producing tall and dwarf F_3 plants in the ratio of 3:1 (Fig. 20.2).

Mendel used this procedure with seven pairs of contrasting characteristics, and in every case one characteristic was found to be dominant, completely excluding the recessive one from the F_1 generation, while the F_2 plants displayed both dominant and recessive characters in a ratio of approximately 3:1. These results are summarized in Table 1.

Mendel was now faced with the task of interpreting his results. Mendel worked entirely alone at a time when nothing was known about chromosomes or cell division, yet he explained his results in a way which ties in almost exactly with modern knowledge of how hereditary mechanisms work.

Interpretation of a monohybrid cross

Remember, Mendel was looking for a law of inheritance which would enable him to predict the outcome of a particular mating, or cross-pollination, and he had greatly simplified this task by studying the inheritance of only one pair of contrasting characteristics. In fact, by this technique he had achieved his aim, for it had revealed a predictable pattern of events from among the seemingly chaotic process of inheritance. This pattern occurred whenever he crossed plants bearing a dominant character with plants bearing a recessive character: the F_1 plants displayed only the dominant character, while the F_2 plants displayed dominant and recessive characters in the ratio of 3:1.

This means that when two contrasting characters are brought together, as in a monohybrid cross, they do not fuse or blend together and so produce offspring with many intermediate forms, such as medium-sized stems or children with one brown eye and one blue. The contrasting characters retain their individual identity and separate, unchanged, in the F_2 generation in a certain predictable order. Mendel summarized all this in a simple statement now known as **Mendel's First Law**. He said: 'In a cross between plants bearing contrasting characters, the characters segregate (separate) in the second filial generation'. But how is it possible for a character to survive unchanged from a cross with its opposite character? To answer this question Mendel proposed this theory of **hereditary factors**.

Mendel suggested that the body of an organism contains a number of microscopic particles which he called factors, and that these factors control the appearance of the organism's hereditary characteristics. According to Mendel, each contrasting characteristic has its own separate factor. For example, the ability to produce a tall stem is controlled by one factor, and the ability to produce a dwarf stem is controlled by another, separate factor. Furthermore, in order to account for the appearance of F_1 generations and the observed ratios of F_2 generations Mendel proposed that factors must operate in pairs. The reasoning behind this assumption can be explained by employing Mendel's system of using different letters of the alphabet to represent the hereditary factors. In the following explanation 'T' represents the dominant factor controlling tallness, and 't' represents the recessive factor controlling dwarfness. Figure 20.3 summarizes this explanation diagrammatically.

Starting with the parent (P_1) plants in the monohybrid cross described above, Mendel argued that the pure line tall plants must contain pairs of factors for tallness – written in symbols as TT – whereas the pure line dwarf plants must have pairs of factors for dwarfness – written tt. Mendel assumed that when P_1 plants produced gametes (pollen and ovules) the factors in each pair must separate so that each gamete receives *one* factor. Thus, when tall plants produce gametes, members of their TT pairs will separate and each of their pollen grains and ovules will receive one T factor. Similarly, all the pollen grains and ovules of the dwarf plants will receive one t factor.

Mendel argued that when a pollen grain from a tall plant fertilizes an ovule from a dwarf plant, and vice versa, a T and a t factor will come together making a Tt zygote. When this zygote grows into an F_1 plant the T will dominate the t and the plant will grow tall. Since all the zygotes from the P_1 cross are Tt this explains why all the F_1 plants are tall. During gamete production in the F_1 plants members of the Tt pairs will separate and each pollen grain and ovule will receive either a T or a t factor. Since there are equal numbers of T and t factors an equal number of T and t pollen grains and ovules will be produced.

Fig. 20.3 Segregation of hereditary factors in a cross between pure line tall and dwarf plants

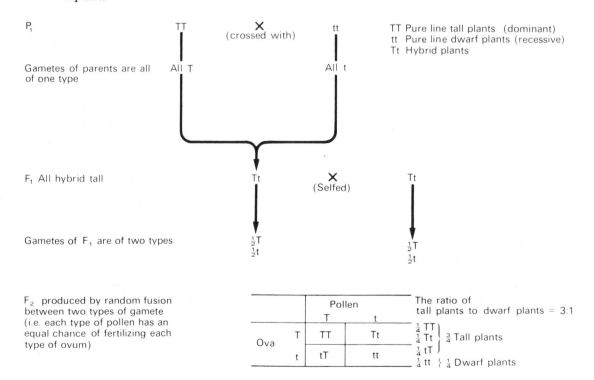

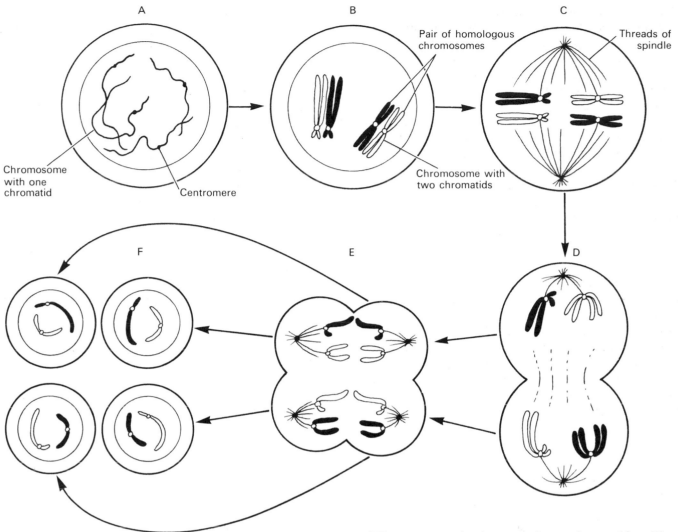

Fig. 20.4 Meiosis (reduction division)

A Chromosomes become shorter and thicker until they are visible as a single strand called a **chromatid**. **B** Chromosomes pair off – they form pairs, the members of which are identical in shape and size. These are called **homologous pairs** of chromosomes (one member of each pair is drawn in black simply to distinguish it from its partner). The cell now manufactures another chromatid alongside the first, so each chromosome now has two chromatids. **C** Homologous pairs of chromosomes now become arranged around the middle (equator) of the cell. The nuclear membrane disappears, and spindle threads are formed. **D** Spindle threads pull the members of each homologous pair away from each other so they move towards opposite ends of the cell. The cell begins dividing in two. **E** Spindle threads now pull the chromatids of each chromosome away from each other. They separate and move in opposite directions. **F** The cell divides into four parts, each containing half the original number of chromosomes, i.e. the **haploid number** of chromosomes. The four cells become four haploid gametes

The next step in the experiment is to self-pollinate the F_1 plants. To understand what happens it must be remembered that pollination and fertilization are purely random events. That is, either of the two types of pollen grain have an equal chance of fertilizing either of the two types of ovule. This fact is extremely important because it accounts for the 3:1 ratio in the F_2 generation. This can be verified in two ways.

First, look at the matrix at the bottom of Figure 20.3. This shows what happens if random pollination actually results in each type of pollen grain fertilizing each type of ovule. In this experiment, three-quarters of the F_2 plants develop from zygotes with a Tt factor (or a tT factor) or from zygotes with a TT factor. These are tall plants. One quarter of the F_2 plants develop from zygotes with a tt factor. These are dwarf plants. This gives a ratio of 3:1.

Second, the probability that random fertilization will produce such results can be demonstrated by means of the coin-tossing test described in exercise A.

To summarize: **Mendel argued that his results can be explained if it is assumed that hereditary characteristics are controlled by pairs of hereditary factors. During reproduction the members of each pair separate and move into different gametes, i.e. a gamete receives one factor from each pair. At fertilization the factors come together again and the pairs are restored. The separation of factors into different gametes and their restoration into pairs as a result of random fertilization is the mechanism behind Mendel's First Law, since it eventually leads to the segregation of contrasting characters in the F_2 generation**.

Segregation has now been demonstrated in all the major groups of organisms which reproduce sexually and has been found to operate more or less as it does in pea plants. However, many new discoveries have been made since Mendel's time, and several have revealed exceptions to his laws of inheritance. Nevertheless, his principles generally hold good even today, and the next section will show how closely they relate to modern discoveries about the cell.

20.2 Chromosomes and heredity

Since Mendel's death the invention of improved microscopes and techniques for staining cells have made it possible to examine the cell nucleus and revealed the presence of objects called **chromosomes**. The study of how chromosomes behave during cell division and reproduction has led to the realization that they must carry the hereditary factors described by Mendel. To understand how this conclusion was reached it is necessary to study a type of cell division called meiosis.

Meiosis
Unlike mitosis, meiosis occurs only within the reproductive organs (except in protists where reproductive organs are absent). Meiosis is concerned with the production of gametes. Another difference is that whereas mitosis produces two daughter cells which have exactly the *same* number of chromosomes as were present in the original parent cell, meiosis produces four daughter cells with *half* the normal number of chromosomes. For this reason meiosis is sometimes called **reduction division**.

Chromosome numbers   A cell with half its normal set of chromosomes is said to have the **haploid number** of chromosomes. The normal set is called the **diploid number** of chromosomes.

Human cells have a diploid number of 46 chromosomes. This means that every body cell has 46 chromosomes. During meiosis sperms and ova are produced with a haploid number of 23 chromosomes. For simplicity body cells are said to be diploid, and gametes haploid.

Clearly, without meiosis young would have twice the number of chromosomes as their parents, and the number of chromosomes would be doubled at each subsequent generation.

The stages of meiosis are described, greatly simplified, in Figure 20.4. This should be studied carefully before reading further.

Mendelian factors and chromosomes
Hereditary factors described by Mendel (i.e. Mendelian factors) are now believed to be part of the structure of chromosomes. The evidence for this comes from observing many parallels between the behaviour of chromosomes during meiosis and fertilization, and Mendel's theory of how factors behave during reproduction.

1. Mendelian factors operate in pairs and, in the first stage of meiosis, chromosomes are seen to form homologous pairs which are composed of chromosomes identical in length and shape. (The 'sex' chromosomes of some animals are an exception, and will be described later.)

2. Mendelian factors separate from each other at gamete formation so that only one from each pair enters each gamete. A gamete therefore contains half the total number of factors. Homologous chromosomes also separate at meiosis, and one from each pair enters each gamete so that a gamete possesses half the normal number of chromosomes.

3. The normal number of both Mendelian factors and chromosomes is restored at fertilization.

These three points show that chromosomes behave in the way that Mendel supposed his factors must behave, and this suggests that chromosomes are factors.

However, more recent investigations have disclosed that there are far more factors than there are chromosomes, and therefore it seemed likely that chromosomes contained a number of different factors.

This has now been confirmed, and the Mendelian factors have been renamed **genes** (which is why the scientific study of heredity is called genetics). Chromosomes are in fact made up of many different genes.

Genes

A gene is a unit of heredity. Each gene controls the development of a set of hereditary characteristics in an organism. In man it is estimated that there are about 10 000 genes contained in the 46 chromosomes of each cell. Among this number there are genes which control the development of all the body organs, and others which control visible characteristics such as the shape of the face, and the colour of the eyes and hair. Characteristics such as height and intelligence are said to be **polygenic**, meaning that they result from the activity of many genes.

Genes affect each other, and in turn are usually affected by the environment of the organism. It has been said of most genes that they provide a 'promise' of a certain result. For example, it is certain that a man who inherits the genes which produce brown eyes will in fact have eyes of that colour, but a man who inherits genes which could produce high intelligence may never develop this characteristic to its full extent unless he is brought up in an intellectually stimulating environment.

Structure of a gene

A gene is a short length of chromosome. In fact, a sequence of genes, all with different functions, are arranged along the length of each chromosome, like beads on a necklace.

In 1944 the American biologist Oswald Avery transferred a substance called **deoxyribonucleic acid**, or DNA, from one type of bacterium to another and, in so doing, discovered that he had transferred the organism's hereditary information as well. It was already known that the cells of all organisms contain DNA, and that in animals and plants DNA is contained in the chromosomes. But as a result of Avery's experiments and numerous others, it became clear that hereditary information is 'stored' in DNA as part of its chemical structure.

DNA can be compared with an architect's plans for a house. Both contain a set of instructions for building something, which in the case of DNA is the body of a living organism. The popular way of describing the building instructions contained in DNA is to call them the **genetic code**, because they are 'translated' into another form when put to use inside a cell.

The genetic code

A gene is a short length of DNA, and so contains a section of genetic code. The section of genetic code in a gene is a set of intructions, in chemical form, for putting amino acids together to make protein molecules.

A cell 'obeys' these instructions by linking together the correct amino acids, in the correct order, directed by a gene's genetic code. Each gene is responsible for assembling a complete protein molecule. DNA therefore controls the type of proteins contained in cells.

In this way, DNA controls the structure of an organism and all the chemical activities within it. This is so because proteins are the building materials out of which cells, tissues, and organs are made. In addition, enzymes and certain hormones are proteins and these control all the chemistry, growth, and development of an organism.

Every cell of an organism contains DNA with a complete set of instructions for building that organism. This set of instructions is passed on to daughter cells during mitosis, and is passed on to the gametes during meiosis. But during meiosis certain changes can occur in the genetic code.

Meiosis and heredity

During the early stages of meiosis, homologous chromosomes grow another chromatid so becoming double structures (stage **B** of Fig. 20.4).

As the homologous chromosomes begin to move apart (stages **C** and **D** of Fig. 20.4) the chromatids of each homologous pair are seen to be in contact at a number of points (Fig. 20.5B). At these points the chromatids break and re-join in such a way that sections of chromatid are exchanged. In other words, there is an exchange of genetic material (DNA) between the chromosomes (Fig. 20.5C). This exchange is called **crossing over**.

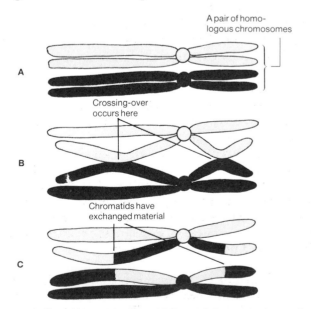

Fig. 20.5 Crossing over – the exchange of genetic material between homologous chromosomes

Each time crossing over occurs, different sections of chromosome are exchanged. Consequently each crossing over is unique.

The result of this exchange of DNA is that the chromosomes which form gametes have a slightly different genetic makeup than the cells of the organism which produced the gametes.

During later stages of meiosis the homologous chromosomes separate and one of each pair enters a gamete. There is no mechanism controlling which member goes into which gamete – it is a completely random process – and will be different each time meiosis occurs. This random movement of chromosomes is called **independent assortment**, because the members of each homologous pair move independently of each other into the gametes.

The importance of crossing over, and independent assortment The importance of these two mechanisms can be seen by following the reproductive process.

When a sperm and an ovum fuse together, the chromosomes from each join up to form the zygote nucleus. For the sake of argument let us call the chromosomes from the sperm the 'male' set of chromosomes, and those from the female the 'female' set.

When the zygote grows into an adult, gametes are produced by meiosis. Then, the descendents of the 'male' and 'female' sets of chromosomes come together again and form homologous pairs. That is, one member of each pair is from the original 'male' set, and the other from the 'female' set which formed the zygote (Fig. 20.6).

When crossing over occurs, therefore, there is an exchange of DNA between the 'male' and 'female' sets of chromosomes. So each crossing over produces a unique mixture of genes from both parents. Then, independent assortment gives each gamete a unique selection of chromosomes as the homologous pairs separate in a random manner. Finally, random fertilization occurs: any male gamete fertilizes any female gamete.

As a result of random crossing over and independent assortment, an organism's gametes contain a never-to-be-repeated combination of its parents hereditary characteristics. Random fertilization then brings together two genetically unique gametes. The result is that successive offspring of the same parents are different.

The combined effect of all these random processes is that sexual reproduction produces far more variation than would otherwise be possible and, as the previous chapter explains, variation is a source of evolutionary change.

But this is still not the end of the story. Section 20.4 explains how a further random event called **mutation** can create completely new genetic material.

20.3 Modern genetics

Having outlined a little of the mechanism of heredity by studying Mendel's First Law, it is now possible to introduce some technical terms and methods used in modern genetics.

Technical terms
The position which a gene occupies on a chromosome is called the **locus** of that gene. On the whole genes remain in the same locus, although there are circumstances in which a gene moves from one locus to another.

Chromosomes occur in pairs (homologous chromosomes) and so genes occur in pairs. In fact the opposite partner of any particular gene occupies the same relative locus on the corresponding homologous chromosome. The genes of such a pair are said to be **alleles** of each other, or an **allelomorphic pair**. The term allele is more often used to refer to all the genes which could occupy a particular locus. The genes which control human blood groups are an example.

There are four main blood groups A, B, AB, and O. These are controlled by three alleles: the genes A, B, and O. Since genes operate in pairs only two of these alleles are present at the same time in a person's cells. Gene O is a recessive to both A and B, and A and B are said to be **co-dominant**, because neither can dominate the other. Table 2 shows how different combinations of these alleles produce various blood groups.

Table 2 Inheritance of human blood groups

Gene pair		Blood group
	Produces	
OO	⟶	O
AA	⟶	A
AO	⟶	A
BB	⟶	B
BO	⟶	B
AB	⟶	AB

The genes of an allelomorphic pair may be identical, in which case they are said to be **homozygous**. For instance, the genes TT in a pure line tall pea plant, and tt in a pure line dwarf plant, are both examples of homozygous genes. Pure lines of any organism are said to be homozygous for the characteristic in question, which is a short way of saying that the allelomorphic pair which controls the characteristic is homozygous.

When genes in an allelomorphic pair are opposite in nature, i.e., when one is dominant and the other recessive, the pair is said to be **heterozygous**. An example of heterozygous genes is the pair Tt found in hybrid pea plants.

The nature and arrangement of genes in an organism is called the **genotype** of that organism. As far as stem length is concerned in pea plants there are three genotypes: TT, tt, and Tt. As far as genotypes are concerned tT is the same as Tt.

It is not always possible to tell the genotype of an organism by looking at its external features. For instance, plants with genotypes TT and Tt look alike, (i.e. tall) but are genetically different. On the other hand, if two plants with the genotype TT are grown on different soils the one on the better soil may grow taller than the other on poorer soil. The word **phenotype** is therefore used in reference to the *visible* characteristics of an organism, as contrasted with genotype (i.e. the genes which it possesses). In a broader sense the term phenotype refers to all the characteristics of an organism which result from the action of the genes, the environment, and the interaction between the two.

A **mutation** is a sudden change in a gene, or a chromosome, which alters the way in which it controls development. The majority of mutations are changes in the DNA of a single gene. The gene will then produce a different type of protein with corresponding effects in the organism. Some mutations affect all or part of a chromosome: parts may break off; parts may break and rejoin in a different way, or become attached to another chromosome; and sometimes a whole chromosome is either lost or gained.

The majority of mutations are harmful. This happens because cells containing a mutation are significantly different in appearance, chemistry, and behaviour from normal types, and may alter or upset the delicately balanced mechanisms of an organism. Very rarely mutations cause vital changes which are lethal. Equally rarely they produce changes which are advantageous. If these spread through a population they may bring about evolutionary change.

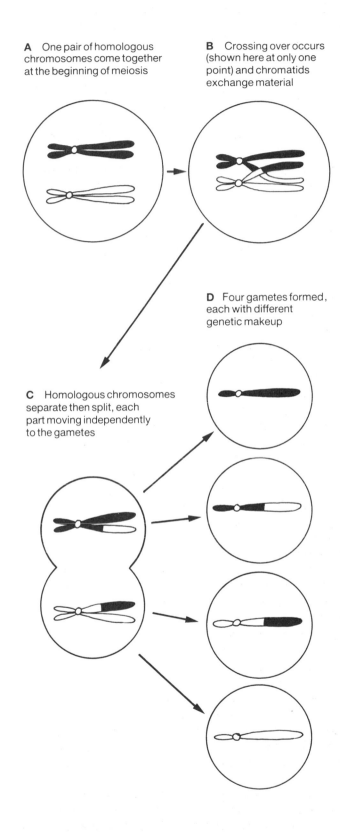

A One pair of homologous chromosomes come together at the beginning of meiosis

B Crossing over occurs (shown here at only one point) and chromatids exchange material

D Four gametes formed, each with different genetic makeup

C Homologous chromosomes separate then split, each part moving independently to the gametes

Fig. 20.6 How crossing over and independent assortment produce uniquely different gametes

An albino and a normal dark-skinned native of the Trobriand Islands (near New Guinea). Albinism results from a mutation in the genes which control skin colouration, so that the colour fails to develop. The skin of an albino is almost transparent and is quickly damaged by exposure to sunlight. The mutation is recessive

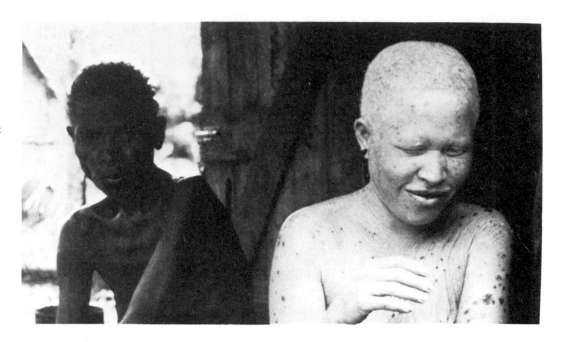

If mutations occur in body cells they may spread by mitosis but are usually restricted to one organism. On the other hand, if mutations occur in the reproductive organs within cells which produce gametes the mutations may be passed on to the next, and subsequent generations.

It is not certain what causes mutations in nature, but several ways have been discovered of artifically increasing the rate at which mutations occur. In some organisms the mutation rate increases as temperature rises: the rate appears to double with each rise of 10°C. Several chemicals cause a rise in mutation rate when applied to living organisms, e.g. mustard gas and formaldehyde. By far the most powerful means of increasing the mutation rate is by exposing an organism to high energy radiation such as X-rays, beta, and gamma rays.

This effect of artificial radiation is a clue to the possible source of natural mutations because radiation is present in all natural environments. The earth is constantly bombarded with high energy radiation in the form of cosmic rays from outer space, and the earth's crust contains radio-active elements such as uranium. Furthermore, in recent years the advance of technology into the use of atomic, or nuclear, energy has increased the amount of artificial radiation. Whether or not this will greatly increase mutation rates remains to be seen. The danger is

there, however, and precautions must be taken to guard against an accidental outflow of radiation into the environment.

Mutations are comparatively rare events because DNA is a stable molecule, and chromosomes are stable structures. For several reasons, most mutations do not reach the next generation, or have much effect even if they are inherited. First, only mutations carried by gametes are inherited, and only a tiny proportion of gametes take part in fertilization to produce offspring. Consequently, inheritance of mutations depends primarily on the proportion of gametes which carry mutated genes. Second, most mutations are recessive and confined to only one gene of an allelomorphic pair. Therefore, the majority of inherited mutations do not appear in the phenotype because they are suppressed by their dominant allele. Third, since mutations are largely harmful, and sometimes lethal, natural selection operates against them when they do appear in the phenotypes, and this limits their spread through the population.

At the same time, it must be remembered that the majority of species consist of millions of organisms producing millions of offspring yearly. Consequently, the rare event of an advantageous mutation appearing in one phenotype can occur a significant number of times in a species as a whole. In other words, a rare event multiplied millions of times can become a common event.

Mutation and evolution

There is a widely held theory that evolution progresses by natural selection of the mutations which occur at random in every species. The theory argues that unfavourable mutations disappear while the rare favourable mutations survive and multiply. This could happen by means of a type of mutation in a species occurring time and time again. These periodic mutations may be unfavourable for generations until suddenly the environment changes, or the species migrates to a different environment where the mutation is favoured by natural selection.

Consider the Peppered Moth (chapter 19). The mutation which produces the dark variety of this moth probably occurred in most generations prior to the industrial revolution but was not favoured by natural selection. Then industrial pollution blackened parts of the moth's environment favouring the dark colouration because of the camouflage which it provided, thereby allowing this variety to multiply.

It seems possible, then, that mutations provide the raw material of evolution in the form of randomly occurring variations, and the mechanism of heredity distributes these variations throughout the species if they are favoured by natural selection.

Use of Drosophila in genetics

Modern geneticists work with many organisms, but the most famous, and one of the most valuable, is a small fly, about 2 mm long, called *Drosophila melanogaster* (Fig. 20.7). *Drosophila* owes its fame to the American geneticist T. H. Morgan and his pupil H. J. Muller, who chose to use this fly in their experiments because it is easily and quickly bred (it has a 10-day life cycle) and gives results in a few weeks which could have taken Mendel at least two or three years using pea plants. Another advantage of *Drosophila* is that when exposed to radiation the wild types yield many distinctly different mutations with, for example, red eyes, white eyes, crumpled wings, and vestigial wings (i.e. wings which are abnormally small).

Figure 20.8 uses modern terms to describe a cross between a mutant *Drosophila* with vestigial wings, and a normal-winged fly. As in Mendel's monohybrid crosses, the F_1 shows only the dominant character, in this case normal wings, and the phenotypes of the F_2 generation show the dominant and recessive characters in a ratio of 3:1. But there remains the problem of distinguishing between homozygous VV flies and heterozygous Vv flies, the phenotypes of which are both normal-winged. At this point Mendel

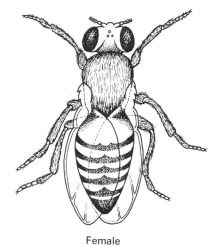

Female

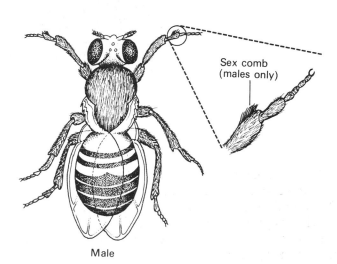

Sex comb (males only)

Male

Fig. 20.7 Comparison between male and female *Drosophila melanogaster*. Females are longer than males, and have a more pointed abdomen with widely spaced black bands. Males have a rounded abdomen with closely spaced bands (giving the abdomen a black tip), and they have a sex comb on each foreleg

self-pollinated his pea plants (Fig. 20.3) but this is impossible with *Drosophila* (and other animals). The problem is solved by a **test-cross**, or back-cross as it is sometimes called, between F_2 flies with an unknown genotype and homozygous vv flies whose genotype is clearly visible from their vestigial wing shape. Figure 20.9A shows that where the unknown genotype is VV, all the F_1 generation develop normal wings. Figure 20.9B shows that where the unknown genotype is Vv the phenotypes of the F_2 generation show the dominant and recessive characters in a ratio of 1:1.

A Changes of phenotypes

Vestigial-winged male   Normal-winged female
(genotype vv)   (genotype VV)

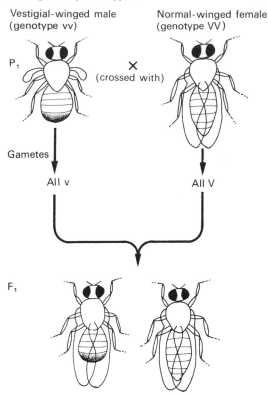

P₁   ×
                (crossed with)

Gametes

All v   All V

F₁

Roughly equal numbers of males and females,
all with normal wings (genotype Vv)

Cross between F₁ flies

Males   ×   Females
V v   V v

Gametes   ½V   ½V
           ½v   ½v

F₂ produced by random fertilization

	Sperms		Phenotype ratio of normal-winged to vestigial-winged flies is 3:1
	V	v
Ova V	VV	Vv
v	vV	vv

¼ VV homozygous normal-winged   ⎫
¼ Vv heterozygous normal-winged ⎬ ¾
¼ vV heterozygous normal-winged ⎪
¼ vv homozygous vestigial-winged ⎭ ¼

B Movements of chromosomes and genes

Chromosomes of   Chromosomes of
vestigial-winged male   normal-winged female

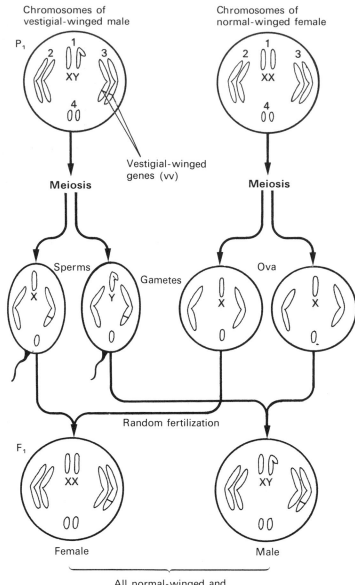

P₁

Vestigial-winged
genes (vv)

Meiosis   Meiosis

Sperms   Gametes   Ova

Random fertilization

F₁

Female   Male

All normal-winged and
heterozygous Vv

Fig. 20.8 Cross between normal-winged and vestigial-winged
Drosophila

A illustrates the phenotypes up to F₁ flies, together with
segregation of genes in the F₂ flies. **B** illustrates the segregation
of chromosomes and vestigial genes between parents and F₁ flies

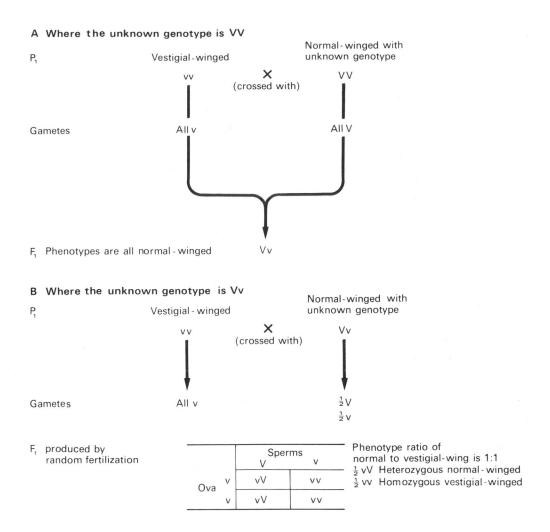

A Where the unknown genotype is VV

P₁   Vestigial-winged   Normal-winged with
                                                 unknown genotype

                          vv   **X**   VV
                                  (crossed with)

Gametes   All v   All V

F₁ Phenotypes are all normal-winged   Vv

B Where the unknown genotype is Vv

P₁   Vestigial-winged   Normal-winged with
                                                 unknown genotype

                          vv   **X**   Vv
                                  (crossed with)

Gametes   All v   ½ V
                                                  ½ v

F₁ produced by   Sperms   Phenotype ratio of
    random fertilization   V   v   normal to vestigial-wing is 1:1
                                                  ½ vV Heterozygous normal-winged
                     v   vV   vv   ½ vv Homozygous vestigial-winged
              Ova
                     v   vV   vv

Fig. 20.9 Use of a test cross to distinguish homozygous dominant VV flies from those with a Vv genotype

Figure 20.8B illustrates the segregation of chromosomes and vestigial-wing genes from parent flies to the F₁ generation. In addition, this figure illustrates the presence of XX chromosomes in female flies, and XY chromosomes in males. These are sometimes called the **sex chromosomes** because they are concerned with determining the sex of the organism.

Sex determination

In *Drosophila*, and many other animals including humans, if sexual reproduction produces a zygote with an X and a Y chromosome that zygote will always develop into a male. But zygotes with an X and an X chromosome always develop into females. The two types of zygote are produced in the following way.

Sex chromosomes separate at meiosis (like all other chromosome pairs) and only one goes into each gamete. Therefore, all female gametes carry one X chromosome, while half the male gametes carry a Y chromosome and the other half an X chromosome. Since there is an equal chance of both X and Y male gametes fertilizing a female gamete there should be an equal number of male and female offspring. For some unknown reason, however, there are usually more males than females. In humans, for instance, the ratio of female to male babies born alive is about 100:106.

It is not clear why the XX combination produces females while the XY combination produces males, since the genes which control the development of sexual characteristics are not restricted to the sex chromosomes but are distributed throughout all homologous pairs. In fact, a zygote appears to have all the genes for producing *both* male and female characteristics. The final outcome as to which sex develops seems to depend on the Y chromosome: when the Y is present only the male characteristics develop, and when it is absent female characteristics develop.

20.4 Genetic engineering

Genetic engineering is the technique of altering an organism's genotype by inserting genes from another organism into its DNA. Once inserted the foreign genes work as if they were still in the organism they were taken from.

This technique allows genetic engineers to 'program' an organism to make useful substances. An example is the way in which a harmless bacterium called *Escherichia coli* (*E. coli* for short) found in the human gut, can be changed into a tiny factory to mass-produce insulin for diabetics.

Insulin from bacteria

The first step is to isolate and cut out the single gene in human cells which controls insulin production. This seemingly impossible task can be accomplished because of the discovery of **restriction enzymes**. These enzymes are naturally occurring DNA 'cutters' which body cells use to dispose of damaged sections of DNA.

By carefully working out which restriction enzyme cuts which section of DNA, scientists can cut out the exact piece of DNA containing the insulin gene.

The gene is inserted into an *E. coli* cell where it combines with the microbe's DNA. Bacteria transformed in this way are cultured in huge vats of nutrient until ready to have the insulin extracted from them (Fig. 20.10).

This work is worthwhile because it provides pure human insulin which is more effective than that at present extracted from cattle and pig pancreas.

Other examples of genetic engineering

Gene transplant techniques make it possible to 'engineer' bacteria capable of producing all manner of useful substances.

E. coli has already been programmed to make human growth hormones for the treatment of growth deficiencies in children. Blood clotting chemicals for haemophiliacs, vitamins, and vaccines can all be tailor-made for the human body as they are made by human genes. Further examples of genetic engineering include bacteria which consume oil spills, and others which give high yields of antibiotics.

Theoretically it should be possible to implant insulin genes into the body cells of diabetics so that they can produce their own insulin. Genetic diseases such as haemophilia could be cured in the same way.

Fig. 20.10 Transplanting genes from humans to bacteria can change these microbes into factories which mass-produce many useful substances

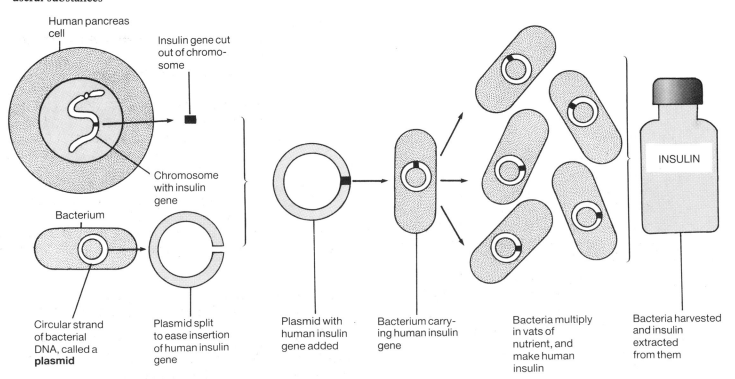

Plant–plant, and animal–plant hybrids

The latest techniques of genetic engineering involve transplanting genes from plants to plants, and even from animals to plants. It is hoped that this will create new strains with built-in resistance to disease; crop plants with resistance to weed killers, and hitherto unheard of plants with, for example, the nutritional value of two or more crop plants (see section 3.10).

In one experiment, genetic material from protein-rich soya plants was combined with carbohydrate-rich corn, barley, peas, and carrots. It was hoped to make a single crop to provide a fully balanced meal, but so far scientists have only succeeded in growing formless clumps of cells and not fully formed plants.

Human genes can be combined with plants. While this conjures up horrific images of human–plant monsters the technique could have beneficial results. Such a transplant would not give rise to a plant with hands and feet! It could produce a hybrid which tells us many things about how human genes operate, and may even produce drugs, hormones etc., which could be harvested from the plant's leaves.

Investigations

A *To demonstrate that random fertilization between equal numbers of dominant and recessive genes produces a phenotype ratio of 3:1*

1. If animals with the genotype Aa are crossed, the F_1 offspring will be produced from random fertilization between equal numbers of A and a sperms and A and a ova. Thus:

	Sperms	
	A	a
A	AA	Aa
a	aA	aa

Ova

$\frac{1}{4}$ AA
$\frac{1}{4}$ Aa $\Big\}$ $\frac{3}{4}$
$\frac{1}{4}$ aA
$\frac{1}{4}$ aa $\Big\}$ $\frac{1}{4}$

The phenotypes occur in the ratio of 3:1

It is possible to show that the 3:1 ratio results from random fertilization in the following way.

2. Take two coins of the same denomination (2p or 10p coins are suitable). Use white sticky paper to label one side of one coin Sperm A, and the opposite side Sperm a; then label one side of the other coin Ovum A, and the opposite side Ovum a. Copy this chart.

Sperm	Ovum	Tally
A	A	
a	A	
A	a	
a	a	

Total =

Total =

3. Working in pairs spin an 'ovum' coin and a 'sperm' coin simultaneously noting how they fall, and enter the result in the appropriate part of the tally column. Repeat this at least 50 times.

4. Does the ratio of one total to the other amount to roughly 3:1?

5. Explain why coin tossing can be used in place of fertilization in this experiment.

6. How does this experiment show that, to be successful, all genetics experiments must involve large numbers of offspring?

B *Variation*

1. Every member of the class should record the following information:

a) The length, to the nearest millimetre, of one index finger.

b) The number of times the heart beats in one minute (after sitting at rest for at least five minutes).

Use this information to prepare histograms or line graphs showing variation in finger length and heart-beat rate in the whole class.

2. Collect at least 50 leaves from one tree.

a) Measure either the length of each leaf (including the petiole) or the width.

b) Use these measurements to construct a histogram or line graph showing variation in leaf size in the plant.

C *A monohybrid cross with maize plants*

1. Maize is useful in genetic experiments for two main reasons.

a) Pure line varieties are available with clearly visible differences such as dark and light coloured grains, and smooth and wrinkled grains.

b) One maize cob contains several hundred grains, each of which results from a single fertilization. Therefore, after a cross has been made, it is only necessary to count the types of grains on the cobs to obtain F_1 and F_2 ratios.

2. Sow seeds of a pure line dark grained variety close to seeds of a light grained variety.

a) Cover the male and female flower heads with plastic bags to prevent self-pollination and to prevent pollination by plants not included in the experiment.

b) When the male flowers are ripe, cross-pollinate the plants by transferring the plastic bags and pollen which has been shed into them from the flowers of the dark variety to the flowers of the light variety, and from the light flowers to the dark flowers.

c) What colour are the grains of the F_1 plants?

d) Which colour is dominant and which is recessive?

e) Sow F_1 grains, but this time ensure that only self-pollination occurs. What is the ratio of dark to light coloured grains in the F_2 plants?

D   *Investigating the effects of environment on phenotype*

1. Obtain rooted cuttings of coleus, geranium, *Tradescantia*, etc. Cuttings must, as far as possible, be the same size, and taken from the same plant so that they are genetically identical.

2. The object is to study the effects of different environments on phenotype. This can be achieved by growing some plants under 'ideal' conditions, and comparing them with plants grown under conditions where *one* environmental factor is unsatisfactory, such as: cold temperatures, poor nutrient supplies, and dim light.

3. After a few weeks compare plants in the ideal environment with those in unsatisfactory environments and note any differences.

Questions

1. Mendel crossed pure line pea plants with green pods, and pure line plants with yellow pods. All the F_1 plants had green pods. Out of 580 F_2 plants, 428 had green pods and 152 had yellow pods. Which characteristic is dominant and which recessive? In the F_2 plants how many are: homozygous recessive, homozygous dominant, and heterozygous? Using G to represent the dominant gene, and g to represent the recessive gene, write out a plan showing the segregation of genes from the parents to the F_2 plants.

2. If two *Drosophila* flies, heterozygous for genes of one allelomorphic pair, were bred together and had 200 offspring, about how many would have the dominant phenotype? Of these offspring some will be homozygous dominant and some heterozygous. How is it possible to establish which is which?

3. Mendel believed that hereditary factors (genes) were always either dominant or recessive. How might he have altered this view had he performed the following crosses? When pure line sweet peas with red flowers are crossed with pure lines having white flowers, all the F_1 plants have pink flowers. When pure line shorthorn cattle with red coats are crossed with pure lines having white coats their offspring have coats with a mixture of both red and white hairs (this is called the **roan** condition).

4. Live bacteria which were all red in colour were placed under an ultra-violet lamp. After several days groups of white bacteria began to appear among the red. What conclusions, if any, can be made at this stage? What further experiments should be performed?

5. The diagram below represents a cross between a heterozygous black male mammal, and a white female. In these animals coat colour is controlled by the alleles B and b. The gene for black coat is dominant. Copy the diagram and answer the questions.

a) In the circles write the genotypes of the parents, the gametes, and the F_1 generation.

b) What is the ratio of young with black coats to those with white coats in the F_1 generation.

c) Which of the animals numbered in the diagram are homozygous?

d) If a black female from the F_1 generation was crossed with its black male parent, what ratio of black to white young would you expect to be produced?

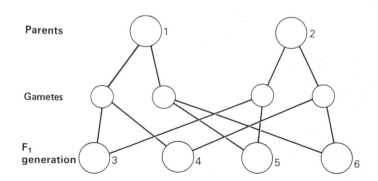

6. The drawing below shows fruit flies produced in a genetics experiment.

a) State the way in which the two types of fly are different, and count the numbers of each type.

b) Which characteristic is dominant and which is recessive?

c) The numbers of each type represent the ratio resulting from crossing two types of fly. Assume that **A** and **a** represent the alleles involved in the cross. Which of the following crosses would produce this ratio:

AA × **aa**; **Aa** × **Aa**; **AA** × **AA**; or **aa** × **aa**?

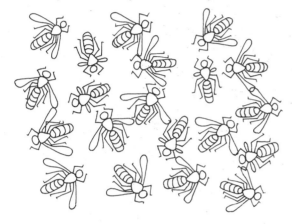

Summary and factual recall test

Mendel was looking for the (1) which govern the way in which (2) characteristics pass from parents to (3). To do this he chose characteristics with (4) variation, which means (5). His basic method involved hybridization which means (6). Study of *one* characteristic by this method is called a (7) cross.

Mendel crossed tall plants with dwarf plants and produced F_1 plants which were (8). From this he concluded that tallness must be the (9) characteristic because (10). Mendel now (11) the F_1 plants and produced F_2 plants in which three-quarters were (12) and one quarter (13), giving a ratio of (14). From this he concluded that dwarfness must be a (15) character because (16).

Mendel suggested that hereditary (17) are controlled by factors that operate together in (18), the members of which (19) during gamete formation so that each gamete receives (20). But at fertilization (21–describe what happens to the factors).

By comparing how factors behave during reproduction with the behaviour of (22) during (23) or reduction division it has become clear that (24). The behaviour of the two are similar in three ways which are: (25).

Factors are now called genes. Each gene has

7. In humans the gene for blue eyes (**b**), is recessive to the gene for brown eyes (**B**). The diagram below represents part of a family tree in which some have brown eyes and some have blue eyes.

a) Using the symbols given above, write the genotype of the mother in the space provided on a copy of the diagram.

b) What is the phenotype of the Grandfather?

c) What is the ratio of individuals with brown eyes to those with blue eyes in the F_2 generation?

d) Write down the genotype of an individual in the F_2 generation who is homozygous, and one who is heterozygous.

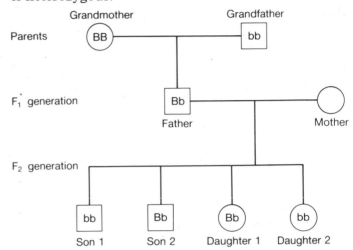

a partner which occupies the same (26) on the corresponding (27) chromosome. Genes of such a pair are said to be (28) of each other. When the members of these pairs are identical they are said to be (29), and when opposite they are said to be (30). The (31) of an organism is the nature and arrangement of its genes, and is contrasted with its phenotype which is (32).

A mutation is (33). Mutation rate has been artificially increased by (34–describe three ways). Only mutations carried in the (35) are inherited, and most of these do not appear in the phenotype because (36), but if they do, (37) selection usually operates against them because most of them are (38). Because of this (39) mutations disappear from a species while (40) ones survive and are multiplied. In time this process can lead to (41) change.

Genes control the types of (42) made in a cell, and thus control the structure and functions of the whole organism because (43).

Genetic engineers alter an organism's (44) by inserting (45) from another organism into its (46). Bacteria can be 'programmed' to produce useful substances such as (47–name four).

21

Parasites, disease, and immunity

A healthy organism is one in which all the physical and chemical processes of life are working in harmony. Apart from accidental damage and starvation, one of the chief causes of disharmony, i.e. disease, in an organism is the presence on or inside its body of another organism which lives as a **parasite**.

21.1 Parasites

A parasite is an organism which lives on or in the body of another living organism called the **host**. Parasites gain food and sometimes shelter from their association with a host, but the host gains no benefit, and is usually harmed in some way.

Some parasites inflict very little damage on their host. Others cause damage which results in disease. This type of parasite is said to be **pathogenic**, and **pathology** is the study of the diseases which they cause.

Types of parasite
Parasites such as fleas, lice, mosquitoes, and certain fungi which live on the outside of their host are examples of **ectoparasites** (Fig. 21.1). Parasites which live inside their host's body are called **endoparasites**. Viruses, protozoa, and parasitic worms are endoparasites (Figs 21.2 and 21.3).

Pathogenic micro-organisms, or microbes, are commonly known as **germs**. Examples are the viruses which cause common colds, bacteria which cause food poisoning, protozoa which cause malaria, and fungi which cause ringworm.

Features of the parasitic way of life
Finding a host Ectoparasites such as mosquitoes, fleas, and lice fly, crawl, or jump from host to host using sense organs to find their way. But endoparasites, such as tapeworms, once inside their

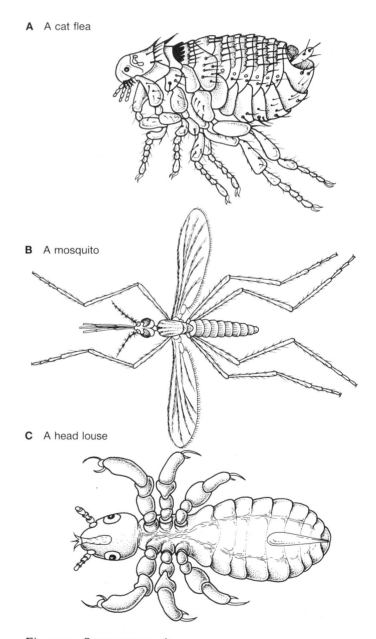

A A cat flea

B A mosquito

C A head louse

Fig. 21.1 Some ectoparasites

host are unable to leave it to infect another host. They have special mechanisms for passing on their eggs to other hosts.

Tapeworms, for example, lay huge numbers of eggs which pass out of the host's body in the faeces, so there is a chance that some will re-infect another host (this is explained in more detail in section 21.6).

Many germs pass from host to host in the bodies of blood-sucking insects such as fleas and mosquitoes. Insects which carry germs are called **vectors**. Female *Anopheles* mosquitoes are vectors of the germs which cause malaria (Fig. 21.1). Other methods by which germs spread from host to host are described in section 21.7.

Feeding without killing the host Unlike predators, a parasite does not deliberately kill its host, since this would destroy its source of food. Sometimes a host may die as a result of the parasite's feeding activities, or from poisons which it releases into the host's body.

Mosquitoes, fleas, and lice hold onto their hosts with clawed feet and feed with mouth parts which allow them to bite through their host's skin and suck juices from its body (Fig. 21.1). Endoparasites, such as germs and parasitic worms, obtain food directly from host tissues, and may even absorb digested food from the host's gut.

How parasites cause disease Many endoparasites release poisonous chemicals called **toxins** into their host's body. Toxins can cause disease symptoms such as high temperature, headache, and vomiting. Symptoms do not appear immediately a pathogen enters a host. There is an interval called the **incubation period** before symptoms appear, during which time germs multiply rapidly and larger parasites develop to full size.

Certain parasitic worms and insects bore through host tissues causing wounds which may then become infected with germs. This is called **secondary infection**.

The advantages of parasitism

Parasites have a plentiful food supply while their host lives. Endoparasites in particular do not have to search for their food, and do not compete with other organisms for it. Endoparasites are protected by their host's body from drought, heat, cold, etc., and from predators, except those which attack the host.

Disadvantages of parasitism

Parasites which live on or in one species of host depend on it exclusively for food. Consequently, they will suffer if their host's numbers are reduced, or if the host becomes extinct.

Some parasites have become completely dependent on their host so that they cannot survive without it. Tapeworms, for example, have no digestive system, no locomotory organs (legs, etc.), and their muscular system is extremely weak. They have a poorly developed nervous system and hardly any sensitivity. An adult tapeworm will die almost immediately if it is removed from its host. Tapeworms can be as long as 8 metres!

21.2 Viruses and disease

In 1935 the biochemist W. Stanley produced a few grams of needle-like crystals by refining juice extracted from tens of thousands of diseased tobacco plants.

Amazingly, this apparently dead, dry powder produced disease when dissolved in water and smeared on healthy tobacco plants, even after it had been stored for months, and sometimes years, in a bottle. This powder was the first pure sample of a *crystallized virus*, now called tobacco mosaic virus, because of the mottled pattern it produces on the leaves of tobacco plants.

Preparation of this powder started a controversy which continues to this day about whether viruses are alive, or just a collection of non-living chemicals with the power to reproduce.

It is certain, however, that viruses are not cells. Viruses are, however, very small. A row of a million average-sized viruses would be only 5 mm long.

Viruses are totally parasitic. They show signs of life only when inside the living cells of another organism. They exist in a dormant state when outside their host.

A virus consists of DNA enclosed in an envelope of protein and fat. It attacks host cells, taking them over and turning them into virus factories.

A virus attaches itself to a cell membrane and injects its DNA into the cell's cytoplasm. Virus DNA makes use of host DNA and other cell chemicals to make replicas of itself. After as little as 30 minutes the host cell bursts open releasing hundreds of new viruses which infect other cells.

Chicken pox, measles, poliomyelitis, the common cold, and influenza are diseases caused by viruses. Section 9.6 describes colds and 'flu in more detail.

Viruses (× 10 000)

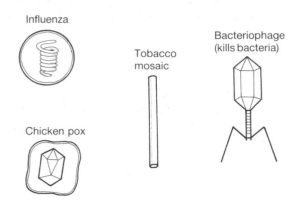

Fig. 21.2   Some examples of viruses

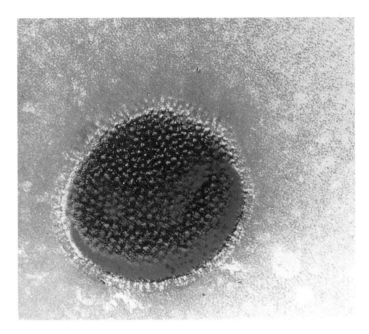

The influenza virus greatly magnified

21.3   Bacteria and disease

Bacteria are unicellular organisms between 0.0005 mm and 0.005 mm long. Figures 21.3 and 21.4 describe the features of bacteria.

Under favourable conditions bacteria reproduce by dividing in two every 20 or 30 minutes. Starting with one bacterium this rate of division could produce 16 million bacteria in 8 hours.

When conditions are unfavourable for growth some bacteria form spores with a protective wall. Spores can survive immersion in boiling water or disinfectant for several hours, and can exist for many years as dust blown about in the wind.

Leprosy, cholera, venereal diseases, and dental caries (decay) are caused by bacteria. Venereal diseases are described in detail in section 17.8 and dental decay in section 4.3.

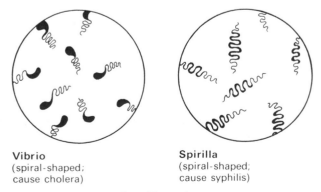

Vibrio
(spiral-shaped;
cause cholera)

Spirilla
(spiral-shaped;
cause syphilis)

Fig. 21.3   Some examples of bacteria

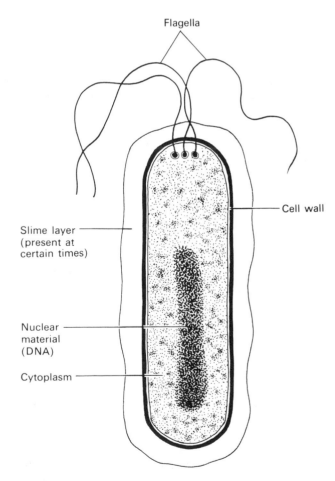

Fig. 21.4   Diagram of a bacterial cell, showing most of the structures found in bacteria

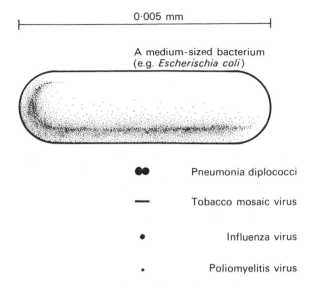

0·005 mm

A medium-sized bacterium
(e.g. *Escherischia coli*)

oo   Pneumonia diplococci

—   Tobacco mosaic virus

•   Influenza virus

·   Poliomyelitis virus

Fig. 21.5   Comparative sizes of bacteria and viruses

21.4   Protozoa and disease

Protozoa are complex unicellular organisms. Very few cause diseases in humans. Amoebic dysentery and malaria are examples.

Malaria is a disease caused by a protozoan called *Plasmodium*. It lives in the blood and liver of an infected person. When a female *Anopheles* mosquito (Fig. 21.1) sucks blood from an infected person it absorbes *Plasmodium* into its stomach along with blood. Here, the *Plasmodium* multiplies and then passes to the insect's salivary glands. When the insect bites other people it squirts saliva, and *Plasmodium*, into their blood stream, and they become infected with malaria.

Malaria can kill, but more often it simply makes the victims unable to work because they suffer from high fever, vomiting, and severe headaches. Drugs such as quinine, and the more modern daraprim, mepacrine, and paludrine are used to kill malarial parasites in humans.

Malaria is controlled by killing the mosquitoes which carry *Plasmodium*.

Whenever possible, mosquito breeding places are removed, or made inaccessible, by draining swamps or covering water tanks and ensuring that tin cans, broken pots, etc. are not left where they can collect rainwater and encourage mosquitoes to breed in them.

Ponds and marshes can be sprayed with chemicals which kill eggs, larvae, and pupae, or they can be stocked with fish such as minnows which eat these stages of the mosquito life cycle.

Mosquitoes hide in dark places during the day, so adults can be killed by spraying interiors of houses, sheds, storerooms, etc. with long-lasting insecticides.

21.5   Fungi and disease

Very few fungi are parasites of humans. One of the commonest fungal diseases is called **ringworm**, because it can produce a circular swelling on the skin. Ringworm fungi attack the scalp, and the soft skin of the groins.

Another very similar fungus attacks the soft skin of the feet, especially between the toes, causing a disease called **athlete's foot**.

Fungal diseases are spread by airborne spores (Fig. 21.6), by contact with infected people and, in the case of athlete's foot, by infected floors and mats on which people walk bare-foot.

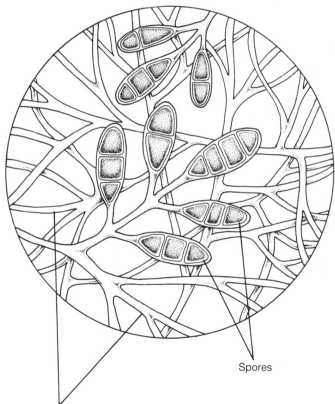

Spores

Hyphae (threads which make up the 'body' of a fungus)

Fig. 21.6   The spore-producing organs of a microscopic fungus which causes ringworm

21.6 Tapeworms and disease

Tapeworms live in the intestines of vertebrates, including humans. A tapeworm consists of a head with hooks and suckers which attach it to the host's intestine wall, and a long flat tape (Fig. 21.7).

Tapeworms do not have a mouth or a digestive system and do not produce digestive enzymes. They are bathed in their host's digested food which is absorbed through the worm's body surface.

Tapeworms produce large numbers of eggs. The beef tapeworm, for example, produces 100 million eggs a year and can continue doing so for 25 years. The eggs pass out of the host's body in its faeces.

The life cycle of a tapeworm involves not one, but two hosts. The first, or **primary host**, is the one in which the adult tapeworm lives and reproduces. The **secondary host** is an animal which is eaten by the primary host. For example the beef tapeworm's primary host is humans, and its secondary host is cattle.

If sewage disposal conditions are primitive, tapeworm eggs may be accidentally eaten by a cow. Inside the cow the egg opens releasing an embryo tapeworm which eventually lodges in muscle tissue. There it can remain unchanged for several years. If the cow is killed and its meat eaten raw, or partly cooked, an embryo can enter a human and develop into an adult tapeworm.

21.7 The spread and prevention of infection

Pathogenic bacteria and viruses can spread from person to person in many ways, the most important of which are: contact with infected droplets from someone with a disease; consumption of contaminated water and food; contact with infected people and contaminated objects; and contact with animals (Fig. 21.8).

Droplet infection

Disease organisms present in the mouth, nose, and lungs can be carried out of the body in droplets of moisture whenever a person breathes out, talks, coughs, or sneezes.

These droplets may spread the disease organism directly to another person, or they may infect water and food which could later be consumed by others.

Viral infections such as the common cold, influenza, and pneumonia are spread by droplets, as are bacilli which cause diseases such as whooping cough and diphtheria.

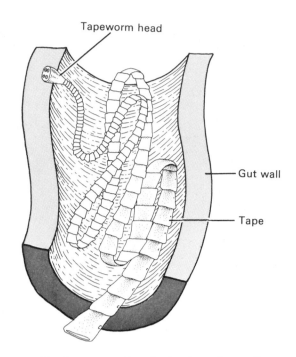

Fig. 21.7   Beef tapeworm (attached to intestine wall)

Contaminated water and food

Food and water can be contaminated with disease organisms by sewage, by contact with people suffering from disease, and by insects, birds, and many other animals. Typhoid fever, cholera, and food poisoning are spread in this way.

Direct contact

Diseases spread by contact with infected people and objects are said to be **contagious**. Infections can be spread by direct bodily contact, or indirectly by touching objects such as handkerchiefs, books, or coins previously handled by infected people. Germs may then be transferred from the hands to the mouth as, for instance, when eating hand-held foods such as sandwiches. Smallpox, measles, and tuberculosis are spread in this way.

Animals

Animal vectors carry disease organisms either inside or outside their bodies. House-flies, for example, carry typhoid, cholera, and dysentery germs both on their bodies and in their faeces. The yellow fever virus is carried from person to person, and from animals to people, inside the bodies of certain mosquitoes, which spread the virus when they suck blood.

Prevention of infection

The most effective ways of controlling the spread of disease have been found by studying the life cycles of pathogenic organisms and discovering how they

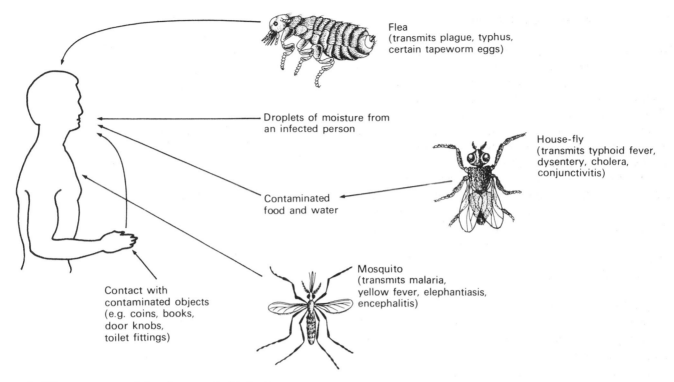

Flea
(transmits plague, typhus,
certain tapeworm eggs)

House-fly
(transmits typhoid fever,
dysentery, cholera,
conjunctivitis)

Droplets of moisture from
an infected person

Contaminated
food and water

Mosquito
(transmits malaria,
yellow fever, elephantiasis,
encephalitis)

Contact with
contaminated objects
(e.g. coins, books,
door knobs,
toilet fittings)

Fig. 21.8 Diagram summarizing the spread of infection

enter the body. This research has led to the development of insecticides, more efficient sewage disposal methods, and methods for decontaminating water supplies and cleansing cities. Personal cleanliness is also necessary if the spread of infection is to be avoided. The skin and hair must be washed regularly to avoid the accumulation of bacteria, and the dirt in which they grow. It is especially important that the hands are kept clean to avoid contamination from, and contamination of, all the articles and objects which are handled daily by different people.

21.8 Food preservation

If food is not quickly preserved in some way after it is harvested it soon 'goes bad' or decomposes, owing to the action of microbes such as bacteria and fungi which begin to grow and multiply in it. Apart from giving food an unpleasant taste, these microbes can be a danger to health. Bacteria of the genera *Clostridium* and *Salmonella* release poisons (toxins) into the food in which they live. These poisons can cause serious illness or death, and can reach dangerous levels even before the food develops an unpleasant taste or smell.

Preserved food can be transported long distances from where it is produced, and stored for long periods

in warehouses or homes without risk of its decomposing or endangering health.

To live in food microbes require moisture, warmth, and usually oxygen. Food can be preserved by storing it in conditions where these requirements are absent.

Drying or dehydration

This is one of the oldest methods of food preservation. In tropical climates, many foods are dried by exposure to air and hot sunlight. Foods such as cereals, rice, beans, coffee, vegetables, fruit, fish, and meat can be placed out of doors on a tray made of wire gauze or reeds to allow free air movement. In addition to removing water, sunlight also kills certain bacteria in food.

Grapes can be dried to form raisins, sultanas, and currants, and eggs and milk can be machine-dried to a powder and stored indefinitely.

Food can now be dried without heat by placing it in a vacuum for a short time. The advantage of vacuum drying is that, unlike drying in heat, it does not alter the flavour of food or destroy any of its vitamins.

Freezing

Domestic refrigerators keep food at about 4°C. At this temperature microbes reproduce very slowly and so food remains fresh for a few days. At temperatures below freezing point bacteria and fungi are

239

unable to decompose food or multiply. A domestic deep-freeze keeps food at −20°C or below. At this temperature food remains fresh for several months. It is essential to cook food soon after it has thawed out as any microbes in it will quickly resume multiplying.

Canning and bottling

Canning is one of the safest methods of preserving food for long periods. The food is first heated to a temperature which kills microbes. In other words, it is **sterilized**. Then, while still hot, the food is put into sterilized cans. Further heating drives all the air out of the can before it is sealed. Domestic bottling of food is done in a similar way, but it is more difficult to seal a bottle so that it remains airtight for long periods.

Chemical preservatives

Salting   This is a very old method of preserving certain types of fish and meat. The food is placed in common salt (sodium chloride). The salt draws most of the water from the food, and it kills any microbes present by drawing water out of them also. Salted butter will keep fresh longer than unsalted.

Sugar   Like salt, strong sugar solution kills microbes by drawing water from them. It is sugar which keeps jam fresh for long periods.

Smoking   Certain types of wood smoke contain chemicals called **phenols**. Phenols poison microbes, and so food can be preserved by hanging it in a smoky atmosphere until sufficient phenols have been absorbed. Smoked fish and bacon are prepared in this way.

Other chemical preservatives   Onions, cabbage, cauliflower, and gherkins can be preserved (pickled) in vinegar. Wine is simply fruit juice preserved in ethyl alcohol. Light wines and beer contain insufficient alcohol to preserve them and so sulphurous acid is added to prevent decomposition. Benzoic acid, boric acid, salicylic acid, and sulphur dioxide are often used, in very small quantities, to preserve meat products such as sausages.

21.9 Smoking and ill-health

Smokers enjoy smoking. It helps them relax and they enjoy the taste. But is this pleasure worth having when we know that smoking kills millions of people a year?

Tobacco: what's in it for you?

When smokers inhale they take into their lungs over a thousand chemicals. Together, these chemicals form a thick black tar which lodges in the delicate air passages and alveoli of the lungs.

Tar contains nicotine, a powerful addictive drug; poisons and irritants such as hydrogen cyanide, carbon monoxide, phenol, acetaldehyde, butane, ammonia, and formaldehyde; together with carcinogens (chemicals which cause cancer) including the deadly nitrosamines.

How tobacco smoke harms smokers

Lung cancer; cancer of the mouth, throat, larynx, gullet, bladder and pancreas; emphysema (thinning and weakening of lung tissue); coronary thrombosis (blockage of arteries to the heart), angina pectoris (pain due to narrowing of arteries to the heart), and chronic bronchitis with phlegm, are all illnesses linked with smoking.

Each year in the United Kingdom over 200 000 people die from these diseases and 90% of these people are smokers.

If you smoke five cigarettes a day you are about six times more likely to die of lung cancer than a non-smoker, and if you smoke twenty a day the risk is nineteen times greater. Smokers are between two and three times more likely to die of heart disease than non-smokers.

In addition, smoking appears to delay the healing of stomach ulcers; it reduces the senses of smell and taste; it slows down reflexes (making smokers more prone to accidents); and gives the breath, clothes, and homes of smokers an unpleasant smell.

People who give up smoking greatly reduce their chances of developing the diseases listed above.

Illness related to smoking is very costly. In Britain it results in the loss of fifty million working days a year and treatment costs several hundred thousand pounds a day.

Cancer   Cancer is a disease in which cells start to divide rapidly until growth proceeds out of control (Fig. 21.9). Cancerous growth can be triggered off by chemical carcinogens in tobacco smoke.

By the time lung cancer is visible on an X-ray it is quite advanced. It can produce a growth so big that it blocks a bronchus, making breathing difficult.

One cancer can start cells dividing in another part of the body. These growths are called **secondary cancers**, and by the time they appear the disease is usually incurable.

Heart disease Smoking makes heart disease more likely in at least three ways.

1. Chemicals in tobacco smoke affect blood platelets (section 6.1) making them 'sticky'. They clump together and, along with red blood cells, form a blockage, or **blood clot**, which can slow or stop blood flow. If this happens in the coronary arteries, heart muscle is starved of food and oxygen and stops working. This is called a **heart attack**.

2. Chemicals in tobacco smoke weaken blood vessel walls so they are more likely to burst under pressure. Nicotine accelerates the heart beat but constricts blood vessels. This suddenly increases blood pressure, putting extra strain on weakened blood vessels, with possible disastrous results.

3. Carbon monoxide in tobacco smoke combines with haemoglobin in red blood cells reducing their ability to carry oxygen. This can be dangerous, especially in people already weakened by heart disease.

Emphysema Chemicals in tobacco smoke damage the walls of alveoli in the lungs so that they become thin and weak. Alveoli may expand in size, or break down completely leaving large empty spaces in the lungs which inflate like balloons. This condition is called emphysema (Fig. 21.10).

Emphysema reduces a lung's surface area for oxygen absorption. It also reduces the elasticity of lung tissue so that breathing is painful and difficult.

Most heavy smokers develop emphysema. It cannot be cured because lung tissue is damaged beyond repair. People with advanced emphysema take so long to exhale that they have trouble attempting a simple act like blowing out a candle flame.

Normal lung lining

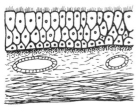

Lung lining with cancer growth

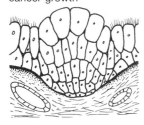

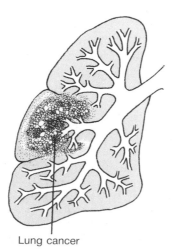

Lung cancer

Fig. 21.9 Growth of lung cancer. A cancer is an uncontrolled growth of cells

A **Normal alveoli**

One air sac

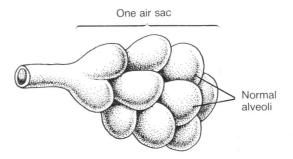

Normal alveoli

B **Alveoli after emphysema has developed**

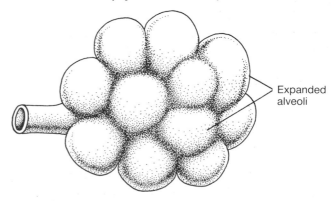

Expanded alveoli

Fig. 21.10 In emphysema alveoli walls are weakened so they expand like balloons

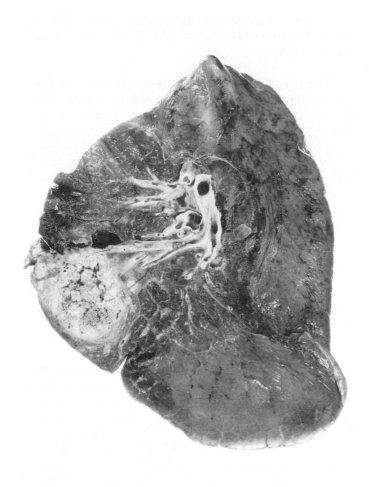

A lung with a cancer (the white part)

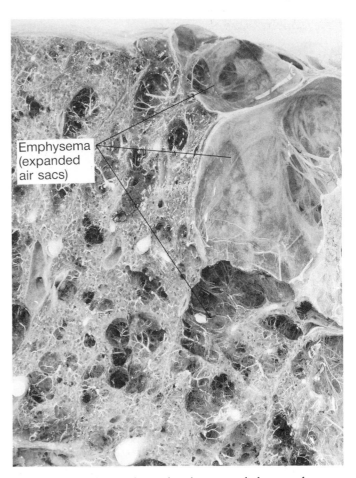

Cross-section of lung-tissue showing expanded areas where emphysema has developed

Bronchitis Bronchitis is inflammation of air passages in the lungs. These passages are lined with cilia (Fig. 9.2) which keep the lungs clean by maintaining a flow of mucus from the lungs to the throat which traps inhaled dust and dirt. Chemicals in tobacco smoke kill cilia, so that this vital cleaning action stops. Mucus and trapped dirt drop down into the lower air passages and alveoli, clogging them and reducing the lung's ability to absorb oxygen.

Continued smoking makes this condition worse. Smoke irritates lung passageways making the walls red and swollen. More mucus is produced to soothe this irritation which only succeeds in clogging the lungs even more.

The result is acute smokers' cough and chronic bronchitis. The chest gets very sore, breathing is difficult, and bronchitis can eventually kill.

How tobacco smoke harms non-smokers

Apart from its unpleasant smell, and the fact that it irritates the eyes, tobacco smoke is dangerous to non-smokers.

The smoke which a smoker inhales has passed through the filter of the cigarette which removes some harmful material. But most of the smoke breathed in by non-smokers comes directly from the burning cigarette end. It has not been filtered when it is breathed in by non-smokers. This is called **passive smoking**, and can theoretically cause all of the smoking-related diseases mentioned so far.

Children and passive smoking

Children are at risk from parents and relatives who smoke. If both parents are heavy smokers their children can inhale as much nicotine and harmful chemicals as if they had smoked 80 cigarettes a year.

Women who smoke during pregnancy tend to have smaller babies than non-smoking women, and their babies are more likely to be born dead, or die shortly after birth.

Smokers who do not smoke during pregnancy can greatly reduce the risk to their baby.

21.10 **Alcohol and ill-health**

Alcohol consumption has risen so much in recent years that doctors believe many people are damaging their health.

How heavy drinking damages health
Heavy drinking can cause a range of diseases. Its effects vary from person to person, but there is clear evidence that alcohol can cause all of the following problems.

Liver damage Regular heavy drinking kills liver cells. Dead cells are first replaced with fatty tissue and later with fibrous tissue. The second of these conditions is called **cirrhosis of the liver**.

Heart disease Heavy drinking over a number of years weakens heart muscle, and this may result in heart failure. It also raises blood pressure, which increases the danger of strokes, especially if the drinker smokes as well (section 21.9).

Brain and nerve damage The brains of heavy drinkers shrink. This has been proved by comparing the brains of identical twins, one of whom is a heavy drinker. Brain shrinkage reduces reasoning powers, and performance in intelligence tests. Alcohol also damages the rest of the nervous system causing loss of sensation, co-ordination, and muscle power.

Fig. 21.11 Standard units of alcoholic drink

Cancer Heavy consumption of alcohol increases the risk of developing cancers of the mouth, throat, and liver.

Obesity and vitamin deficiency Heavy drinkers can easily become overweight. They may also suffer vitamin deficiency if alcohol is replacing food as a source of energy.

Stomach ulcers Alcohol irritates the stomach lining causing it to produce extra gastric juice. This can lead to the development of stomach ulcers.

Pregnant women If a pregnant woman drinks alcohol the effects on her baby are greater than on her own body, because the baby is so small. Heavy alcohol consumption can be very damaging, especially in early pregnancy when most of the baby's development takes place. Its birth weight may be reduced, miscarriage is more likely, and the baby may be deformed or mentally retarded.

How much drink is too much? Different drinks contain different amounts of alcohol. So **standard units** of drink have been established to help people control consumption of alcohol (Fig. 21.11). An important fact to remember is that a woman's health is more easily damaged by alcohol than a man's.
The Health Education Council says that men who drink less than 20 units a week, and women who drink less than 13 units a week, are unlikely to damage their health.

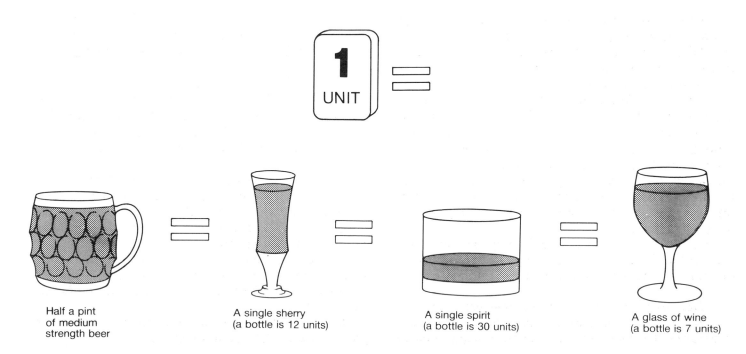

Half a pint
of medium
strength beer

A single sherry
(a bottle is 12 units)

A single spirit
(a bottle is 30 units)

A glass of wine
(a bottle is 7 units)

21.11 **Immunity and immunization**

The body possesses many defences against parasites, and together these give it what is called **natural immunity**. In humans, these natural defences include the skin, the ciliated membranes lining the respiratory system, stomach acid and enzymes of the digestive system, the type of white blood cells called phagocytes, and chemicals called **antibodies** in the blood and tissue fluid. In addition, medical science has developed ways of helping the body to develop additional defences against parasites. Together these defences are referred to as **artificial immunity**.

Natural immunity

The defences described below are summarized diagrammatically in Figure 21.12.

The skin The outer surface of the body is covered with a layer of dead cells known as the **cornified layer** of the skin (Fig. 10.9). As fast as they wear away or are damaged, these dead cells are replaced by a region of live growing cells beneath the surface, called the Malpighian layer. In addition, the dead cornified layer is kept supple, water-repellent, and mildly antiseptic by an oily substance called **sebum**, produced by sebaceous glands in the hair follicles. The human skin therefore forms a waterproof, germ-proof, self-repairing barrier preventing the entry of germs and dirt into the body.

The skin which covers the eyes, called the **conjunctiva** (Fig. 14.7), is extremely thin and delicate, but it is protected by a whole battery of natural defences, including antiseptic tear drops, and the blink reflex of the eyelids.

The respiratory system Dust and germs breathed in through the nose do not usually reach the lungs; they are trapped in sticky mucus which covers the membranes lining the nasal cavity and trachea. Trapped dirt and germs are then carried by cilia (Fig. 9.3) to the oesophagus, where they are swallowed and eventually passed out of the body in the faeces.

The digestive system Many germs are unavoidably consumed with food and drink. Fortunately they are usually harmless, but in any case the majority are killed by stomach acid and digested by enzymes.

The examples of natural immunity described so far may be thought of as the body's first line of defence. But if these defences are broken, as happens when the skin is cut, grazed, or burned, the body's second line of defence comes into operation, and this is controlled by the blood.

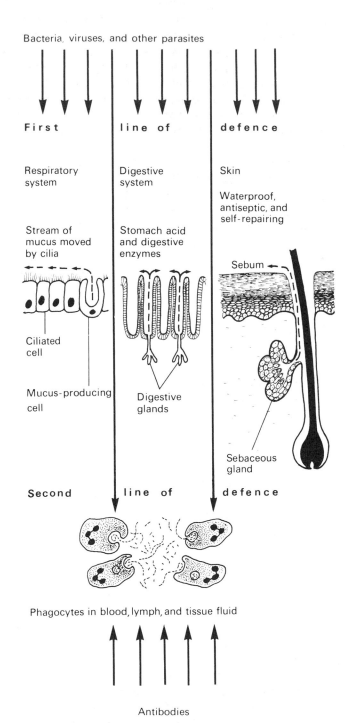

Fig. 21.12 Diagram summarizing natural immunity

Action of phagocytes Phagocytes are white blood cells (Fig. 6.2) which destroy bacteria by engulfing and digesting them in the same way that an amoeba eats its prey (Fig. 4.2). At the same time, some phagocytes are killed by toxins produced by the invading bacteria, and the dead cells resulting from this 'battle' gather beneath the blood clot as a white fluid called pus. But unless the infection is of uncontrollable size, the bacteria are eventually destroyed and the pus absorbed, leaving the wound clean and free to heal.

Sometimes the body is invaded by large numbers of germs which overwhelm its outer defences. Infections of this size are dealt with in three ways. First, bacteria in the blood and tissue fluid are destroyed by free-swimming phagocytes of the type already described. Second, bacteria in the lymph vessels are destroyed by large phagocytes attached to the walls of the lymph nodes (Fig. 6.12). Third, bacteria and their toxins in any part of the body are attacked by chemicals known as antibodies, which are produced by the body during an infection.

Antibodies To understand what antibodies are and what they do, it is first necessary to understand that the substances which make up the bodies of parasites, and the toxins which they produce, are chemically different from the substances which make up the human body. The body is able to detect the presence of parasites and their toxins by 'recognizing' these chemical differences. Whenever this happens specialized cells in the lymph nodes, spleen, liver, and bone marrow react by producing antibodies. Antibodies are chemicals which combine with the 'foreign' chemicals in the bodies of parasites and in their toxins, either destroying these chemicals or rendering them harmless in various ways. Any substance, from whatever source, which stimulates the production of antibodies, is called an **antigen**. Parasites and most of their toxic products are antigens.

Antibodies are known to be specific, which means that each type of antibody can combine with one specific antigen and no other. The antibody against measles virus antigen will destroy this virus alone and have no effect on other disease organisms. The body must be able to produce different antibodies to combat each type of antigen, but it is not yet clear how this is done. Antibodies also vary in their effects on antigens. These effects are summarized below, and in Figure 21.13.

1. **Opsonins** are antibodies which combine with antigen material on the outer surface of bacteria. This seems to make the bacteria more likely to be destroyed by phagocytes. It is as if the antibody makes a bacterium more 'appetizing'.

2. **Lysins** are antibodies which kill bacteria and viruses by dissolving them.

3. **Agglutinins** are antibodies which cause bacteria and viruses to stick together in clumps. In this state the germs can neither penetrate host cells nor reproduce properly.

4. **Anti-toxins** are antibodies which combine with toxins produced by disease organisms, and render them harmless to the body.

Antibodies are produced slowly at first, but if the disease persists for more than three or four days they are produced in much larger quantities. Moreover, the antibodies against certain disease organisms remain in the blood for many years after the infection has disappeared, thereby giving the body a built-in resistance (immunity) to the disease in the event of re-infection. A child who has recovered from chicken-pox or measles for example, is unlikely to suffer from the disease again despite repeated exposure to infection, owing to the presence in his blood of specific antibodies against these germs.

Artificial immunity

The production of antibodies in response to disease organisms is an example of **active immunity**, that is, 'active' in the sense that the antibodies are produced by the body in response to an antigen. It is now possible to make use of the body's capacity for active antibody production in order to produce immunity by artificial means so that the body can be prepared in advance to fight off infections. This is done by introducing **vaccines** to the body.

Vaccines A vaccine can be made either from a suspension of dead disease organisms – a suspension of germs which have been inactivated in various ways so that they no longer cause a disease – or from germs which are very similar to those which cause a serious disease but are actually harmless. When such vaccines are injected into the blood-stream the body responds by producing antibodies as if it were undergoing an attack from the actual disease organism. These antibodies remain in the blood and thereby make the body immune to the disease. The period of immunity so produced varies according to the disease from a few months to several years.

One of the first people to use a vaccine was Edward Jenner (1749–1823), a country doctor in a small village in Gloucestershire. Jenner became interested in a local legend that people who had recovered from a mild disease known as cowpox, because it is caught from cows, never thereafter suffered from the

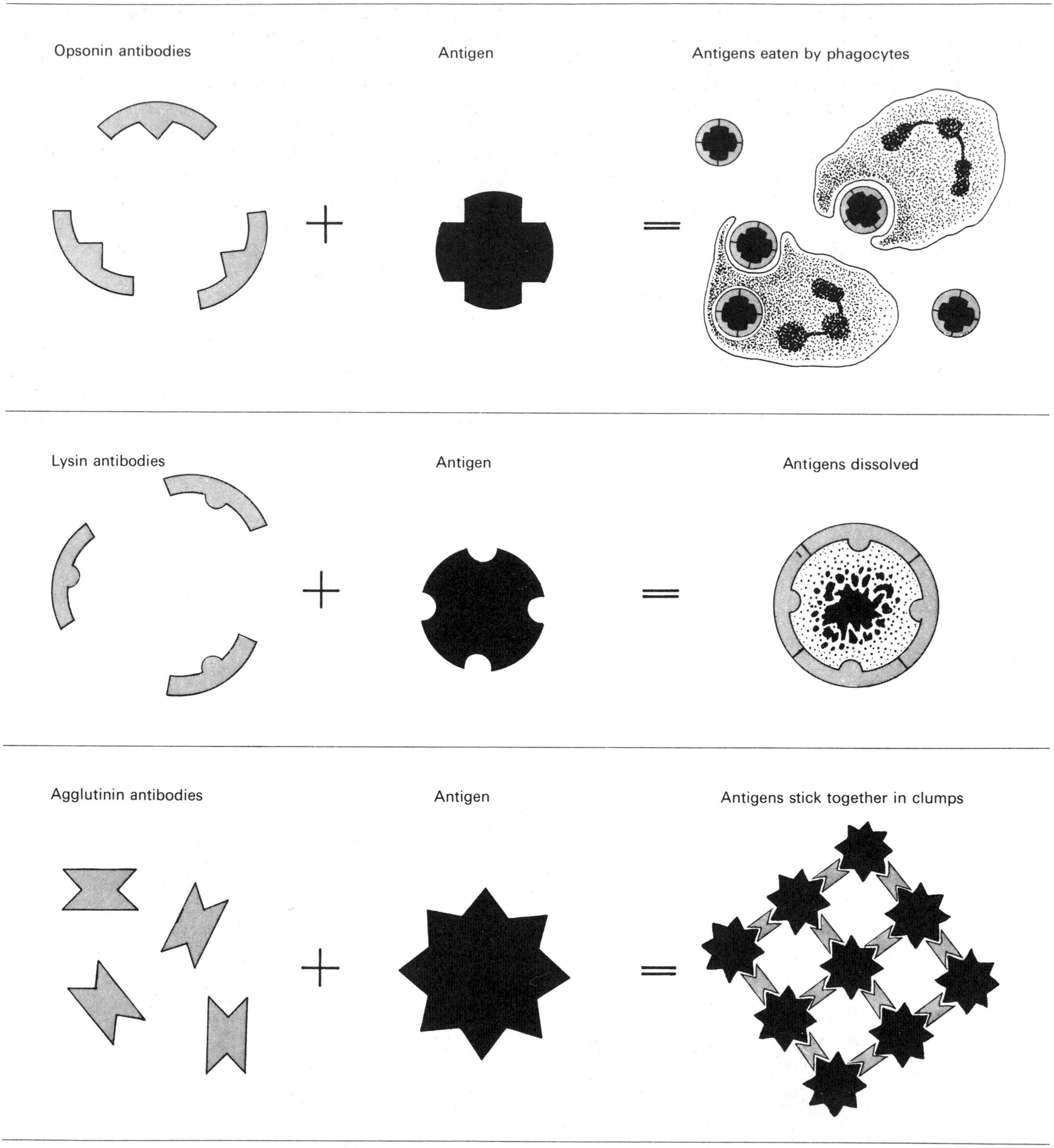

Opsonin antibodies   Antigen   Antigens eaten by phagocytes

Lysin antibodies   Antigen   Antigens dissolved

Agglutinin antibodies   Antigen   Antigens stick together in clumps

Fig. 21.13 Diagram showing how certain antibodies work.

Antibody molecules are not really shaped like those in the
diagram. The shapes indicate that each antibody will react with
only one specific antigen

dreaded disease of smallpox. To test the truth of the legend Jenner performed an extremely dangerous experiment. He scratched the skin of a healthy eight-year old boy and rubbed pus from the hand of a milk-maid suffering from cowpox into the wound. The boy caught cowpox and quickly recovered from it. Jenner then inserted pus from a patient suffering from smallpox into the boy in a similar manner. The boy did not catch smallpox on this occasion, nor on a subsequent occasion when Jenner again attempted to infect him.

Cowpox virus is called vaccinia after the Latin word for cow, and because of its association with Jenner's work it has given rise to the modern terms vaccine and vaccination. There is now vaccines available against many diseases, including typhoid fever, poliomyelitis, cholera, and bubonic plague.

Ready-made antibodies Medical science has devised a way of supplying ready-made antibodies, which are injected into the body to assist in fighting germs before its own defence mechanisms have come into operation. This treatment is an example of **passive immunity**, which is given that name because the body acquires added immunity without doing any work.

Diphtheria antibodies, for example, are produced by injecting horses with toxins produced by diptheria bacilli. The horses then produce antibodies of the antitoxin variety to neutralize the diphtheria poison in their bodies. By taking blood from these horses and extracting serum from it, a supply of anti-toxin is obtained which is extremely effective in treating and helping to prevent the spread of diphtheria in humans.

Horses are also used to produce anti-toxins against tetanus; and an attack of measles may be prevented, or its effects lessened, by injecting serum taken from humans who have recently recovered from this disease.

Gillray's cartoon illustrating the supposed effects of vaccination

Investigations

Safety precautions in microbiology

The following precautions are necessary in order to avoid health hazards when growing bacteria cultures.

1. Treat all bacteria cultures as potentially hazardous.

2. Fasten down the lids or plugs of culture containers with adhesive tape before class examination.

3. Never prepare cultures from faeces, pus, sputum, or material from a lavatory, hand basin, or animal cage.

4. Destroy all cultures before disposal. Glass culture containers should be placed in hypochlorite disinfectant and then opened.

5. Pencils and pens should never be placed in the mouth nor should labels be moistened with the tongue.

6. Protect all cuts with waterproof dressing before handling cultures.

7. Wash the hands thoroughly before leaving the laboratory.

A   Making a culture of live bacteria

1. Boil a few grams of finely chopped meat, carrots, and potatoes in water for 15 minutes, then filter off the solid matter to obtain a fairly clear broth.

2. Leave the broth in open test-tubes for a few hours. Plug the tubes with cotton wool and leave them in a warm place (at approximately 25°C) until the broth has 'gone bad' owing to the growth of bacteria. This is an example of bacteria culture.

B   Stained preparations

1. Smear some saliva on a slide and allow it to dry. Scrape a little food material from between the teeth, mix it with a drop of water on a slide, then smear it and allow it to dry.

2. The dried smears may be stained in either gentian-violet, crystal-violet, or methylene blue stain. Put a drop of stain on the smear and allow it to remain there for 2 to 3 minutes before washing it off under a slow-running tap. Dry the back of the slide and place a cover slip over the stained smear. Observe under low, then under high-power magnification. What shapes are the bacteria?

C   An investigation of the effect of heat, cold, and chemicals on bacteria

1. Heat

a) Plug six test-tubes with non-absorbent cotton wool and sterilize them in an autoclave (pressure cooker) for 15 minutes at 103 kPa.

b) Prepare a quantity of broth as in A above, and pour about 2 cm into each of the sterile tubes, replugging them immediately afterwards. Place all the tubes upright in a beaker half-filled with water and boil them for another 15 minutes.

c) When the tubes are cool put a few drops of live bacteria culture in each from A2 above, replugging them immediately afterwards, and taking great care to wash the hands thoroughly afterwards, and sterilize the dropper used to transfer the bacteria. Leave the tubes in a warm place for a day.

d) Call the next day 'day one' and proceed as follows:

Day one – Boil all six tubes in a beaker of water for 5 minutes and allow to cool. Put them in a warm place.

Day two – Put two tubes on one side in a warm place and label them A and B. Boil the other four tubes for 5 minutes, allow them to cool and put them in a warm place.

Day three – Put two tubes on one side and label them C and D. Boil the other two for 5 minutes, and when they cool label them E and F. Put them with the others in a warm place. A and B are therefore heated once; D and C are heated twice; and E and F are heated three times. Leave the tubes in a warm place for two or three days and examine them, without unplugging, for traces of bacterial growth.

e) Heating to 100°C kills growing and reproducing bacteria but not spores. How does this fact help explain the results of this experiment? Explain how this procedure could be used to sterilize food which is made inedible by pressure cooking (i.e. by temperatures in excess of 100°C).

2. Chemicals and cold

a) Devise experiments making use of bacteria cultures prepared according to C1 above to verify: that strong salt and sugar solutions kill bacteria; and that refrigeration slows down the rate of bacterial growth.

b) Devise other experiments to test the effectiveness of various antiseptics and disinfectants in killing bacteria, when used at different strengths.

c) How do these results show how food may be preserved from decay owing to the action of bacteria, and diseases may be stopped from spreading?

d) Never touch live bacteria cultures with the hands. *Always* wash the hands after handling equipment containing live bacteria. *Always* dispose of bacteria cultures by boiling them and washing the equipment in strong disinfectant.

D *To demonstrate the presence of bacteria throughout the environment*

In the following experiment use either commercially prepared sterile nutrient agar and disposable petri dishes, or proceed as follows.

1. Prepare a quantity of broth as in A above, and while it is hot stir in 1.5 g of agar powder to every 100 cm³ of liquid. Prepare about 12 sterile test-tubes as in C1(a) above, add 2 cm³ of hot agar to each, then replug and boil them in a beaker of water for 15 minutes. Take the tubes from the water and prop them against something so that the agar solidifies at an angle – making what is known as an agar 'slope'. Finally place the tubes upright in a test-tube rack. (*Note*: if absolute sterility is required then the test-tubes must be autoclaved for 15 minutes after adding the agar.)

2. Leave one tube untouched as a control. Each of the remaining tubes should be contaminated in one of the following ways, then replugged and labelled accordingly:

A   with soil
B   with dust from various parts of the room (using more than one tube if necessary)
C   with aquarium or pond water
D   with tap water
E   leave open to the air for 1 hour indoors
F   leave open out of doors
G, H, J, etc.   think of other ways to contaminate these tubes

3. Watch for signs of bacterial growth.

E *Investigating cigarette smoke*

1. Prepare the apparatus in Figure 21.14.

2. Suck the smoke of at least five cigrettes through the apparatus, then look at and smell the wool and water.

Questions

1. An experiment was prepared as illustrated in Figure 21.15. Tube A was left untouched. Tubes B, C, and D were prepared as illustrated, then boiled for 15 minutes in a beaker of water. A went bad in one day, B in two days, C in one week, and D remained fresh indefinitely, or until the bung and 'S' tube were removed. Disprove the following statements:

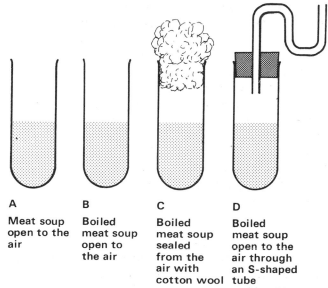

A
Meat soup open to the air

B
Boiled meat soup open to the air

C
Boiled meat soup sealed from the air with cotton wool

D
Boiled meat soup open to the air through an S-shaped tube

Fig. 21.15   Diagram for question 1

a) Meat soup turns into bacteria.

b) The process of boiling meat soup alters it so that it cannot turn into bacteria.

c) Boiling soup alters the air in the tube so that the soup cannot turn into bacteria.

Why does the soup in C eventually go bad, whereas that in D remains fresh? How do these results help to show that bacteria are living creatures which can exist floating in the air around us?

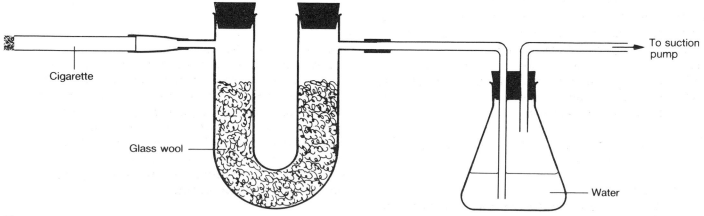

Fig. 21.14   Apparatus for investigation E

2. Make a list of all the features of a tapeworm which enable it to live as it does. In what way do these features make tapeworms completely unfitted to an independent (non-parasitic) existence? What are the advantages and disadvantages of a parasitic way of life?

3. Study Table 1.

a) If a man aged 35 smoked 15 cigarettes a day how many years of life could he lose beause of his habit.

b) If he cut down to one a day how much longer could he live?

c) What could smoking do to his body which is likely to shorten his life?

Table 1 How cigarettes reduce life expectancy

HOW CIGARETTES REDUCE YOUR LIFE EXPECTANCY (from a study of American male cigarette smokers)									
PRESENT AGE	25	30	35	40	45	50	55	60	65
LIFE EXPECTATION (YEARS EXPECTED)									
NON SMOKER	48·6	43·9	39·2	34·5	30·0	25·6	21·4	17·6	14·1
1–9 CIGARETTES PER DAY	44·0	39·3	34·7	30·2	25·9	21·8	17·9	14·5	11·3
10–19 CIGARETTES PER DAY	43·1	38·4	33·8	29·3	25·0	21·0	17·4	14·1	11·2
20–39 CIGARETTES PER DAY	42·4	37·8	33·2	28·7	24·4	20·5	17·0	13·7	11·0
YEARS LOST FOR 20–39 PER DAY SMOKER	6·2	6·1	6·0	5·8	5·6	5·1	4·4	3·9	3·1

Summary and factual recall test

Parasites are organisms which obtain (1) from the (2) body of another organism called the (3). Parasites which cause disease are said to be (4). Germs are (5) which cause disease. Examples are (6–name four).

Viruses cause diseases such as (7–name four). They show signs of life only when (8). Viruses attack hosts by injecting (9) into their cells, which turns them into (10).

Bacteria cause diseases such as (11–name four). Under favourable conditions, bacteria (12) every (13) minutes. Under unfavourable conditions some bacteria form (14) which can survive e.g. (15–name two things).

Protozoa cause diseases such as (16–name two). (17) is caused by *Plasmodium*. It is carried from one person to another by (18), which can be controlled by (19–describe two ways).

Fungi cause diseases such as (20–name two). These are spread by (21–describe two ways).

Tapeworms live in the (22) of vertebrates. They have two hosts, the second of which is (23) by the first.

Diseases spread by droplet infection are (24–name two). Droplet infection occurs in the following way (25). Food and (26) can be contaminated with germs by (27–name two ways). Contagious diseases are those spread by (28). Disease spread in this way are (29–name two). Animals which spread infection are called (30). Examples are (31).

Cigarette smoke contains carcinogens, chemicals which cause (32). Other diseases linked with smoking are (33–name three).

Heavy drinking can cause disease such as (34–name at least three). Men who drink (35) units a week and women who drink (36) units are not likely to harm their health.

Sebum, produced by (37) glands in the (38), make the skin (39–describe three effects). Germs entering the respiratory system are trapped in (40), and moved by (41) to the (42) where they are (43). Most germs in food are killed by (44), and (45).

Germs inside the body are destroyed by a type of (46) blood cell called (47). In addition, antibodies made by cells in the (48–name four places) destroy germs and toxins in four different ways: (49–name the antibodies and give their functions).

Vaccines can be made from (50–name three things). The body reacts to a vaccine by producing (51), which make it (52) to attack from certain germs. Vaccines are available against (53–name four diseases).

Antibody production in response to germs is an example of (54) immunity, which is given this name because (55). Treatment with ready-made antibodies is an example of (56) immunity, so called because (57).

22

Soil

Soil is not just a dead substance. It is an environment which provides both food and shelter for a wide variety of organisms such as bacteria, fungi, worms, and insects, which carry out their whole life cycle within it. Man and many other animals also depend on soil, because it provides support, water, and minerals for the plants which are their food.

This chapter describes the composition of soil, its origin, cultivation, and a number of methods for studying its physical and chemical properties.

22.1 Composition and origin of soil

Shake a few grams of garden soil in a test-tube of water for about thirty seconds, allow it to settle for a few minutes and compare the result with Figure 22.1. This is called a **sedimentation test**, and is a method of separating the major components of soil, which are

Fig. 22.1   Results of a soil sedimentation test

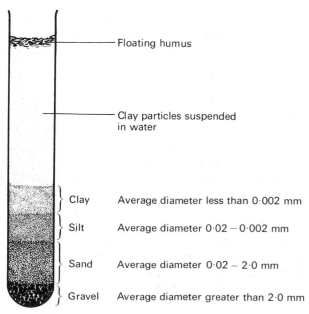

Floating humus

Clay particles suspended in water

Clay — Average diameter less than 0·002 mm

Silt — Average diameter 0·02 — 0·002 mm

Sand — Average diameter 0·02 — 2·0 mm

Gravel — Average diameter greater than 2·0 mm

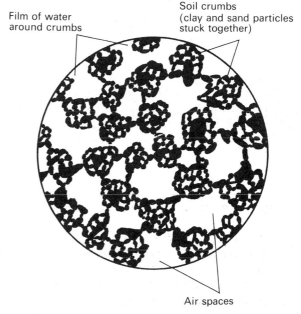

Film of water around crumbs

Soil crumbs (clay and sand particles stuck together)

Air spaces

Fig. 22.2   Magnified (diagrammatic) view of soil crumbs

humus and **mineral particles**. Humus is the black material which floats to the top of the tube, and mineral particles are those which settle to the bottom. They settle in order of weight, the heaviest particles (which are usually the largest) settle first.

Humus

When animals and plants die their remains decompose owing to the saprotrophic activities of bacteria and fungi. One of the end-products of decay is the black fibrous humus material which floats to the surface in a sedimentation test. This fibrous humus comes mainly from cellulose and lignin fibres of plants and the hard parts of animals. The soft tissues of dead organisms decay into a liquid which passes into the soil forming a sticky coating around mineral particles. This glues the particles together into clumps called **soil crumbs** (Fig. 22.2). The soft decaying tissues also form nitrates, phosphates, and ammonium salts. These dissolve in water and are essential for the healthy growth of plants.

The presence of fibrous humus and well developed soil crumbs are characteristics of fertile soil. First, a soil with these features does not usually become water-logged during rain because the air spaces between the crumbs provide adequate drainage. Second, these air spaces allow oxygen to reach plant roots, bacteria, fungi, and other soil organisms, permitting rapid respiration and growth. Third, humus retains moisture which would otherwise drain away after rainfall. Fourth, humus absorbs dissolved minerals from soil water thereby preventing them from being washed out of the soil. Fifth, humus binds the mineral particles together so that they are not easily blown away by high winds.

Mineral particles

Mineral particles originate from rocks which have been broken up by various physical and chemical processes known as **weathering**. These processes are summarized in Figure 22.3.

Physical weathering   There are several ways in which physical weathering can take place. Water in cracks and crevices may shatter the rock as it freezes and expands. Cracks may be widened further by the pressure of plant roots which grow within them. Over many years these processes, together with the removal of soluble substances from rocks by rainwater, can reduce them to boulders and then smaller fragments. When this happens on a slope, loosened rocks roll or are washed downwards, undergoing further wear and tear as they knock and rub against each other. This kind of weathering is called mechanical breakdown. When rock fragments are washed into rivers they can be carried long distances, and during this time mechanical breakdown continues as the rocks are rolled over and pounded together. Eventually, this reduces them to microscopic particles of sand, silt, and clay.

Chemical weathering   Carbon dioxide in the air dissolves in falling rain to make a weak solution of carbonic acid. When this acid falls on rocks, especially

Fig. 22.3   Diagram summarizing the ways in which soil is formed

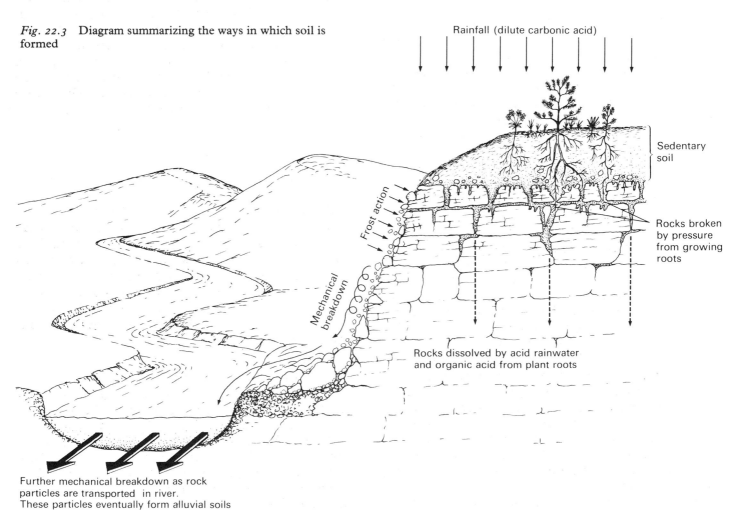

Rainfall (dilute carbonic acid)

Sedentary soil

Frost action

Rocks broken by pressure from growing roots

Mechanical breakdown

Rocks dissolved by acid rainwater and organic acid from plant roots

Further mechanical breakdown as rock particles are transported in river. These particles eventually form alluvial soils

limestone, it reacts chemically forming soluble substances which wash away, leaving insoluble fragments behind which develop into soil. As rainwater percolates through the soil it picks up more carbon dioxide from respiring animals and plant roots. The carbonic acid is therefore stronger by the time water reaches underlying rocks. In addition, plant roots produce weak organic acids which also contribute to the chemical decomposition of rocks below the surface, resulting in a gradual deepening of the soil layer.

Besides humus and minerals, air, water, and living organisms are essential components of fertile soils.

Air

Most soil organisms, including plant roots, are aerobic (i.e. use oxygen in respiration). Consequently no soil is fertile unless oxygen can circulate through it. The amount of air, and therefore oxygen, in soil depends mainly on the size of its particles: the larger the particles the larger the air spaces between them.

Water

Soil water is essential for plant growth both on its own, and because it carries in solution the many different minerals required by plants.

Soil water comes mostly from rainfall. Some of this water evaporates back into the atmosphere, and some drains away. A large amount of water is held in the soil by forces called **adsorption** and **capillarity**. The force of adsorption causes water to gather in a thin film around the surface of the soil particles (Fig. 22.2). Consequently, the volume of water retained by a soil depends on the total surface area of the particles which it contains. The smaller the particles in a soil, the greater its total surface area (question 2). Therefore, in equal volumes of sand and clay, the clay will retain most water by adsorption. This adsorbed water attracts more water by a force called capillarity. The strength of capillarity depends on the distance apart of the soil particles. The closer they are together the stronger the force of capillarity (this is demonstrated in exercises B3 and B4). Owing to their small size, particles of clay are closer together than the larger particles of sand. Consequently, clay soils have greater capillary force than sandy soils. In fact, the capillary force of clay soils is great enough to overcome the force of gravity and so clay actually draws water upwards by capillarity from lower regions as fast as it evaporates from the surface. This is an important feature of clay soils; it means their surface layers are kept moist for a longer period than sandy soils during drought.

Soil which contains all the water it can possibly hold is said to be at **field capacity**. The water which it cannot retain drains away and may eventually form a completely waterlogged layer deeper down, the upper level of which is called the **water table**. Depending on the amount of rainfall and the rate at which water can seep away through underlying rocks, the depth at which the water table occurs may vary from a few centimetres to several hundred metres.

Soil life

Countless small and microscopic organisms inhabit soil. Among the most important as far as soil fertility is concerned are saprotrophic bacteria and fungi which decompose dead organisms, releasing nitrates and other minerals necessary for plant growth (nitrogen cycle, chapter 23, Fig. 23.1).

Earthworms also increase soil fertility. They aerate the soil by burrowing through it, and some worms turn the soil over by taking it into their mouths, passing it through their digestive systems, and finally passing it to the surface as a worm cast. Most worms add to soil humus by dragging leaves down into their burrows as food, and leaving them there partly eaten.

Other burrowing animals which turn the soil over include rabbits, moles, badgers, certain millipedes, and insects. All animals increase soil fertility with their droppings, and ultimately with their dead bodies. Soil also contains many pests which destroy plant life, such as slugs, wire-worms, and parasitic fungi.

22.2 Types of soil

Soil can be classified in many different ways. For example, it can be classified as **alluvial** or **sedentary**, depending on how it was formed; as **topsoil** or **subsoil**, depending on the depth at which it is situated; and as **loam**, **sand**, or **clay**, depending on its composition. These are different ways of classifying what may be the *same* soil; it is quite possible, therefore, for a soil to be described as alluvial, and a topsoil, and at the same time a loam.

Sedentary and alluvial soils

A sedentary soil is one which is situated on top of the rock from which it is developed (by weathering), like that on top of the rock formation in Figure 22.3. An alluvial soil is one composed of mineral particles which have been carried long distances from their place of origin by rivers or glaciers, and then deposited on flat areas such as valley bottoms and coastal plains.

Topsoil and subsoil

Both sedentary and alluvial particles slowly develop into fertile soil as they are colonized by plants, and animals, and as they gather humus from the activities of bacteria and fungi. But these processes take place only in the upper regions of the soil, where they lead to the formation of a layer called topsoil. Good topsoil is dark in colour owing to the presence of humus, and is filled with organisms. Below the topsoil is subsoil, which is lighter in colour owing to the absence of humus, and contains very little life except the deeper roots of plants. Because of its humus content and the presence of organisms, topsoil is more fertile than subsoil. The best topsoil of all is called loam (Fig. 22.5).

Sedentary and alluvial soils can be classified into different types according to the substances of which they are composed, e.g. loam, sand, and clay, and the proportions in which these substances exist. These substances in different proportions give a soil its particular physical and chemical properties.

Loam soils

Loams are the most fertile of all soils, because they possess all the characteristics which promote vigorous growth. The best loams consist of about 50% sand, 30% clay, and 20% humus, with a little lime, all of which are mixed together and formed into well-developed soil crumbs. Soils of this type are well-aerated, drain freely, and yet retain plenty of moisture and dissolved minerals. Loams warm up quickly in the spring, and are easily cultivated by digging or ploughing.

Sandy soils

Sandy soils contain a high proportion of large mineral particles combined with little humus and hardly any clay or silt. Soils of this type have several advantages. There are large spaces between sand particles which give the soil good drainage and aeration, and this enables plant roots to penetrate quickly during germination. Sandy soils are easily cultivated all the year round because their particles do not stick

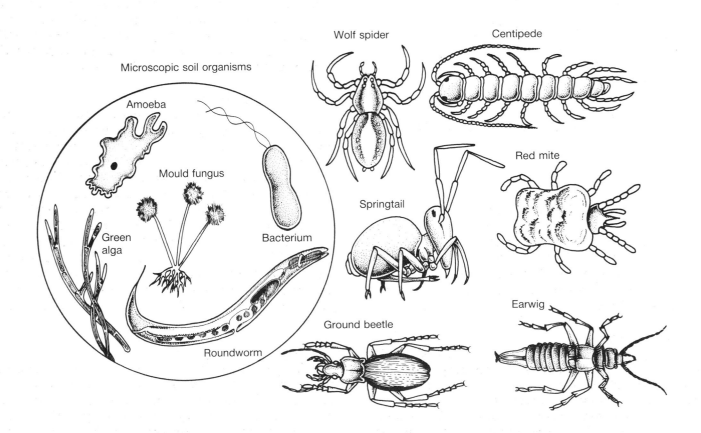

Fig. 22.4 A few of the many creatures found in soil and among dead leaves

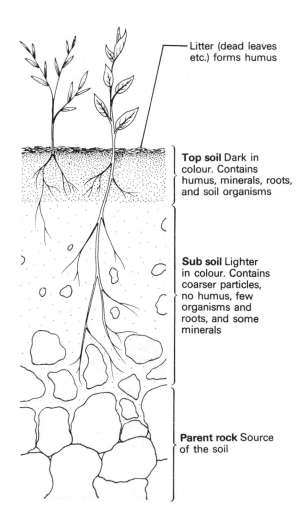

Fig. 22.5 Types of soil found by digging down from the surface to parent rock. This view of soil is called a **soil profile**

The labels in the figure read:

Litter (dead leaves etc.) forms humus

Top soil Dark in colour. Contains humus, minerals, roots, and soil organisms

Sub soil Lighter in colour. Contains coarser particles, no humus, few organisms and roots, and some minerals

Parent rock Source of the soil

together in large clumps, which is why they are generally called 'light' soils. But sandy soil does not retain water for long after rainfall, and this causes plants to wilt and die very quickly during drought. The rapid movement of water through sandy soil washes away its dissolved minerals which quickly reduces its fertility.

Clay soils

Clay soils contain a high proportion of minute mineral particles. Unlike sand, clay retains moisture during drought owing to its capillarity, and it loses minerals at a slow rate owing to its poor drainage. The disadvantages of clay are related to the smallness of clay particles. These are crowded together in compact masses which are sticky and heavy when wet, making the soil extremely difficult to cultivate. Wet

clay is almost completely devoid of air, and in prolonged drought it dries and splits into rock-hard lumps.

22.4 Cultivation of soil

The soils of wild uncultivated areas are more or less balanced and unchanging. As fast as minerals are extracted from them by plants they are replaced by other minerals from animal droppings and the decaying remains of dead organisms. But this is not the case in garden and agricultural soils. Here, the natural sequence of life, death, and decay is replaced by intensive cultivation in which plants are removed from the soil for human consumption. Under these conditions, unless the minerals extracted by these plants are replaced artificially, the soil rapidly deteriorates and soon becomes incapable of supporting plant life.

However, it is possible to go further than simply maintaining soil fertility by replacing lost minerals; soils can be improved artificially so that they are capable of supporting a far larger crop of plants than in their natural state. In general all types of soil improvement involve attempting, as far as possible, to turn poor soils into best quality loam.

Improving clay soil

Heavy clay soils can be made lighter by deep ploughing or digging in the autumn. This produces large clods which are broken down by winter frosts. Drainage of clay can be improved by adding the substances which it lacks in order to turn it into loam, such as sand, and humus in the form of peat or well rotted manure. An especially effective method of improving clay is to add lime. This reacts chemically with clay particles making them join together in clumps like soil crumbs. This is called **flocculation**, and is investigated in exercise C2. Clay also tends to be rather acid, a condition which reduces the growth of certain crop plants. The acid is neutralized by the addition of lime.

Improving sandy soil

Sandy soil can be improved by adding humus in the form of peat and manure. When this is ploughed into the soil it improves water-holding capacity and mineral content. In dry weather, evaporation from the surface of sandy soil can be reduced by laying wet rotting manure or peat on the soil surface. This process is called **mulching**.

Fertilizers

Of the many mineral elements present in soils, nitrogen, phosphorus, potassium, magnesium, calcium, and sulphur are required by plants in the greatest quantities. Consequently, these elements must be added to garden and agricultural soils at regular intervals if fertility is to be maintained. A **fertilizer** is a substance which supplies one or more of these essential elements.

An **organic fertilizer** is one which is derived from animals and plants, whereas an **inorganic fertilizer** is a factory-produced chemical.

Organic fertilizers These include farm yard manure, well-rotted compost, dried blood, and bone meal. They are slow-acting fertilizers because they must first decompose in the soil before releasing their mineral elements. They add humus as well as minerals to soil and are 'natural' in the sense that they replace the natural decomposition by which uncultivated soils maintain fertility.

Inorganic fertilizers These are chemicals which dissolve immediately in soil water and so are immediately available to plants. They are useful for quick results and can be formulated to contain balanced proportions of minerals to exactly match the needs of crop plants.

However, long term use of chemical fertilizers can be harmful to soils. They can lead to a break down of soil crumbs and an accumulation of organic acids in soils. If used in excess these processes are accelerated. Moreover, before plants can absorb them, the excess minerals are washed into rivers and lakes where they cause pollution problems (section 25.2), and may enter drinking water.

When a soil's crumb structure is destroyed, it becomes a fine powder which is easily blown away by the wind, and washed away by water. This type of soil erosion is increasing in Britain and Europe, and is especially damaging in the intensively cultivated corn belt of the U.S.A.

22.4 Soil erosion

Soil erosion is the removal of fertile topsoil from a region by heavy rain and wind. Erosion exposes the infertile subsoil, or bare rock, and can change once fertile regions into deserts.

The main causes of soil erosion are the excessive use of inorganic fertilizers described above, over-grazing, and deforestation.

Over-grazing

This is a severe problem in tropical countries. If herds of cattle are grazed for long periods on one area of land they eat most of the vegetation, thus exposing the soil. Their feet compress the earth, forming a hard surface layer, or **hardpan**, which will not allow water to soak through.

During heavy rainfall these two factors cause water to run off hillsides in a continuous torrent. This quickly loosens soil particles and within a few days a sheet of topsoil can be removed from a region, and deep channels, or **gulleys** can form.

Intensive farming and weathering has caused soil erosion

Deforestation

If forest trees are removed, the thin soil underneath is exposed to tropical sun and rain. Soil temperature immediately rises by up to twenty degrees Celsius and, unless carefully cultivated, soil is washed away by the next rainstorm.

Refugees from Tigray in Ethiopia on a long march, in search of food, through what once was a fertile land

This leaves a hard infertile residue that is of little use to farmers, and which cannot be re-colonized by forest plants. So deforestation quickly leads to desertification.

The disastrous famines which have killed millions of Africans in recent years have shown the world the tragic consequences of chopping down trees. Not long ago the deserts of Sudan and Ethiopia were covered in forests. In the rainy season these held on to water and provided a regular flow of water to the plains below.

The forests have been cut down and used for fuel and wood products by an ever-expanding population. The consequent soil erosion, deteriortion of grass-lands, and drying up of water courses has devastated the land.

Investigations

A *An investigation into the composition of various soils*

Carry out the following tests on soils collected from different environments, e.g. meadows, marshes, woods, moorland. If this is not possible at least ensure that samples of clay, sand, and good quality loam soil are available. Relate the results of the following tests to the type of vegetation found on each type of soil. Finally, summarize the results in a chart so that the various soils can be compared at a glance.

1. *Sedimentation test*

Repeat the sedimentation test on equal volumes of a variety of soils, but use a measuring cylinder instead of a test-tube. After the sediments have settled roughly estimate the proportions of humus and each type of mineral particle.

2. *An estimation of water content*

a) Weigh a soil sample in an evaporating basin, heat it for 30 minutes at approximately 100°C and weigh it when cool. (The soil may be heated in an oven, or over a beaker of boiling water, but it must not be heated to a temperature at which its humus content begins to burn.) Repeat until no further weight loss is observed. The difference between the original weight and the weight after heating rep-resents the amount of water that the soil contained. Calculate the result as a percentage of the original weight.

b) Devise a way of discovering the field capacit-ies of various soils. Try to account for any differences discovered.

3. *An estimation of humus content*

a) Weigh a quantity of dry soil from exercise A2 above in a crucible, and heat it strongly for about 20 minutes to burn away all its humus content.

b) Weigh again when cool and calculate the humus content as a percentage.

c) Find the difference between the humus content of topsoil and subsoil.

4. *An estimation of air content*

a) Find the volume of an empty tin (which has had one end removed cleanly with a can opener) by filling it with water and emptying it into a large measuring cylinder.

b) Punch one or two holes in the bottom of the tin and push it, open end first, into the soil until its bottom is level with the surface. Dig the tin out of the soil without disturbing its contents, turn it the right way up and smooth its surface level.

c) Empty *all* the soil from the tin into a large measuring cylinder containing a known volume of water. After bubbles have stopped rising from the soil note the amount by which the water level has risen up the measuring cylinder. This rise represents the volume of solid matter in the soil.

d) Calculate the volume of air in the soil sample. Calculate the result as a percentage.

B   *An investigation into the physical properties of various soils*

Carry out the following tests on the same variety of soils and add the results to the chart.

1. *To compare the rate at which air percolates through various soils*

It is extremely difficult to measure air percolation through undisturbed soils. Nevertheless, useful comparisons can be made using dried powdered samples of soil.

a) Place a known volume of soil in the funnel of the apparatus illustrated in Figure 22.6. Tap the funnel to make the soil settle.

b) Start a stop-watch and simultaneously open the clip. Water will drain from the tube, pulling air through the soil sample. Stop the watch when the tube is empty.

c) Repeat, using an equal volume of a different soil.

d) What factors affect the rate at which air percolates through soil?

e) In which type of soil will roots receive the most air.

2. *To compare the water-retaining capacities of soils, and the rates at which water percolates through them*

a) Place a known volume of dry powdered soil in the funnel of the apparatus illustrated in Figure 22.7. Pour a known volume of water gently on to the soil, and note the volume of water which eventually drains into the measuring cylinder. The difference between the two volumes represent the amount of water retained by the soil.

b) Start a stop-watch, and simultaneously pour another known volume of water on to the wet soil. Note the length of time taken for this water to drain through the soil.

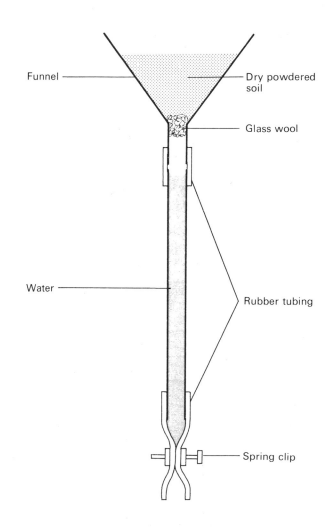

Fig. 22.6   Materials and apparatus for exercise B1

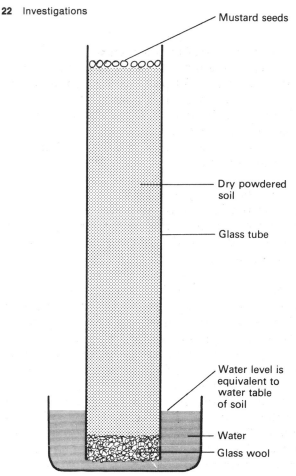

c) Repeat these two tests on different types of soil.

d) What factors affect the water-retaining and drainage characteristics of soil?

e) What type(s) of soil: retain so much water and drain so slowly that they are likely to flood during heavy rainfall; drain so readily that they are likely to have the soluble minerals washed out of them?

3. *To compare the ability of various soils to take up water by capillarity*

Capillarity results in water moving upwards against the pull of gravity from the water table. The apparatus illustrated in Figure 22.8 can be used to measure this.

a) Obtain a number of wide glass tubes about 20 cm long and prepare them as illustrated using a different type of dry powdered soil in each.

b) Note the date on which the bases are immersed in water, and the dates on which the seeds germinate.

c) Account for any differences observed.

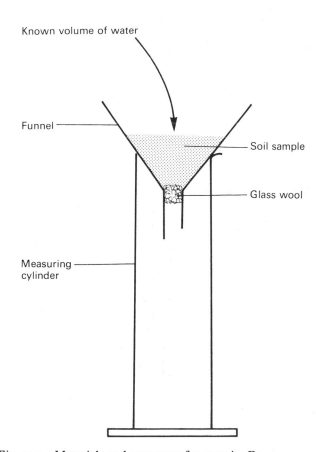

Fig. 22.7   Materials and apparatus for exercise B2

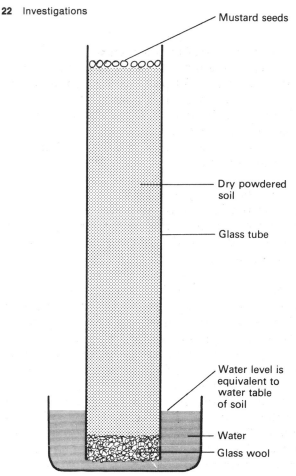

Fig. 22.8   Materials and apparatus for exercise B3

4. *To demonstrate the relationship between the height to which water rises by capillarity and the width of the channel through which it passes*

a) Strap two pieces of glass together with elastic bands, then slip a length of matchstick between them to hold two edges apart (Fig. 22.9).

b) Lower this apparatus into a shallow dish of coloured water. Tap the glass a few times, and observe the way in which water rises between the two sheets.

c) What is the relationship between channel width and the height of capillary rise?

d) How does the result of this test help to explain why water rises to a greater height by capillarity in clay soil than in sandy soil.

C   *An investigation into the chemical properties of soil*

1. *To discover the degree of acidity or alkalinity of a soil sample (i.e. its pH value)*

Obtain a quantity of universal indiator, such as BDH or as prepared by a biological supplier. (An indictor is a substance which changes colour according to the pH of chemicals which are mixed with it.)

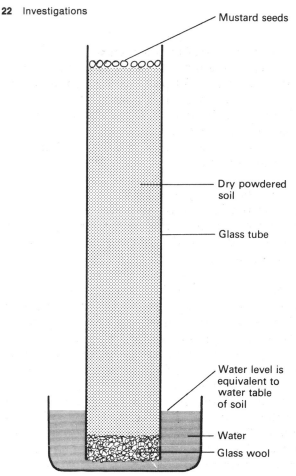

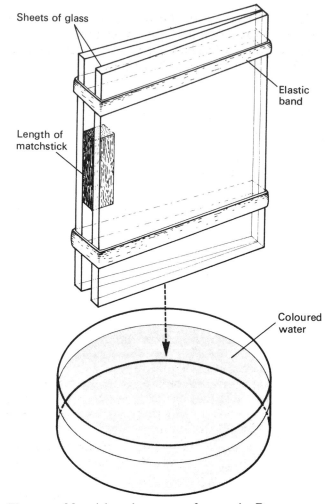

Sheets of glass

Elastic band

Length of matchstick

Coloured water

Fig. 22.9   Materials and apparatus for exercise B4

a) Place a few drops of indicator on a white tile, mix a little soil with it, then tilt the tile to make indicator run out of the soil.

b) Observe the indicator's colour and read off the soil's pH value (use a colour chart). To be of any use this test should be related to the type of vegetation found growing on a soil.

2. *To investigate the effects of lime on clay*

a) Two-thirds fill two large test-tubes with distilled water and add an equal volume of dry powdered clay to each. Shake both tubes thoroughly, add lime water to one tube and an equal volume of distilled water to the other. Shake the tubes again and put them in a test-tube rack to settle. What are the effects of flocculation?

b) Fill two funnels with wet clay soil and place each on a measuring cylinder (Fig. 22.7). Pour a known volume of lime water into one funnel and an equal volume of distilled water into the other. Compare the rates at which they drain through clay, and explain any difference observed.

D   *An investigation of the living things found in soil*

1. *To verify that soil contains living organisms*
a) Hang a cloth bag containing a sample of moist garden soil inside a flask of lime water (as illustrated in Figure 8.5).
b) As a control, prepare another flask containing a bag of baked soil. Note any changes in the lime water, and explain.

2. *To verify the presence of bacteria and fungi in soil*
Prepare tubes of agar as described in exercise D of chapter 21, and sprinkle a little soil from different localities into each. Observe the growth of bacteria and fungi.

Questions

1. A bottle was filled with water, sealed with a screw cap and placed in the freezer compartment of a refrigerator. A few hours later the bottle shattered. Explain how this result illustrates one of the factors responsible for soil formation.

2. A box 10 cm square will hold 1000 glass beads 1 cm in diameter, or 125 glass beads 2 cm in diameter. The surface area of a sphere $= 4\pi r^2$. If $\pi = 3$, what is the total surface area (a) of all the 1 cm beads, and (b) all the 2 cm beads? How does the difference help to explain the different water-holding capacities of sand and clay?

3. *a)* How are the rock particles of soil formed?
b) What is humus and how is it formed?
c) In what ways do plants benefit from humus?

4. How do earthworms increase soil fertility?

5. *a)* What are sedentary and alluvial soils?
b) What are the main differences between topsoil and subsoil, and which is the most fertile?

6. Which of the statements below describe: loam, sandy, and clay soil:
a) 50% sand, 30% clay, 20% humus
b) contains a high proportion of minute mineral particles
c) contains a high proportion of large mineral particles
d) does not detain water for long after rainfall
e) the most fertile of all soils?

7. How can sandy and clay soils be improved

8. *a)* Describe the differences between sheet, rill, and gully erosion.
b) Why does over-grazing often result in soil erosion?
c) Describe three ways in which soil erosion can be prevented.

9. Study the apparatus illustrated below. At the beginning of the experiment the level of coloured liquid was the same in both arms of the 'U' tube.

a) What is the aim of the experiment?

b) What process in the unsterilized soil causes the levels of the coloured liquid to change?

c) If salt solution had been used instead of soda lime what would the result have been?

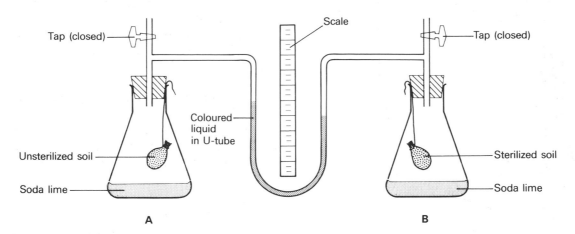

Summary and factual recall test

When organisms die their remains (1) owing to the (2) action of bacteria and (3) in the soil. One end-product of this process is a black fibrous material called (4) which comes mainly from (5–name three sources). The soft parts of decaying organisms form a (6) which sticks soil particles together into clumps called soil (7). Soft decaying tissues also form chemicals such as (8–name three) which are essential for (9) growth. Humus and crumbs are an important part of the soil because (10).

In order of increasing size the types of mineral particle in soil are: (11). These particles originate from rock by a number of (12) and (13) processes called weathering. Cracks in rocks are widened by (14) which freezes and (15); and by pressure from the growth of plant (16). Rocks are also broken down by mechanical weathering such as: (17–name two).

(18) dissolves in falling rain to make weak (19) acid. This reacts with rocks, especially (20), forming (21) substances which (22) away leaving small fragments that develop into (23). More of this acid is formed from (24), together with (25) acid from plant roots. These deepen the soil by (26).

Air is an essential part of soil because (27). The amount of air in soil depends on (28).

Water is an essential part of soil because it carries (29) in solution. A force called (30) retains water around soil particles. This water attracts more water by a force called (31), which gets (32–stronger/weaker) the smaller the distance between soil particles. This explains why (33) soils can draw water upwards from the (34) table against the force of grav-

ity. The field capacity of soil is (35).

Bacteria and fungi increase soil fertility mainly by (36). Worms (37) the soil by burrowing through it. Some worms turn the soil over by (38), and others add humus to the soil by (39).

A sedentary soil is one which is situated (40). An (41) soil is formed from particles carried long distances by (42) and (43), and then deposited in places such as (44–name two). Topsoil differs from subsoil in at least three ways: (45).

Loam soils consist of about 50% (46), 30% (47), and 20% (48), with a little (49). All these are formed into well-developed (50). Loam is the most fertile of all soils for a least four reasons: (51).

Sandy soils consist mostly of (52–large/small) particles with little or no (53–name two other components). The agricultural advantages of sandy soil are (54–describe four). Its disadvantages are (55–describe two).

Clay soils consist mostly of (56–large/small) particles. The advantages of clay are (57–describe two). Its disadvantages are (58–describe three).

Uncultivated soils can be described as 'balanced' because (59). Garden and agricultural soils must have (60) added to them regularly because (61). Heavy clay soils can be improved by (62–describe four ways). Sandy soils can be improved by (63).

Three examples of organic fertilizer are (64). These release minerals to the soil (65–slowly/quickly) because (66). Inorganic fertilizers release their minerals (67–slowly/quickly). Two examples of inorganic fertilizers are (68).

23

The balance of nature

The planet Earth is a ball of rock hurtling through space. The rock has depressions in it filled with water and flat areas covered with soil, and it is entirely contained in an envelope of air. But this is only the physical world. The Earth is inhabited by countless millions of living organisms which make up the biological world.

This chapter is concerned with the continuous exchanges of material and energy which take place between and within these two worlds. These exchanges occur in cycles, which involve a continuous circulation of substances between organisms and their physical environment. Substances such as oxygen, carbon dioxide, water, and minerals are constantly absorbed by organisms, but as fast as these substances are lost from the physical world they are replaced by the natural processes of photosynthesis, respiration, excretion, and decay in the biological world. In other words, there is a **balance of nature** in which losses equal replacements, and in which materials are used and re-used over and over again.

The **nitrogen cycle** is an excellent example of how one element circulates around and through the physical and biological worlds (Fig 23.1).

Fig. 23.1   The nitrogen cycle

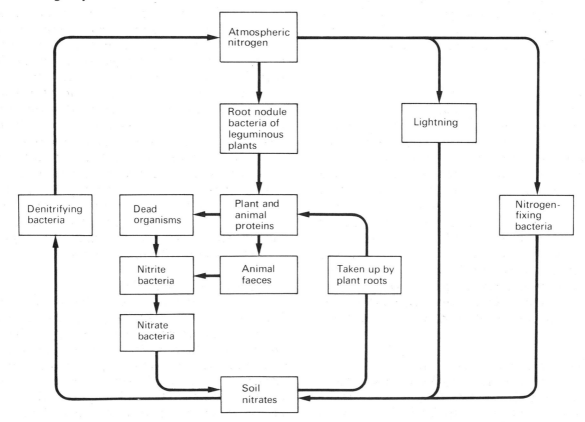

23.1   The nitrogen cycle

Life cannot exist without nitrogen. It is an essential component of all proteins. Nitrogen gas forms approximately four-fifths of the atmosphere but neither plants nor animals can make use of it in this form. Plants can take in nitrogen only in the form of nitrates which they absorb from the soil, and the only way animals can obtain nitrogen is by eating plants or animals which eat plants. Consequently, both plants and animals depend, directly or indirectly, on mechanisms which replenish the soil with nitrates as fast as they are removed from the soil by plant roots. These mechanisms which replace nitrates in the soil may be divided into two types: those which transform atmospheric nitrogen into soil nitrates, and those which transform plant and animal protein into nitrates.

Transformation of atmospheric nitrogen
Nitrogen is transformed into nitrates by lightning and by certain soil bacteria. (Nitrates are compounds of nitrogen and oxygen.)

Lightning   During lightning flashes an extremely high temperature is generated. This results in nitrogen combining with oxygen to form nitrous and nitric oxide gases. These gases dissolve in rainwater and form nitrous and nitric acid. These acids soak into the soil where they react with other chemicals to form nitrates.

Nitrogen-fixing bacteria   Bacteria of this type use carbohydrate and atmospheric nitrogen to make compounds which are eventually released into the soil as nitrates. This process is called **nitrogen-fixation**. Some of these bacteria obtain carbohydrate from soil humus. Others live inside the root cells of leguminous plants (e.g. peas, beans, clover, and vetches) where they cause tiny swellings called **root nodules** (Fig. 23.2). Here the bacteria obtain both protection and carbohydrate from the plant cells, and in return they release nitrates into the plant tissues and the soil. (An association of this kind in which two different organisms benefit from living together is called **symbiosis**.) Leguminous plants are often cultivated in order to increase the nitrate content of agricultural soils.

Transformation of plant and animal proteins
Saprotrophic, or putrefying bacteria and fungi decompose proteins in dead animals and plants. In this process ammonia is released and immediately dissolves in soil water. Here, the ammonia combines

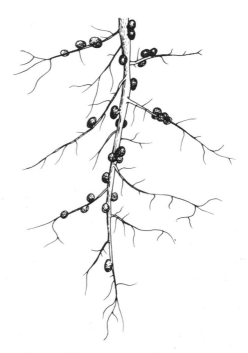

Fig. 23.2   Root nodules on a bean plant

with soil chemicals to form various ammonium compounds which are converted into nitrates by nitrifying bacteria.

Nitrifying bacteria   There are at least two types of nitrifying bacteria in the soil. First, **nitrite bacteria** turn ammonium compounds into nitrites by combining them with oxygen. (Nitrites are chemicals with less oxygen in their molecules than nitrates.) Second, **nitrate bacteria** combine nitrites with more oxygen to form nitrates. By this sequence nitrogen in proteins is changed into a form which can be absorbed by plant roots. The same series of bacteria also form nitrates out of the nitrogen-containing compounds in animal droppings and urine.

Lightning and the nitrifying bacteria replace nitrates removed from soils by plant roots. But there are some bacteria which remove nitrogen from soil nitrates and return it to the atmosphere as gas. These are called **denitrifying bacteria**.

Denitrifying bacteria
Waterlogged and heavy clay soils contain very little oxygen and therefore from a biological point of view they are anaerobic. Dentrifying bacteria are anaerobic, whereas nitrifying bacteria are aerobic. Consequently, in waterlogged conditions and in heavy clay soils denitrifying bacteria are the more active of the two. Denitrifying bacteria obtain their energy by breaking down nitrites and nitrates in the soil into nitrogen and oxygen gases. These gases pass

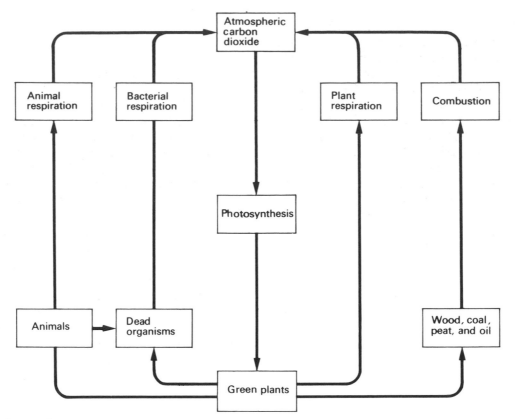

Fig. 23.3 The carbon cycle

into the atmosphere where the nitrogen is no longer available to plants. For this reason it is good agricultural practice to improve drainage of soils where flooding is likely, and to break up clay soils by the methods described in section 22.4.

23.2 The carbon cycle

The carbon cycle is another example of how an element circulates continuously within the physical and biological worlds (Fig 23.3). Carbon dioxide is absorbed by green plants during daylight hours as a raw material of photosynthesis. Despite this process the amount of carbon dioxide in the atmosphere remains constant at about 0.033%. This is because it is replaced in a number of ways as fast as it is absorbed by plants.

Respiration

Some of the carbohydrate produced by photosynthesis is used to build up the plant body, but the remainder is eventually respired for energy. This releases carbon dioxide back into the atmosphere.

Carbon atoms in the carbohydrates, fats, and proteins of plants are transferred to the bodies of her-

bivores when these animals feed on plant tissues. Later these carbon atoms may be transferred again if the herbivores are eaten by carnivores or omnivores. As these animals respire some of the carbon atoms are released to the atmosphere as carbon dioxide.

Decay

After death, the bodies of organisms are decomposed and absorbed as food by saprotrophic bacteria and fungi. Carbon atoms in this absorbed material are released to the atmosphere as the saprotrophs respire.

Combustion

Combustion, or burning, of inflammable materials results in the release of carbon atoms as carbon dioxide. Combustion can form part of the carbon cycle in the following ways.

Carbon absorbed by a tree during photosynthesis and used to build woody tissue in its trunk will be returned to the atmosphere if the tree is chopped down and burned as fuel. Over millions of years, the bodies of dead organisms have produced fossil fuels such as coal, oil, and natural gas. Some of these organisms were plants which took carbon from the air during photosynthesis, and some were animals which fed on the plants, or on animals. Therefore, when these fossil fuels are burned today, they release carbon atoms which were trapped by photosynthesis in plants that lived millions of years ago.

The circulation of carbon in each of these ways can give individual carbon atoms a long and varied history. For instance, carbon atoms in the breath of a gardener may be absorbed by photosynthesis into the leaves of her rose bushes where they are incorporated into sugar. Later, the same carbon may enter the bodies of greenfly as they suck sugar from rose phloem. It could then be transferred to ladybirds which eat greenfly; to insect-eating birds which feed on ladybirds; to hawks which feed on insect-eating birds; and then returned to the atmosphere by hawk respiration.

This sequence of events illustrates that the raw materials of life are never lost or created, but re-used endlessly as they circulate between organisms and their environment. But circulation is not restricted to carbon and nitrogen; there is a continuous circulation of all the chemicals which make up living things, including water.

23.3 The water cycle

The world has an enormous water supply – more than 600 million cubic kilometers. But at any one time 97% of this water lies in the salty seas, and about 2% is trapped in glaciers and polar ice caps. This leaves less that 1% available to us as fresh water in lakes, rivers, streams, and porus underground rocks.

If this meagre supply were not continually renewed rivers and lakes would go dry, plants would wither, and all life on land would come to an end.

Except during severe droughts, this does not happen because water is constantly on the move. It circulates between the oceans, atmosphere, and land in an unending **water cycle** (Fig. 23.4).

Heat from the sun keeps the water cycle operating. It warms the surface of the sea, lakes, and land, causing water to evaporate into water vapour.

Water also evaporates from plants as they transpire, and from all living things as they respire.

Fig. 23.4 The water cycle

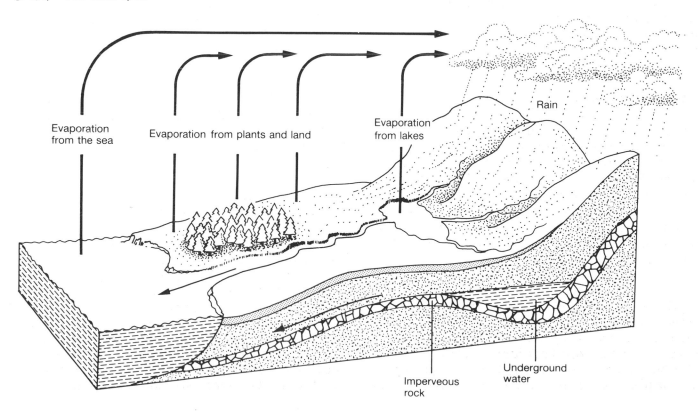

Evaporation from the sea

Evaporation from plants and land

Evaporation from lakes

Rain

Imperveous rock

Underground water

Warm, moist air from all these sources rises into the atmosphere. As it rises it cools and water vapour condenses into billions of water droplets, many of which collect to form clouds.

At first, these droplets are so light they float on air currents, but some collide, join together, and form larger and larger drops. In time they become heavy enough to fall as rain drops, which starts the cycle all over again.

Fig. 23.5   The parts of a food chain

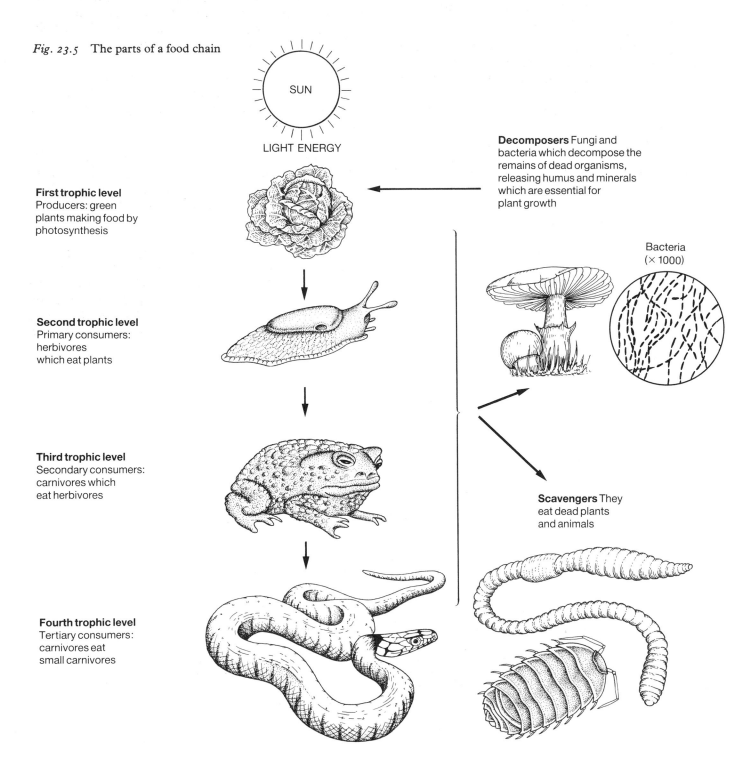

SUN

LIGHT ENERGY

Decomposers Fungi and bacteria which decompose the remains of dead organisms, releasing humus and minerals which are essential for plant growth

First trophic level
Producers: green plants making food by photosynthesis

Bacteria (× 1000)

Second trophic level
Primary consumers: herbivores which eat plants

Third trophic level
Secondary consumers: carnivores which eat herbivores

Scavengers They eat dead plants and animals

Fourth trophic level
Tertiary consumers: carnivores eat small carnivores

23.4 Food chains

Think again about the sequence of events starting with photosynthesis in rose bushes, and ending in hawks. (Section 23.2):

1. rose leaves are eaten by greenfly
2. which are eaten by lady birds
3. which are eaten by insect-eating birds
4. which are eaten by hawks.

This sequence illustrates another important biological principle, that the materials and energy needed for life pass from one organism to another along what are called **food chains**.

Like every other food chain this example begins with **producer organisms**, which are always green plants producing food by photosynthesis.

Subsequent 'links' in a food chain are made up of **consumer organisms**. They consume plants and each other.

Trophic levels of a food chain

The position that an organism occupies on a food chain is called its **trophic level**. This position depends on whether it is a plant or an animal and, in the case of an animal, on what it eats. Figure 23.5 illustrates these points using a simple food chain.

Green plants occupy the first trophic level of any food chain, since they produce the food which supports the whole chain. The second trophic level is occupied by herbivores. These are called **primary consumers** because they eat producers. The third trophic level is occupied by **secondary consumers**. These are carnivores which eat herbivores. The fourth trophic level is occupied by **tertiary consumers**. These are carnivores which eat smaller carnivores. Some food chains have fifth trophic levels, usually occupied by large carnivores.

Parasites, scavengers, and decomposers

Parasites (e.g. germs, tapeworms etc.) feed on all members of a food chain. So they do not occupy any special trophic level.

Scavengers such as certain beetles, earthworms, millipedes, and wood lice, feed on dead plants, dead animals, and animal droppings. This never-ending supply of dead material is known as **detritus**.

Decomposers, such as bacteria and certain fungi, also attack detritus. Together with scavengers, they perform the vital task of breaking detritus down into humus and minerals essential for the healthy growth of plants. Without them soils would become infertile, plants would die, and all food chains would collapse.

Fig. 23.6 Flow of energy through food chains

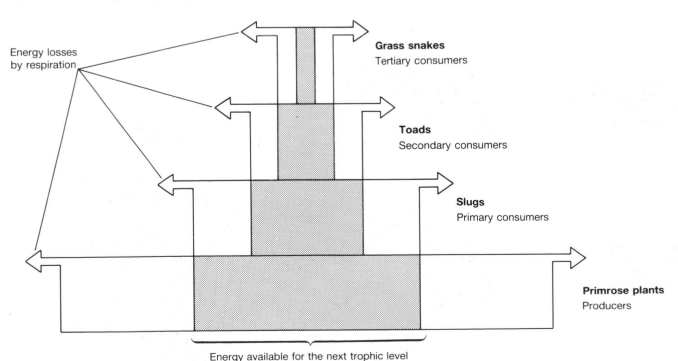

Energy losses by respiration

Grass snakes
Tertiary consumers

Toads
Secondary consumers

Slugs
Primary consumers

Primrose plants
Producers

Energy available for the next trophic level

Flow of energy through food chains

Few food chains have more than four trophic levels. Why? The answer is linked with the fact that some energy is lost at each trophic level, and so cannot be passed on to the next, higher level.

When, for example, a slug eats a primrose plant, it is not obtaining *all* the energy which the plant soaked up from the sun. First, some of that energy was lost as heat as the primrose respired. Second, the slug does not eat the entire plant – roots, leaves, flowers, seeds and all. Third, it cannot digest all the plant material it does eat. Some indigestible material passes through its body and out as faeces. A slug, therefore, gets only a fraction of the energy which went into making the primrose plant. The same things happen when a slug is eaten by a toad, and when a toad is eaten by a grass snake.

This means that each level of a food chain has to make do with what is left over after deducting energy lost by respiration in the level below, and after deducting indigestible material from what can be eaten (Fig. 23.6).

The limit is reached by about the fourth trophic level of a food chain. At this level, these deductions have reduced available energy to such an extent that there is rarely enough left to support another population of consumers.

Fig. 23.7 A pyramid of numbers

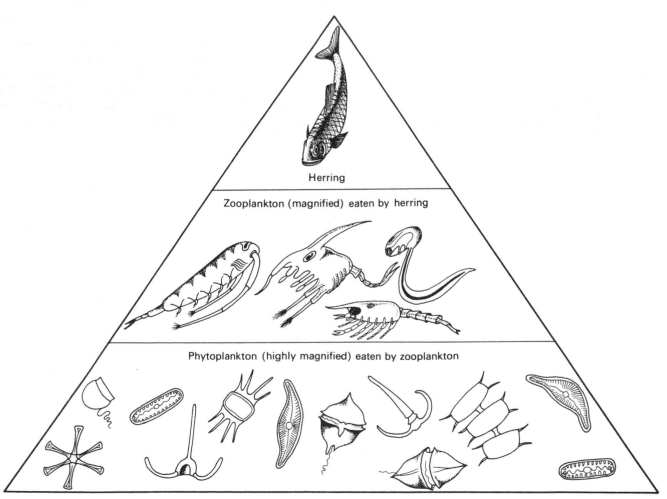

Herring

Zooplankton (magnified) eaten by herring

Phytoplankton (highly magnified) eaten by zooplankton

23.5 Biological pyramids

The energy lost at each link in a food chain means that each trophic level can pass on only a fraction of its energy to the next trophic level.

It follows, therefore, that a certain number of producers can support fewer primary consumers, which can support fewer secondary consumers, and so on. This gives rise to a characteristic of certain food chains called a **pyramid of numbers**.

Pyramids of numbers

Figure 23.7 shows how a marine food chain exists as a pyramid of numbers. A huge number of microscopic plants called phytoplankton support a much smaller number of zooplankton, which support only one herring.

The number of organisms at each trophic level of a food chain does not always give us a true picture of its structure. A single oak tree, for instance, can support thousands of insects, which would give us an upside-down pyramid.

A more accurate understanding of food chain structure is obtained by considering not the total number of organisms at each tropic level, but their total weight, or **biomass**.

Pyramids of biomass

An oak tree with a certain biomass can support a population of tortrix moth caterpillars with a much

smaller biomass. These, in turn, support a smaller biomass of hawks (Fig. 23.8).

23.6 Food webs

In nature, food chains are not isolated from one another. Consumer organisms rarely depend on only one type of food, and often a particular food is eaten by more than one consumer.

Zooplankton, for instance, are eaten by herring and by many other consumers including prawns, shellfish, marine worms, and whalebone whales.

In consequence, food chains are interconnected at many points forming what is best described as a **food web**.

Figure 23.9 illustrates a woodland food web. It shows that many different organisms interact with one another in such a complicated way that each type comes to depend on many other organisms for its existence.

Food chains and webs are other examples of the balance of nature, since they depend upon a delicate balance between losses and gains. If anything happens to disturb this balance, such as disease or pollution, which destroys a link in even one food chain, a whole food web can be affected. Exercises at the end of this chapter examine some of the problems caused by such disturbances.

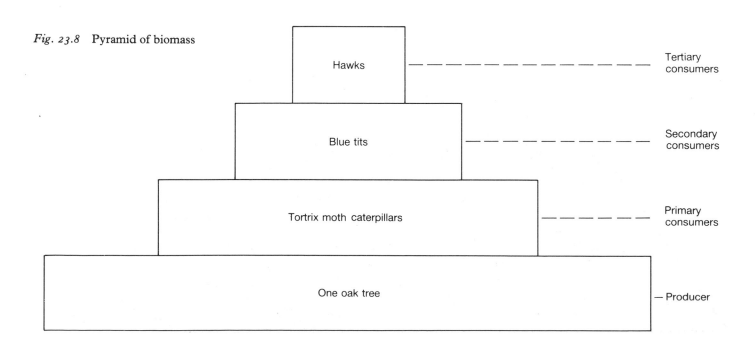

Fig. 23.8 Pyramid of biomass

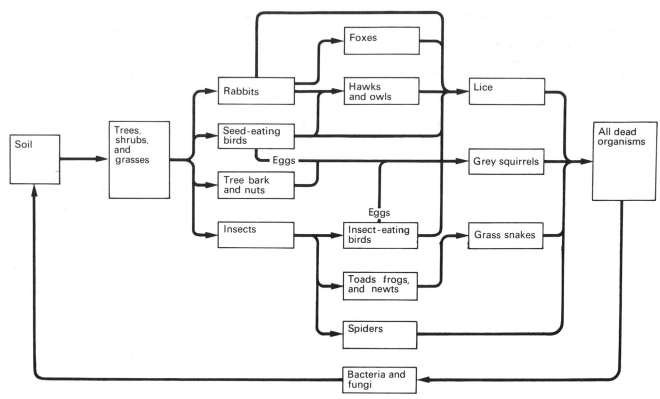

Fig. 23.9 Part of a food web in an area of woodland surrounding a small pond

Questions

1. *a*) Why are green plants known as producers, and why are most other organisms known as consumers?

b) Sort the following into producers, primary and secondary consumers, parasites, scavengers, and decomposers:

cows	grasses
ticks	bread moulds
lions	earthworms
spiders	gazelles
hyenas	tapeworms.

2. Study Figure 23.9 and answer the following questions.

a) Name the producer organisms.

b) How many complete food chains are to be found in the diagram?

c) Describe what might happen to the other organisms if a disease killed all the rabbits.

d) Why might the removal of all the grey squirrels improve the living conditions for the other animals?

e) If, after removing the grey squirrels, all hawks and owls were to be killed, why might this harm the living conditions of other animals?

f) Name those animals whose food supply would be reduced (both directly and indirectly) if the area were to be sprayed with insecticide.

g) What part is played by bacteria and fungi in maintaining this food web?

3. Opposite there is a diagram of the nitrogen cycle, with certain parts omitted.

a) Describe the two ways in which atmospheric nitrogen is converted into nitrates (labels **A** and **B**).

b) Mechanism **A** involves an example of symbiosis. Explain symbiosis and describe this particular example.

c) Name two organisms which produce ammonia from dead organisms (label **C**). What is the name of this process?

d) What are the organisms called which change ammonia into nitrates (label **D**)?

e) Explain how some soil nitrates are converted back into atmospheric nitrogen (label **E**).

f) Which substances released by animals form ammonia (label **F**)?

4. Opposite there is a diagram showing parts of two important natural cycles.

a) Name the two cycles.

b) Name the process represented by **A**, and write a chemical equation (in words only) for this process.

c) Name the process represented by **B**.

d Name the process represented by **C**, and write a chemical equation (in words only) for this process.

e) Name the process represented by **D**.

f) What substance is represented by **E**?

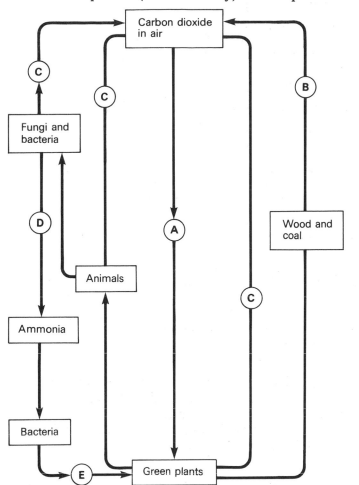

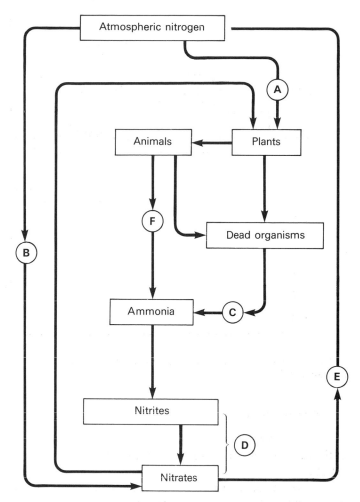

Summary and factual recall test

Nitrogen is an essential component of (1) and so is vital to all life. Plants cannot absorb nitrogen from the (2) but only in the form of (3) which they absorb from the (4). Some of the nitrogen absorbed by plant roots is replaced by nitrogen-(5) bacteria which use (6) nitrogen and (7) to make compounds which eventually release (8) into the soil.

Dead organisms are decomposed by (9). The process of decay releases (10) which is first converted into (11) by one set of bacteria and then into (12) by another set. Together these organisms are called (13) bacteria. The events described above are called the nitrogen cycle because (14).

(15) removed from the air by photosynthesis is replaced in three ways as follows (16). This process is called the (17) cycle because (18).

(19) from the sun (20) water causing it to evaporate into the atmosphere and form (21). Here, (22) drop-

lets get larger and larger until they fall as (23). This is called the (24) cycle.

All food chains begin with producer organisms which are (25). Next in line are (26) consumers which are (27) that eat (28), and next are the (29) consumers, which eat (30). Each position in a food chain is called a (31) level. Scavengers such as (32–name at least two) and decomposers such as (33–name at least two) are essential because they break down detritus into (34) and (35) needed for (36).

Few food chains have more than (37) links because some (38) is lost at each level. Therefore a certain number of producers support a (39) number of consumers. This gives rise to a pyramid of (40) or, if the total weight of each link is measured, a pyramid of (41).

Food chains are linked at many points to form a food (42). This linkage happens because (43).

24

Ecology and ecosystems

24.1 What is ecology?

Ecology is the study of the complex inter-relationships which exist between organisms; and also between the non-living world and the organisms which it supports.

Every living thing is linked to every other living thing as they eat and are eaten in the food webs of the living world.

Living things are also linked to the non-living world of sunlight, rocks, soil, water, and air. They depend on it for energy, support, shelter, and the basic raw materials of life.

Ecologists study these complex interrelationships to see how living systems work. They look at how energy is distributed via food webs; how organisms are adapted to particular ways of life; how climate, soil, landscape, and pollution affect living things; and also how we can conserve the teeming millions of creatures with which we share this planet.

24.2 Communities, habitats, and ecosystems

Communities and habitats

A group of organisms which live in a particular place, such as a forest, sea shore, or pond, is called a **community**.

The actual place which a community occupies is called its **habitat**. A rock pool, for instance, is a habitat with a community made up of sea weeds, crabs, sea anemones, etc.

Communities consist of many different species linked together in a food web. The members of each species make up a **population**. In a forest, for example, there will be a population of each type of tree, bird, mammal, and ant, just as a town has populations of people, cats, and dogs.

The **density** of a population is the number of individuals living in a particular habitat. Density depends mainly on birth rate, death rate, fertility, and migration in and out of the habitat.

Ecosystems

An **ecosystem** is a community of organisms together with the habitat in which it lives. This means that an ecosystem is made up of all the producers and consumers in a community; the parasites, scavengers, and decomposers; the rocks, soil, water, and air of the physical environment; and the circulation between this environment and the community of materials such as nitrogen, carbon, water, and oxygen.

Terrestrial ecosystems include grasslands, forests, and tundra. Aquatic ecosystems include the seas, sea shores, estuaries, rivers, and lakes.

The living or **biotic** part of an ecosystem obtains its energy from sunlight, and its raw materials from the non-living, or **abiotic** part of the ecosystem.

Sunlight energy trapped by plants during photosynthesis is passed via food webs to the entire biotic component of an ecosystem, and is eventually lost as heat. On the other hand, raw materials of life such as nitrogen and carbon are not lost. Partly through the action of decomposers, they are used and re-used over and over as they circulate between the biotic and abiotic components of an ecosystem.

24.3 Niches and adaptations

A niche is a way of life which enables a species to occupy a particular place within a community. This 'way of life' includes all the things a species does to survive, such as the type of food it eats, how it finds its food, how it has babies, avoids predators, survives the winter, etc.

A niche can be compared with a human profession. To survive as a bus driver you must know how to operate the controls of a bus, manoeuvre it through traffic without crashing into things, and find your way to the terminus.

The 'profession' of being a swallow, for example, is far more complex. They must be able to catch insects on the wing, build nests in sheltered places, feed their young and, as winter approaches, find their

A woodland habitat

A sea-shore habitat

way over thousands of kilometres of land and sea to warmer climates in the south.

To survive in a niche, a species must be very good at many different tasks. But there must also be opportunities for it to use its skills. A swallow must have plenty of insects to catch, sheltered places to nest, and natural predators must not be too plentiful. When we destroy habitats we remove these opportunities and thereby remove niches. This can lead to the extinction of species.

Each ecosystem has its own set of niches and, over millions of years, each species has evolved adaptations which allow it to survive in one of them.

Adaptations

Adaptation is a feature of an organism which improves its chances of surviving in a particular environment. All organisms, including the reader of this book, are marvellously adapted for life on earth. There are both general and specific adaptations.

General adaptations These include the more obvious features of an organism, such as lungs for breathing air, and gills for breathing in water; exoskeletons, endoskeletons, and muscles for support and movement; digestive systems by which animals obtain energy and food materials for their metabolism; and green leaves by which plants obtain energy and food.

Specific adaptations These are the adaptations which enable an organism to occupy a particular niche. For example, specialized teeth which enable some mammals to eat flesh, and others to eat plants. Special insect mouth parts which enable some to suck blood, some to drink nectar, and others to eat leaves. Special features in plants which allow some to survive in deserts, some to survive in rain forests, and others to survive constantly submerged in water (Fig. 24.1). The study of adaptations is one of the most important aspects of ecology, because they are the key to understanding the interrelationships between organisms, and between organisms and their environment.

Duck
The beak can hold slippery food like frogs and fish. Ducks filter small animals from water and mud by squirting it through grooves in the edges of the beak

Flamingo
Feeds on algae by filtering water through tiny grooves inside the beak

Humming bird
Sucks nectar from flowers, using its long beak and a tongue with a tubular tip

Parrot
Uses its beak to crack open hard seeds and nuts

Eagle
Uses its beak to kill prey and tear flesh from bones

Swallow
Catches insects by flying with its beak held wide open

Fig. 24.1 Examples of adaptation. Each beak is adapted to enable the bird to occupy a particular niche

24.4 Colonization and succession

From time to time new habitats become available and existing habitats are cleared of their original communities. This allows a new community to **colonize** the area. Man-made examples of new habitats are the heaps of powdered rock and earth dug out of mines, and natural examples include banks of mud deposited where rivers flow into the sea, and areas cleared by fire.

At first these new habitats are lifeless. Soon, however, a number of plants and animals colonize the area, but it may be many years before a stable unchanging communitity is established. One set of organisms gives way to another, and another, in what is called an **ecological succession**. Finally, this succession of organisms ceases and then the area is described as having a **climax community**. Figure 24.2 shows an example of the sequence of events which occurs after fire clears woodland.

24.5 Ecological factors

To most people the word 'environment' means their surroundings. In ecology, however, environment means the sum total of all the factors which affect living things.

Ecological factors can be classified into four groups: **climatic factors**, **physiographic factors**, **edaphic** (soil) **factors**, and **biotic factors**, the first three of these can be grouped together as **abiotic factors**.

Climatic factors
The main climatic factors which affect life on land (terrestrial habitats) are rainfall, humidity, temperature, wind, and light. The climatic factors of water (aquatic habitats) include the dissolved salts which give water a certain salinity; the suspended particles which cause cloudiness or turbidity, and affect the depth to which light can penetrate; the temperature, and water movement.

Rainfall, humidity, and temperature are responsible for the main vegetative zones of the world, such as rain forest, grassland, and deserts.

Rainfall affects vegetation because it supplies the moisture essential for growth and transport within a plant. Temperature affects the rate of metabolism (e.g. respiration) and therefore growth in plants, and humidity (the amount of water vapour in the air) affects the rate at which plants lose water by transpiration (evaporation from leaves).

Winds are important because they control cloud movement and, thereby, the distribution of rainfall. Wind also affects temperature, and it increases the rate of transpiration. These factors affect the type of plant life which can grow on land exposed to strong winds. Winds can also cause serious erosion, and alter the physical structure of a habitat.

Light is an important ecological factor because it is essential for photosynthesis. Light intensity and day length vary according to latitude and the season of the year.

Physiographic factors

Physiographic factors include features of the landscape such as coastlines, lakes, rivers, valleys, hills, and mountains.

Landscape has a profound effect on living things, mainly because it produces local changes in the overall climate of a region. A deep river valley has a different climate from a nearby mountain top, and a coastline has a different climate from adjoining inland areas.

Fig. 24.2 Succession in woodland

A Fire drives away mammals, flying insects, and birds. It kills most non-flying insects, larvae, and nymphs, and most plant life

B Grass begins to grow. This provides food for primary consumers: grass-eating insects, (e.g. grasshoppers), seed-eating birds, and grazing mammals (e.g. rabbits). Secondary consumers also begin to colonize the area: spiders, foxes etc.

C Bushes and shrubs appear.

D Trees gradually replace shrubs. Small trees encourage leaf-eating insects, and fruit-eating birds

E A fully established forest ecosystem (climax community). This has trees at various heights, or canopy levels. Each canopy will have its own communities and food webs.

275

Edaphic factors

Edaphic factors are those associated with the soil. The type of soil and its fertility have a great influence on the plants which will grow in any habitat, and the plants in turn influence animal populations.

Sandy soil and clay, for instance, are likely to support entirely different plant and animal populations because they have very different characteristics. Sandy soil dries out quickly after rain and has large air spaces between its particles. Clay, however, is almost airless, and frequently becomes waterlogged in the rainy season. In the dry season clay splits into rock-hard lumps.

Biotic factors

Biotic factors include all the ways in which the organisms of a community affect each other.

Section 23.6 describes how food webs form as organisms feed on each other, but feeding is only one of many biotic factors. Plants compete with each other for light and water. Frequently, as in forests, the taller plants cause extensive shade which restricts the type of vegetation that can grow beneath them. Some trees release substances into the soil which poison other plants, leaving it free for them alone to exploit.

Animals affect plants in several ways. Bees, butterflies, and moths pollinate flowers; termites and worms aerate the soil and help create humus; and many animals take part in seed dispersal. People, however, are the most powerful biotic factor on earth. They dig up trees and burn away bush to clear land for farming, they build roads, cities, and reservoirs, hunt animals for food and pleasure, and they pollute (poison) the environment with the wastes of modern civilization. Pollution is described in chapter 25.

24.6 Practical ecology

Beginners should remember two things before making an ecological study. First, it is very important that you study only a small area with one clearly defined community. Examples are suggested at the end of this chapter. If too large an area is studied the work involved becomes too great to justify the end results, and you soon become discouraged. Second, do not destroy the habitat you have come to study. To avoid this, you must try to identify plants without uprooting them, and when animals have been identified and studied return them to their natural environment. Do not trample all over a habitat. A class of enthusiastic students can quickly do enormous damage to a small pond or shallow stream.

Ecological studies should have definite aims, such as the following.

1. To compare plants and animals in two different habitats. In particular compare the ways in which organisms are adapted for survival in each habitat.

2. To study the animals and plants in one habitat in relation to their environmental conditions. Again, this means studying adaptations to the environment.

3. To study a single animal or plant species in relation to local ecological factors.

These aims can be achieved by measuring ecological factors such as temperature, humidity, rainfall, etc.; by collecting and identifying specimens, and estimating their numbers; and by mapping a habitat in various ways. Sometimes it is necessary to design and carry out simple experiments to solve problems encountered during an ecological study.

Measurements

Temperature Ordinary thermometers can be used to measure temperature differences between shaded and open areas of a habitat. Daily temperature variations are recorded with a maximum-minimum thermometer (Fig. 24.3).

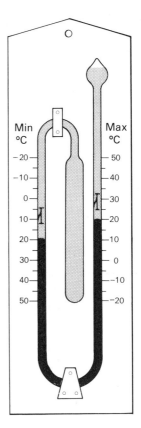

Fig. 24.3 A maximum-minimum thermometer

Humidity The amount of water vapour in the air (humidity) can be calculated by using a paper hygrometer (Fig. 24.4).

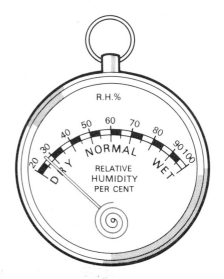

Fig. 24.4 A paper hygrometer

Rainfall A rainfall gauge can be made from a tin can, a measuring cylinder and a funnel (Fig. 24.5). Rainfall is calculated using the formula d^2/D^2h where $d =$ the diameter of the measuring cylinder, $h =$ the height of rainwater in the measuring cylinder, and $D =$ the diameter of the mouth of the funnel.

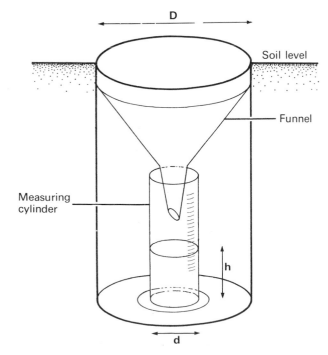

Fig. 24.5 A rainfall gauge

Fig. 24.6 A Secchi dish

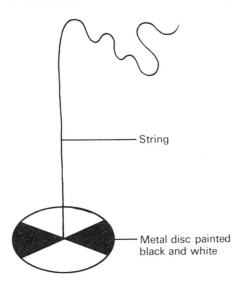

Turbidity The transparency of water is important because it affects the depth to which light can penetrate, and this affects the growth of algae and other submerged plants. Turbid water contains suspended particles. Turbidity can be measured by lowering a **Secchi disc** into the water (Fig. 24.6) and noting the depth at which it is no longer visible.

Water movement The rate of water movement along a stream can be measured by timing a floating object as it travels a known distance along the stream.

Light intensity It is possible to compare the intensity of light reaching various parts of a habitat using a photographic exposure metre. Readings should be made at the same time (e.g. noon) on several consecutive days, and at different seasons. Average light intensities at each season can then be calculated. You should note the plants and animals found in deep shade, moderate shade, and in bright light.

Wind It is important to know the direction of the prevailing wind for a habitat. This is the direction from which wind blows during each season. This is important because wind direction affects rainfall and temperature. Actual wind speed is not important. Note which parts of a habitat are exposed to prevailing winds, and which are sheltered. Note how animal and plant populations differ in exposed and sheltered places.

Measurement of pH pH is a measure of the acidity or alkalinity of an environment (pH 7 is neutral; that is, neither acid nor alkaline). The pH of water is found using litmus indicator paper. The pH of soil is found using a universal liquid pH indicator. Place a few drops of liquid indicator on a white tile, and mix a little soil with it. Tilt the tile to make the indicator run out of the soil across the tile. Observe the indicator's colour and, from the supplier's colour chart, read off the soil's pH value. Now you can relate the soil's pH with vegetation growing on it. You will find that some plants prefer acid soils, whereas others prefer neutral, or alkaline soils.

Other properties of soil in a habitat can be investigated using the methods described in chapter 22. Usually, however, it is enough to note whether the soil is sand, clay, or loam.

Collecting

The aim of collecting is to obtain specimens for study. Many organisms do not have to be collected: plants and many animals can be studied where they live. Only collect specimens which must be taken back to the school laboratory for study, or specimens which cannot be studied at once because they live in places such as water or under soil. Never collect more than one example of each species and if possible return it to the place where it was found. This will ensure that little damage is done to the habitat being studied.

Plants can usually be collected in plastic bags. Sometimes it is necessary to make simple drawings as well, to show what the parts which cannot be collected, or which will quickly fade or wilt, look like.

Animals are more difficult to collect because they often hide in inaccessible places. A number of different nets can be made from broom handles, strong wire, and netting.

Small animals can be sucked out of crevices using a **pooter** (Fig. 24.7), and a **Tüllgren funnel** can be used to coax small animals from soil and dead leaves (Fig. 24.8).

Place specimens in separate numbered containers (tubes, jars, etc.) and write the same numbers in a notebook, giving details of where the specimens were found.

Identification

Identifying every specimen as far as its genus and species is both time consuming and unnecessary. Books are available which help to identify most flowering plants to their genus, but it is sufficient to identify only the families of other plants such as ferns and mosses.

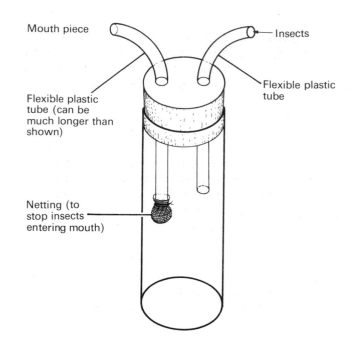

Fig. 24.7 A pooter

Fig. 24.8 A Tüllgren funnel

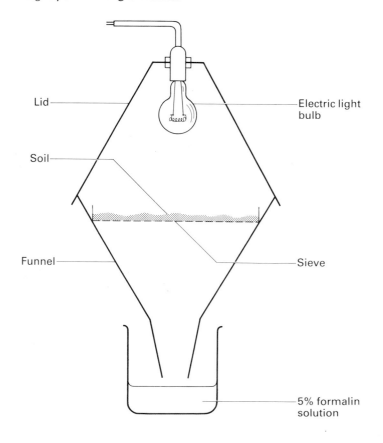

It is particularly difficult to identify certain invertebrates further than order or class, and to save time it is often wise to invent a temporary name, like 'snail with brown and yellow stripes'.

Recording the organisms in a habitat

Any serious ecological study must involve recording information about the animal and plant communities in a habitat. This information can be qualitative or quantitative. **Qualitative** information is about the names of species found in a habitat, and their adaptations. **Quantitative** information is about the numbers of each species in a habitat.

Recording qualitative information

The simplest way of recording qualitative information is to list the species found in a habitat. Far more interesting and informative records are **transects** of a habitat.

Making a line, or profile transect

A line transect shows the types of vegetation which exist on a line across a habitat (Fig. 24.9). This is a quick method of showing the influence on vegetation of factors such as different soils, waterlogging, trampling of feet along footpaths, etc.

Mark a length of rope at 1 m intervals and stretch it between two pegs across the habitat so that it crosses different types of vegetation, or prominent physiographic features.

Draw an outline (profile) of the transect on graph paper to show features such as the slope of the land, trenches, logs of wood, large stones, pools, streams, etc.

Starting at one end of the transect record the plants which touch the rope at certain intervals (e.g. 10, 50, or 100 cm depending on the density of the vegetation). Using symbols to represent the different species, draw these plants on the graph paper. It is also possible to indicate areas of shade by drawing a thick black line at ground level, and different types of soil by shading beneath the line which indicates ground level.

Recording quantitative information

The methods used in quantitative recording depend on the size of the area to be studied. In small areas (1 m^2 or less) you may be able to count all the members of certain species.

It is not practical to count all the organisms in a large area, unless they are large, like trees, and widely spaced. Instead, study a number of small areas chosen at random inside the habitat. This is called **random sampling**. Random samples can be obtained using a piece of apparatus called a **quadrat**.

Fig. 24.9   An example of a transect

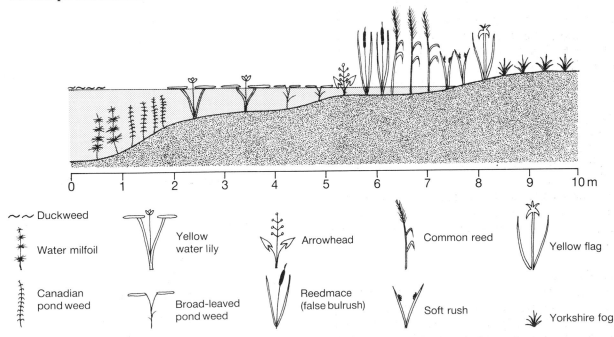

~ ~ Duckweed

Water milfoil

Canadian pond weed

Yellow water lily

Broad-leaved pond weed

Arrowhead

Reedmace (false bulrush)

Common reed

Soft rush

Yellow flag

Yorkshire fog

Using a quadrat A quadrat is a square or rectangular frame made of metal or wood, or is pegged out with string on the ground. A fixed, wood or metal quadrat can be made up to 1 m², but it is often more convenient to use a smaller quadrat 0.25 m². Quadrats should be divided into smaller areas, called mini-quadrats, with string or thin wire (Fig. 24.10).

Many plants grow in circular clumps. A square quadrat would cover the clumps and little else, so it is preferable to make a rectangular quadrat.

Quadrats must be placed at random throughout a habitat. This is done by throwing one over your shoulder (but first ensure that it will land on vegetation and not on another ecologist!). Do not deliberately throw a quadrat towards vegetation which looks interesting.

Quadrats are best suited to sampling vegetation. Sampling animal populations is very difficult and should be avoided by beginners. What you do after the quadrat lands depends on the type of information you require.

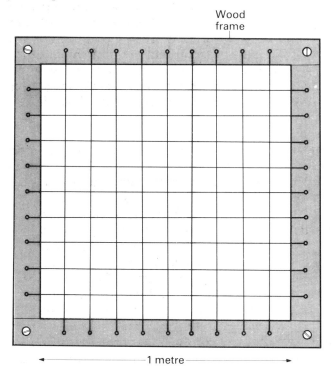

Wood frame

Fig. 24.10 A fixed quadrat (divided into mini-quadrats with string)

1 metre

Types of quantitative information
The main types of quantitative information are density, frequency, and percentage cover.

Density This is the number of organisms per unit area of habitat (e.g. the number per m²). To obtain the density of a species within a habitat you count the number present within a quadrat each time it lands. Continue with this procedure until the quadrat has been cast throughout the habitat. Then calculate the average number of this species per unit area.

Frequency This is the number of times a particular species is found when a quadrat is thrown a certain number of times (e.g. the number of times it is found during, say, a hundred throws). To calculate frequency you count the number of different species within the quadrat each time it lands, and make a note of their names. If, for example, you throw it a hundred times, you then count the number of times each species was found, and express each result out of a hundred. In this way you can discover which is the commonest (most frequent) species in a habitat, the next commonest, down to the rarest.

Percentage cover This is the percentage of the total area of a habitat which each species covers. Percentage cover can only be calculated with species which grow in clumps or large patches. You can calculate percentage cover of the species in a small area by drawing a grid map. If a large area is to be studied you must estimate the percentage cover of the species found inside a quadrat each time it lands. Then calculate the average cover, over a number of throws.

Investigations

A   *Studying communities in the laboratory*

1. *A rotting log community* If you can find a well-rotted tree trunk, cut a length from it which will fit in an aquarium or other large container. Put a layer of soil and rotting leaves on the bottom of the container and lay the log on this. Snails, ants, spiders, centipedes, millipedes, beetles, moss, fungi, etc., live in rotting wood. A glass lid, or a ring of vaseline around the upper edge of the container will stop insects escaping. Feed ants and other insects with bread crumbs spread on a damp sponge moistened with sweetened water.

2. *A pond community* Spread washed sand and gravel on the bottom of an aquarium sloping it up to the back. Fill the aquarium with clean pond or river water, then plant as many different types of pond weed as you can find, including floating varieties. Wait a week for the plants to begin growing before introducing pond animals. Introduce snails, *Cyclops*, *Daphnia*, flatworms, and perhaps a minnow or other

small fish. If you introduce dragonfly larvae or large water beetles they will eat other small animals. Do not put the aquarium in direct sunlight as this will over-heat the water and encourage growth of algae, which will turn the water green.

B *Ecological projects*

Use the collecting and sampling methods described in this chapter in the following ecological projects.

1. *Colonization* If land is cleared by burning or digging it will soon be colonized by plants and animals. With time, early colonizers will be replaced by others, and these too will be replaced, until a stable community is established. These changes are called **succession**, and the final stage is called a **climax community**.

a) Colonization of cleared land. Remove all the vegetation from a square metre of grassland. Dig out as many roots as you can, but leave as much soil as possible for new plants to colonize. At regular intervals, for example once every month, map the plants growing in the square.

b) Colonization of a stream bed. Make a strong wooden frame (0.25 m^2), and cover it with fine metal gauze strengthened with thick wire netting. You now have a tray with a wire gauze bottom. Fill this tray with washed sand and pebbles to make it look as similar to the bottom of a stream as possible. Leave it on the bed of a stream for about a month. It can then be lifted out and examined to discover which animals have colonized the area. Try to estimate the numbers of each species present.

c) Colonization by micro-organisms. Hang several clean glass microscope slides in a well established aquarium. After a week remove the slides, clean and dry one side, and examine the other side under a microscope. You will see many different types of algae and protozoa attached to the glass. Replace the slides and examine them at weekly intervals. Is there any evidence of succession?

d) Colonization of bare rock. Look for areas of rock on a sea shore which are thickly covered with sea weed. Clear a small area (0.25 m^2) down to the bare rock. Examine this area at monthly intervals and record the stages of colonization.

2. *Study of a pond or stream*

a) Animals can be collected from fresh water using nets. Aquatic plants can either be studied where they grow, or by dragging some from the water with metal hooks on the end of a pole or length of rope. It is important not to excessively damage the habitat by

this method.

b) Study the effects on vegetation and animals of: shade from overhanging trees; distance from the bank (i.e. depth of water); movement of water at various distances from the bank of a stream; and turbidity if any.

c) If there are signs of water pollution make a special study of its effect on organisms by comparing polluted and unpolluted areas.

d) Look for adaptations which help plants and animals survive in water. Compare the roots, stems, leaves, and flowers of land and water plants. Examine animals for adaptations which allow them to move in water and avoid being swept away by water currents. Examine the various types of respiratory organs in aquatic animals.

3. *Studies of special habitats*

a) Animals in a compost heap. Take small, measured samples of compost from different levels of a compost heap. The most conspicuous animals can be collected with a pooter after spreading each sample on newspaper. Less conspicuous animals can be obtained by putting samples in a Tüllgren funnel. Collect animals from each sample in separate numbered specimen jars. Count the numbers of each species found at different levels. Does the frequency of each species vary according to the level investigated?

b) Animals under stones. Examine stones, planks of wood etc., in woodland, grassland, and on the seashore. In what ways does this habitat differ from the surrounding habitat? What are the special features of animals which live in this habitat?

c) Animals associated with food crops. Look for animals which eat the crops, and their predators.

d) Animals associated with one type of plant. A surprising number of different animals can be found living on one type of plant. Nettles, for example, are associated with over a hundred different insects, including beetles, plant bugs, plant hoppers, butterflies etc., as well as snails, spiders, and many other animals; and a food web made up of about 20 different animals has been found in thistle flower heads. Observe parasites, especially solitary wasps which lay eggs inside caterpillars, and observe carnivores.

e) There are many different special habitats which are worth special study: bird's nests, holes in tree walls and river banks, parasites of various animals, etc. Can you think of more examples?

Questions

1. *a*) 250 g of soil was placed in an oven and heated at 100°C until its weight became constant. Its weight was then 165 g. What percentage of the original soil was water?

b) Describe how you would make an ecological study of the organisms in a sample of soil and dead leaves taken from a forest floor.

2. Study the bird feet illustrated below. How is each adapted to a particular way of life?

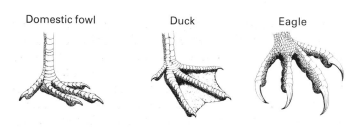

Domestic fowl   Duck   Eagle

3. A group of students studied two areas of grassland: one lightly trampled and the other heavily trampled. The histograms below show the numbers of plants of five different species found in random samples taken within each region.

a) How many of species E were found in each region?

b) What is the effect of *increased* trampling on species G and H?

c) Which species is most affected by trampling?

d) Which species is least affected by trampling?

e) How would you obtain random samples of plants in these areas?

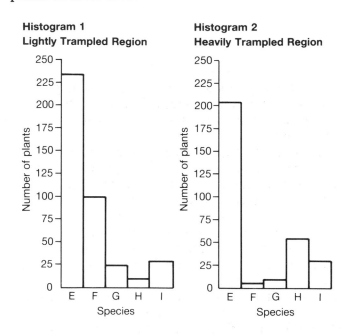

Summary and factual recall test

A (1) is a group of organisms which live in a particular place. The place the group occupies is its (2), such as (3–name three examples). Together, the members of a species make up a (4). An ecosystem is made up of (5–name its parts). The living part of an ecosystem is called the (6) part, and the non-living part is the (7) part. The way of life of a species is its (8).

General adaptations include (9–list three examples). Specific adaptations allow a species to occupy its particular (10). Examples of such adaptations are (11–name two).

The way in which one set of organisms gives way to another during colonization is called (12). The final stage is called a (13).

Climatic factors include (14–name at least three). Physiographic factors include (15–name at least three). Edaphic factors are associated with (16). Biotic factors include the ways (17) affect each other.

25

Pollution and conservation

A substance is called a **pollutant** when its presence causes harm to living things. This means that a substance may be a pollutant or not according to circumstances.

Carbon dioxide, for example, is harmless to animals and useful to plants when present in the air at its normal level of 0.033%. But at much higher concentrations carbon dioxide is classed as a pollutant owing to its harmful effects on animals. Similarly, nitrates are vital to the healthy growth of plants, but they become pollutants when they accumulate in lakes and cause such a rapid growth of green algae that the water looks like pea soup and other organisms become endangered. Some substances are both useful and harmful at the same time. Examples are the chemicals applied to seeds such as wheat to prevent fungal parasites attacking the seedlings. These chemicals are useful insofar as they increase crop yields, but are harmful (pollutants) when they enter food chains starting with seed-eating birds and ending with hawks and owls.

In view of these facts a pollutant can also be defined as a substance present in the wrong amount, at the wrong place, at the wrong time.

25.1 Air pollution and acid rain

Life on earth is supported by a layer of air called the **troposphere** which extends to a height of approximately 10 km. It is usual to describe pollution of the troposphere by stating the concentration of pollutants in parts per million (ppm), which means the number of cubic centimetres of pollutant in a cubic metre of air. The main cause of air pollution is the burning of fossil fuels such as coal and oil.

When fossil fuels are burned in factories, power stations, vehicle and aeroplane engines etc., they produce smoke and fumes containing gases such as sulphur dioxide, carbon dioxide, carbon monoxide, and oxides of nitrogen.

At first, smoke and fumes are hot and rise into the air, especially if they come from high chimney stacks. Some of these substances remain in the air for long periods and may be carried hundreds of kilometres. While in the air they can enter clouds, dissolve in water vapour and form acid solutions which fall back to earth as rain. This type of pollution is called **acid rain**.

Some smoke and fumes fall back to earth without dissolving in water vapour. This type of pollution is called **dry deposition**.

Air pollution from a coal by-products plant

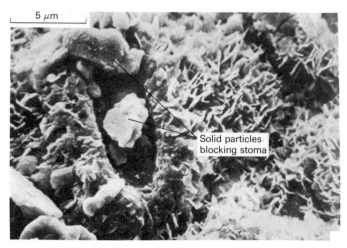

An oak leaf stoma blocked by particles of dirt caused by air pollution

The effects of air pollution on medieval stonework at York Minster

Dry deposition

Smoke This contains particles of various sizes which settle out of the atmosphere as dust and soot.

They settle on cities, blackening paintwork, masonry, and washing on the line. They settle on the leaves of plants and cut off their light thus limiting photosynthesis. The smallest particles block plant stomata and slow down transpiration and the exchange of gases with the atmosphere. Dust is inhaled by people and, if present in high concentrations, may cause serious respiratory ailments.

Sulphur dioxide In concentrations as low as 0.1 ppm this gas damages plants by chemically attacking their leaves. It erodes their waxy cuticle and prevents the formation of chlorophyll. There is evidence that crop yields can be reduced by sulphur dioxide even when levels are too low to produce visible defects.

Sulphur dioxide corrodes metal railings, balustrades, bridges etc., and eats away stonework, especially limestone. Many famous buildings like St Paul's Cathedral in London have been damaged. Sulphur dioxide forms acid vapour when breathed into the lungs. This causes little harm in concentrations below 1 ppm, except that it aggravates ailments like acute bronchitis.

Plant-like organisms called lichens are useful indicators of sulphur dioxide pollution, because certain varieties are killed by concentrations of only 0.05 ppm. If these sensitive lichens disappear from an industrial area, but are still present in surrounding rural areas it indicates that sulphur dioxide pollution is having some biological effects.

Oxides of nitrogen These come mainly from the exhausts of cars, diesel engined lorries, buses, and trains. They corrode metal and stone, and can have harmful effects on people, especially those with anaemia or lung disease. These gases are also extremely unpleasant to inhale and can irritate the lungs and eyes.

Under certain conditions, oxides of nitrogen are transformed by sunlight into peroxacyl nitrates, which gather in the atmosphere as visible clouds called **photochemical fog**. This can damage plants by attacking the spongy mesophyll layer of their leaves. At high concentrations it causes severe eye irritation.

Photochemical smog is particularly dense over the Los Angeles area of the United States where it has reduced the yield of citrus fruits and damaged many other plants. Fortunately, this type of fog is rare in Britain, largely because there is much less sunshine.

Lead poisoning Depending on the fuel used, exhaust gases can contain certain lead compounds. These can be inhaled directly, or fall on crops where they contribute to lead in food. There is evidence that lead can damage the developing brain of a foetus, or a young child. Air pollution is not the only source of lead poisoning. It is found dissolved in water from homes with lead piping; certain types of paint (especially those made before 1983); and cosmetics such as kohls and surmas imported from India.

A Pollutants disperse upwards

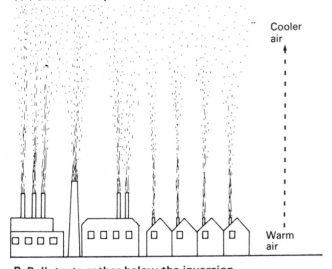

B Pollutants gather below the inversion

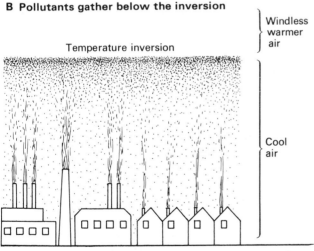

Fig. 25.1 Comparison between normal conditions and those which lead to smog formation

Temperature inversions Under normal atmospheric conditions wind disperses smoke and fumes before they can seriously harm people. But these pollutants are not dispersed when there is a condition called a **temperature inversion**. This happens when still, cold air is trapped under a layer of warmer air (Fig. 25.1).

From the 5th to the 9th of December 1952 there was a temperature inversion over London. During this time sulphur dioxide levels increased to 0.75 ppm, and smoke gathered in the atmosphere along with sulphuric and nitric acids. All of these compounds gave rise to a thick yellow fog, or 'smog'. Over this period the death rate in London increased rapidly; in one area there were about 4000 more deaths than would normally be expected at that time of year.

Since 1952, legislation in Britain such as the Clean Air Act, combined with the increased use of oil instead of coal as a source of power in industry and the railways, has done a great deal to reduce the risk of smog formation, and air pollution in general.

Acid rain (wet deposition)

Emissions of sulphur dioxide and oxides of nitrogen from power stations, factories, and motor vehicles cause the formation of sulphuric and nitric acids in rain clouds. If rain falls through polluted air it picks up more of these gases and increases its acidity (Fig. 25.2).

It is estimated that, in Europe, about 90% of this acidity comes from power stations, and the remaining 10% from factories and motor vehicles.

Acid rain damage

Many people are convinced that acid rain is causing serious damage to plants, soils, and freshwater life around the world.

Severe acid rain damage to conifers at Brantridge park

Damage to plants In West Germany acid rain is blamed for damaging 34% of that country's forests. Old exposed trees are most affected. There is progressive death of young shoots; leaves or needles turn yellow and fall off; fine root structure is damaged and the whole tree eventually dies. The same thing is happening to forests in Central Europe, Scandinavia, Britain, and the United States.

There is also serious damage to crop plants in farms near cities. Recent evidence suggests that yield losses in British cereals, root crops, and vegetables amount to £200 million a year.

Damage to soils A soil rich in lime will neutralize acids it receives from rainfall. Peaty soils, and those formed on acid rocks like granite could be more affected because they are already acidic. Acid rain may do more than simply make a soil more acid. It may cause mineral nutrients to be washed away and can release toxic chemicals into a soil such as aluminium and mercury. Under normal conditions these would be safely locked up in an inactive state. Thus, acid rain could make soils less fertile, and even poisonous to plants.

Acid rain may also wash aluminium and mercury out of soil into lakes and rivers where they poison aquatic life.

Damage to fresh water life In Southern Norway and in Sweden stocks of atlantic salmon have declined so far that they are now almost extinct. It is believed that this is due to acidification and mineral poisoning of their river breeding grounds.

Thousands of lakes in these two countries have suffered a decline in their fish populations. Fish are thought to be killed when acidity releases aluminium into the water. This builds up as a layer of aluminium hydroxide on their gills. Winds can spread air pollution at a rate of 500 km a day. Therefore, under the influence of westerly winds, acid rain formed over European power stations could be falling on Scandinavian lakes three days later.

Is acid rain harmful?

Many people believe that the answer to this question is yes. Others believe that while acid rain is important, it may be one of many factors responsible for the damage described above. Perhaps severe drought, frost, disease and other factors weaken wildlife over the years so that the extra stress caused by acid rain is too much. It tips the ecological balance the wrong way, which leads to disaster.

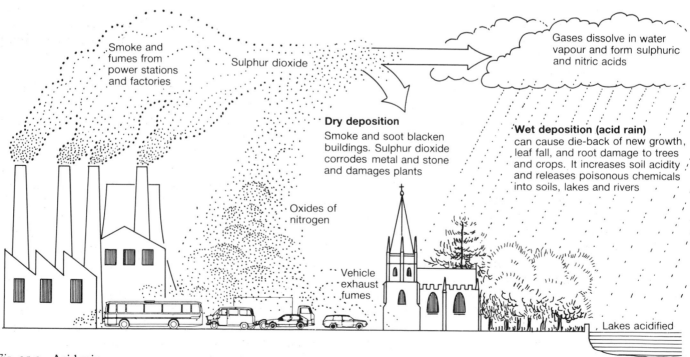

Fig. 25.2 Acid rain

Can acid rain be cured?

There are several ways in which sulphur dioxide emissions can be reduced, but all of them are expensive.

1. Coal can be crushed and washed before it is burned. This would remove 10% of its sulphur.

2. Oil can also be treated to remove sulphur.

3. Cleaning systems can be fitted into chimneys to remove sulphur dioxide before it can be released into the atmosphere.

4. Improved furnaces can be built which burn fuel more effectively and produce less pollution.

5. Vehicle exhausts can be fitted with a device which removes pollutants from engine emissions.

West Germany has reduced the emission of oxides of nitrogen from vehicles by introducing lower speed-limits on motorways.

25.2 Water pollution

At present British rivers are being polluted by domestic, industrial, and agricultural waste to such an extent that in certain regions purification processes cannot make the water fit for human consumption or, in severe cases, even for industrial use. This pollution has caused great damage to the animal life of inland waters. In England and Wales, for example, about 5% of rivers are totally without fish.

Domestic waste Domestic waste can be made relatively harmless at sewage treatment plants, but in many areas population growth has temporarily overloaded these plants so that much untreated sewage is discharged into rivers and the sea. When solid organic matter in sewage enters a river it is decomposed by bacteria which multiply within it and begin to use oxygen at a rate that quickly deoxygenates the water, making it uninhabitable for most types of fish, insects, crustaceans, and acquatic worms.

The treatment of sewage removes the bulk of its organic matter. But this process creates a problem by releasing chemicals such as phosphate from the sewage, which is added to an even greater amount of phosphate released from the use of synthetic detergents. When sewage rich in phosphate is discharged into rivers it may be joined by nitrate which drains out of agricultural land that has been treated with quick-acting inorganic fertilizers. As all this phosphate and nitrate gathers in water it gives rise to a condition known as **eutrophication**, which is the accumulation of minerals that promote plant growth. When eutrophication occurs in rivers, lakes, and reservoirs the algae that live in the water multiply at a much faster rate than normal. In most cases the algae gather at the surface where they form a thick green mat that prevents light from reaching plants living lower down. Consequently, these submerged plants die and the water, deprived of their photosynthetic oxygen, quickly becomes uninhabitable by animal life.

Industrial waste One of the main problems with industrial waste is that it often contains highly poisonous and long-lasting pollutants such as compounds of cyanide, lead, mercury, copper, and zinc. These chemicals are dangerous even in low concentrations because they can accumulate in fish and other aquatic creatures and then spread along food chains to other animals such as water birds, otters, and even man.

Agricultural waste The traditional agricultural methods in which animal manure is spread on land used to grow crops cause little or no pollution. However, many modern farmers keep poultry, cattle, and pigs in buildings and have no other land on which to use their manure. All too often, owing to labour shortages and the high cost of transport, this manure is disposed of by methods which pollute local streams and rivers when it could have been utilized as a valuable fertilizer. Once in the water untreated animal manure decomposes causing reduced oxygen levels as described above.

Other pollutants resulting from modern agricultural practice include chemicals in crop sprays used to kill insects and fungal pests. If these chemicals are washed into local rivers and ponds they disrupt food chains in the same way as industrial waste.

Pollution of the sea The oceans of the world contain nearly 600 million cubic kilometres of water, but this does not mean they are large enough to absorb all the waste that man is pouring into them without becoming polluted.

Until oil supplies run out, or until science discovers an alternative and cleaner source of energy, there will be a continual risk of oil spillage into the sea from tankers and offshore oil rigs. When this happens the oil kills sea birds either by poisoning them or by sticking their feathers together so that they cannot fly in search of food, and when washed ashore oil kills all forms of life on rocks and in sand.

Nevertheless, living things have a remarkable capacity for recovery. Provided these disasters do not occur too often sea bird colonies can replace their lost members, and shore life has been known to return to normal within only a year of heavy oil pollution. Unfortunately, in using detergents to clear oil

Drax power station

Eggborough power station
Ferrybridge power station

Pollution of the river Thames

pollution man has often caused more damage than if he had left the oil to be decomposed naturally by bacteria. It has been shown that these detergents, in concentrations of only 1 ppm, kill most sea-shore life, and that where they are used life takes at least two years to restore itself to normal. A less harmful alternative is to cover floating oil with powdered chalk. The oil then sinks to the bottom of the sea and is decomposed by bacteria.

A large quantity of untreated sewage is discharged into the sea, and this can cause pollution, especially where it gathers in restricted areas such as bays and inlets. Otherwise it is carried by tides and coastal currents into deeper water where it quickly decomposes and is then relatively harmless. A more serious problem arises from the discharge into the sea of poisonous industrial wastes, like those described above. When these gather in shallow-water fisheries they may cause a danger to human health by accumulating in fish and other sea creatures which are consumed as food.

Drax, Ferrybridge, and Eggborough emit 2000 tonnes of sulphur dioxide a day, causing pollution in Scotland, Norway, and even the arctic

25.3 Other types of pollution

Radiation

Radiation such as X-rays, beta, and gamma rays can cause various types of cancer, a blood disorder called leukaemia, and damage to sperms and ova which can lead to deformed babies.

Natural radiation comes from outer space in the form of cosmic rays, and from igneous rocks such as granite. Artificial radiation comes from certain medical and industrial processes. Little if any harm comes from these sources, but there is increasing concern about radiation from the testing of nuclear weapons, from the generation of nuclear power, and from the re-processing of spent nuclear fuel.

There is always some risk, however negligible, that radiation will escape from a nuclear power station, and from a nuclear re-processing plant. A long-term problem comes from the disposal of nuclear waste. It can be processed to extract re-usable uranium, and the residue stored in metal containers, fused into glass, or encased in concrete. It can then be stored on the surface in secure buildings, or put in caverns deep below the surface in geological formations free from earthquakes. Unfortunately, it remains radio-active for long periods, sometimes thousands of years. So we are passing on the problem to generations yet to come. Remember that other types of fuel cause pollution problems too.

Noise

Noise from cars, motorcycles, aeroplanes, dogs, children, radios, televisions etc., is sometimes described as pollution because it can cause mental and physical harm.

Prolonged loud noise from sources such as industrial machinery and discotheque music can damage the inner ear and cause partial deafness. Noises can also cause sleeplessness and mental depression. In some cases it is possible to take legal action against people who disturb others by creating noise over a long period.

25.4 Control of pollution

Pollution control can be enforced by law, or it can be achieved voluntarily.

Legislation

Many acts of Parliament have been devised to control pollution. These include The Clean Air Act of 1968; Rivers (Prevention of Pollution) Act 1961; Noise Abatement Act 1960; and the Control of Pollution Act 1974.

Technical

Engines can be designed to reduce exhaust emissions. Smokeless fuels can be used in homes, and factory and power station chimneys can be fitted with pollution reduction devices. Plastic containers can be made from bio-degradable materials which decompose naturally.

Recycling

Recycling involves the reprocessing of waste back to the state from which new items can be produced; 76% of what we throw away could be recycled: paper can be pulped and made into new paper goods; glass and metals can be melted down and re-used; certain types of plastics can be used again, and kitchen refuse can be composted and used as garden fertilizer.

Recycling does not simply control pollution by reducing the size of rubbish tips, it slows the rate at which raw materials are used up, and it saves energy and money.

Pollution is a problem because man is clever but not always wise. He has been clever in learning how to exploit the world's resources to his own advantage, but not wise enough to foresee the environmental damage which results from this exploitation. Before it is too late man must learn how to reduce his interference with the balance of nature, otherwise he may endanger all forms of life on this planet, including his own.

25.5 The destruction and conservation of wildlife

It is estimated that there are at least 10 million different species alive to-day, but only 1.6 million have been named so far. At present, because of human activities, the world may be losing up to three species a day, and the rate is increasing, so inevitably some species are being lost before we even know they exist.

Extinction is forever; because it is not simply death – it is an end to birth.

How and why is wildlife being destroyed

There are three major reasons for wildlife destruction: we are destroying habitats for our own use; we are poisoning wildlife with our wastes; and we are killing wildlife for fun and for luxury goods we could do without.

Most of this destruction is the inevitable price we must pay if we are to feed, clothe, and find homes for an ever-increasing population. But the picture is not all gloom and despair. This section describes not only how wildlife is being destroyed, but how it can be saved.

Urbanization

The growth of cities and road systems gobbles up land. Every 1.5 km of motorway covers an area the size of 35 football pitches and requires 10 000 tonnes of aggregates, sand, and hardcore. This means extending quarries, which often eats into wild countryside. Once abandoned, however, a quarry makes an excellent wildlife habitat, and can even be a place of beauty.

Conservation area at Cwmcarn, Gwent, once the site of a colliery tip

Most urban growth covers top grade agricultural land, and this puts pressure on farmers to spread into wildlife habitats. However, this spread can be limited by improving productivity of existing farmland.

There is one way in which wildlife can benefit from urban growth. City parks and gardens can be made much richer in wildlife than an equivalent area of modern farmland. Roadside verges can provide a vast refuge for wildlife if they are not sprayed with herbicides. Many derelict industrial sites have been transformed into flourishing nature reserves (see photograph).

Agriculture versus wildlife

Traditional farming methods created an environment in which our ancestors lived in harmony with wildlife. But modern methods can pose the most serious of all threats to nature.

In a day, farmers can clear woodland which has

A habitat being destroyed by fire

taken hundreds of years to grow. They can drain marshes, fill in ponds, dig up hedges, and spray pesticides over huge areas killing pests as well as harmless insects, which starves or poisons insect-eating birds and animals.

Farmers can add fertilizers to heaths, downlands, and ancient meadows to 'improve' them. This treatment makes a few vigorous species of grass grow quickly and smothers dozens of slow-growing plants that become rare or extinct.

The Nature Conservancy Council presides over 167 National Nature Reserves and nearly 4000 Sites of Special Scientific Interest (S.S.S.I's). Even these are not safe. Every year 4% of them are damaged or completely destroyed, and this rate is probably higher for the best potential farm lands like ancient meadows and marshes.

Farmer conservationists

There is bound to be some conflict between agriculture and wildlife conservation but it can be kept to a minimum.

Avoiding needless destruction Some farmers plough up unique habitats out of ignorance. There must be better communication between conservationists and farmers.

Crop protection without chemicals Pesticide should be used to kill pests and not as a means of wiping out everything that moves. It is wasteful and unnecessary to spray crops so regularly that every pest is destroyed, and frequent spraying can lead to pests developing resistance to chemicals used.

It is possible to forecast pest outbreaks. Insect traps are used to estimate the number of pests in an area. Farmers can be warned if the numbers of a pest increase to danger levels so they can spray at the right time.

Creating new habitats Some farmers plant trees and shrubs, and even make ponds in areas where cultivation is difficult.

Killing for fun and profit

There are still people who depend on hunting for some of their livelihood, but more often the killing is for pleasure and profit. As the resource becomes more scarce, the price asked gets higher, the temptations greater – extinction becomes more likely.

Tigers and vicunas (a South American relative of the camel) are on the verge of extinction, and yet tiger skin coats and vicuna cloth are still on sale in London. Many rare species of snake, lizard, and crocodile are killed to make handbags, belts, and shoes.

Asian and African elephants are hunted for their tusks and skin. These big mammals could disappear from many areas to satisfy people's desire for ivory ornaments, piano keys, billiard balls, and trinkets.

Perhaps sheer scarcity of these products will eventually price them out of the market, but by then the animals may have been hunted to a point from which they can never recover.

Protection of Wildlife

Legislation In Britain, the Wildlife and Countryside Act of 1981 was passed to protect animals and plants. Some Governments help conserve wildlife by creating game reserves and national parks, where hunting and building are restricted or forbidden.

Voluntary Organizations There are scores of organizations throughout the world whose aim is to protect and conserve wildlife.

Some are campaigning organizations which put pressure on governments, landowners, industrialists etc., to stop habitat destruction and reduce pollution. Friends of the Earth and Greenpeace are well-known examples.

The main aim of some organizations is to raise money for the conservation of threatened wildlife. The World Wildlife Fund is famous for this kind of work.

Other organizations preserve wildlife habitats by purchasing them. These 'nature reserves' are looked after by local wardens and volunteers, to ensure that they remain undamaged for all time. The Royal Society for Nature Conservation and its many local trusts is the largest organization of this kind in the United Kingdom. Their junior branch, called Watch, encourages concern for the environment among young people.

These few examples of conservation are intended to show that it is not simply the opposite of exploitation. Conservation can be described as an attempt by man to develop a way of life that does not disturb the balance of nature, and yet allows him to maintain all the advantages of modern civilization. Unfortunately man is a very long way from reaching this goal, for to do so he must solve some extremely difficult problems: he must control the growth of population so the number of people alive at any one time remains more or less constant; and he must learn how to satisfy his needs without polluting his environment, and wasting the world's resources. Man must find ways of using and re-using his resources over and over again, as does every other living thing except man.

Investigations

A   *Pollution*

School groups can help control pollution by careful observation of the environment, without the use of expensive equipment.

1. *Water Pollution* Organize a survey of local ponds, canals, rivers, streams, and beaches. Look for new sources of pollution such as oil, chemicals, and dumped rubbish. Look especially for dead animals such as fish and birds.

By looking at various points along a river bank it is often possible to find the source of pollution. If it appears to come from a particular factory, garage, farm etc., report it to the police or Regional Water Authority.

The state of a piece of water can be judged by studying 'indicator animals'. If a stream contains stonefly and mayfly larvae it is clean and healthy. If only sludgeworms and rat-tailed maggots are to be found it is badly polluted.

2. *Air pollution* A simple method of assessing air pollution is to place white painted boards in a number of locations from the centre of a city to surrounding countryside. Compare the accumulation of soot and dust on each board after a fixed period of exposure.

B Recycling

Set up a recycling scheme. This one of the most popular, and profitable, forms of conservation.

1. First, and most important of all, decide what to collect and where each type of article can be sold. You can find local buyers of waste paper, plastic, aluminium foil etc., in the Yellow Pages or a 'phone book. Find out the minimum quantities they will take, and if items must be clean.

2. Produce a short leaflet stating what you aim to collect and why.

3. It is best to have collections once a week, or even monthly. Remind everyone the night before. Take care sorting inflammables, like paper. You may create a fire risk.

C Conservation

1. *Protecting a threatened habitat* Find out if your area has untouched habitats such as meadows, woods, ponds, marshes etc. Make a survey of plants and animals found in the habitat. Make regular checks for signs of pollution, and industrial, housing or agricultural developments. Take action by reporting any threat to the habitat to your local wildlife trust.

2. *Create a nature reserve* It is possible to create a small, but interesting, nature reserve in a garden or, with permission, in your school grounds or even a local park.

a) A pond. Pond making is quite easy. Dig a shallow hole and line it with heavy duty polythene sheeting. Hold this in place with large stones around the edges and small stones and soil at the bottom. Leave it full of water for a week before inserting water plants, and let these become established before starting to think about animal life. Don't put goldfish into your pond. They will eat all the invertebrates. In fact it is interesting just to leave the pond alone for a year. You will be surprised at how quickly wildlife colonizes the water without your help.

b) Log piles. If you leave a few logs to rot in an unused corner of a garden they will attract beetles, woodlice, centipedes, millipedes, digger wasps, hover flies, and birds. Occasionally brightly coloured fungi will grow out of the rotting wood.

c) Wild patches. A tidy, well-weeded, and well-sprayed garden is not a good place for wildlife. A bit of wilderness if far better. You can stop a wilderness spreading by surrounding it with bricks or paving stones. Let nettles, ragworts, dandelions, thistles, clover, hogweed, wild carrot, and a wide variety of grasses grow in your wild patch. Each will attract its own horde of invertebrates, as well as bird visitors.

d) Insect gardens. If your garden must be neat and tidy at least grow plants which attract butterflies and other insects. Such plants include: verbena, alyssum, phlox, ageratum, candytuft, sweet william, wallflower, golden rod, thyme, and especially buddleia.

Questions

1. *a)* List some of the pollutants which are released into air and water from homes and farms.

b) In what ways could these pollutants endanger wildlife and humans?

2. *a)* What are the main sources of sulphur dioxide and oxides of nitrogen which are released into the atmosphere?

b) What damage can they cause when first released into the environment?

c) What damage can they cause if they are absorbed into rainclouds and fall as rain?

3. Over the last few years the amount of lead in petrol has been reduced. Why was this done? (Find out.)

4. Explain with examples what recycling is and why it is important.

5. List some of the ways in which governments, voluntary organizations, farmers, and you personally can help preserve wildlife.

6. Study the graphs on this page. They show some of the effects of discharging untreated sewage into a stream.

a) What causes the effect shown in graph A? In what ways will the effects shown in graph A harm animals?

b) What happens to the oxygen concentration as water flows downstream?

c) How will the effects shown in graph B harm submerged water plants?

d) What does graph C tell you about the effect of sewage disposal on water animals?

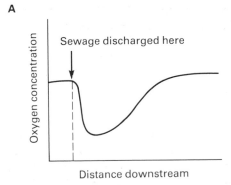

A

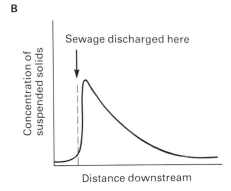

B

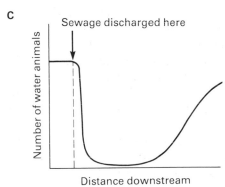

C

Summary and factual recall test

A substance is called a pollutant when it (1). Air pollution comes from the burning of (2) and (3). These produce smoke and fumes containing gases such as (4–name four). Dust in smoke settles on plants harming them in two ways (5). Sulphur dioxide damages plants by (6), and damages buildings by corroding (7). Photochemical fog is formed by the action of (8) on oxides of (9). Lead compounds in vehicle (10) gases can damage the (11) of (12) children.

When fumes from (13) and (14) enter clouds they form (15) and (16) acids which fall as acid rain. It is claimed that acid rain damages the (17) and (18) of trees, makes soil more (19) and releases poisonous (20) such as (21) and (22) into soils. These can be washed into lakes where they (23). Acid rain could be prevented by (24–list at least three methods).

Untreated sewage entering rivers is (25) by bacteria, which remove (26) from water making it uninhabitable for (27–name four organisms).

Industrial waste often contains long-lasting pollutants such as (28–name four). These, and poisons used in agricultural (29) sprays can disrupt food chains in the following way (30).

Radiation such as (31) can cause diseases such as (32). Noise from (33–three sources) can cause (34–two effects).

The three major reasons for wildlife destruction are (35), (36), and (37). Wildlife can be conserved when farmers create new habitats by (38), and reduce the use of (39) sprays. Gardens and parks can be made attractive to (40), and game reserves created where (41) is prohibited.

Index

Page numbers in **bold**, indicate major references in the text

Answers to factual recall tests

The following are provided as one set of possible answers for the chapter summaries. It is not suggested that these are the only correct answers; in many cases there are other equally appropriate words, phrases, or sentences,

Chapter 1

(1) fins, wings, legs (2) plants do not move about from place to place (3) leaves turning towards the light, roots growing into the soil (4) light, sound, touch, smell, taste (5) co-ordinate (6) light, gravity, water (7) energy (8) growth (9) repair (10) autotrophic (11) carbon dioxide (12) water (13) sunlight (14) photosynthesis (15) herbivores, carnivores, omnivores, parasites, saprotrophs (16) oxygen (17) energy (18) carbon dioxide (19) gills, lungs (20) excretion (21) egestion (22) other living organisms (23) spontaneous (24) living matter can arise spontaneously out of non-living matter (25) animal growth is limited, whereas plants can grow throughout their whole life span (26) all the chemical and physical processes of life (27) respiration (28) photosynthesis (29) balanced (30) speed up (31) catalyst (32) *Homo sapiens* (33) genus (34) species (35) they help prevent confusion when naming organisms (36) fertile

Chapter 2

(1) cell membrane (2) selectively (3) cytoplasm (4) enzymes, oil droplets, glycogen granules, starch grains (5) organelles (6) mitochondria, chloroplasts (7) chromosomes (8) deoxyribonucleic acid – DNA (9) the development of hereditary features (10) plant cells are enclosed in a cellulose cell wall; they contain vacuoles and chloroplasts (11) tissues (12) labour (13) muscles, nerves (14) organ (15) the heart (16) surface area (17) volume (18) all cells absorb food and oxygen, and remove waste products through their surface (19) number (20) size (21) growth occurs all over an animal's body until it reaches adult size; plants have restricted growth regions (22) chromosomes (23) hereditary (24) diffusion (25) water (26) semi-permeable (27) weak (28) strong (29) cellulose (30) turgor (31) wall (32) turgid (33) wall pressure prevents further entry of water molecules (34) seedling, herbaceous plants, leaves, flowers (35) lose (36) plasmolysis

Chapter 3

(1) sweet fruit, jam, treacle, bread, potato, rice (2) energy (3) seventeen (4) glycogen (5) muscles (6) liver (7) fat (8) butter, lard, suet, dripping, olive oil (9) thirty-eight (10) adipose (11) under the skin, around muscles, around the heart (12) skin (13) heat (14) it has double the energy yield, and is less bulky (15) meat, liver, kidney, eggs, fish, soya (16) growth and repair (17) they contain all the essential amino acids (18) energy (19) beri-beri, pernicious anaemia, scurvy, rickets (20) fifteen (21) eggs, milk, green vegetables, fruit (22) constipation, heart disease, diabetes, cancer of the bowel (23) food eaten should provide only enough energy for daily needs (24) the correct proportions of carbohydrate, proteins, and fats should be eaten (25) age, sex, occupation, body size, pregnancy (26) it contributes to heart disease, high blood pressure, diabetes, gall bladder disease, cancer of the bowel (27) they may give rise to allergies, and behaviour problems in children (28) unequal (29) deforestation and over grazing (30) harnessing biological processes to produce useful substances (31) oil, weak mineral solutions, maize starch

Chapter 4

(1) simpler (2) soluble (3) insoluble (4) absorbed (5) enzymes (6) starchy (7) sugars (8) glucose (9) fats (10) oils (11) fatty acids (12) glycerol (13) proteins (14) amino acids (15) inside (16) *Amoeba* (17) outside (18) intestine (19) mouth (20) anus (21) it breaks down food into easily swallowed pieces; it increases the surface area of food and so speeds digestion; it mixes food with saliva (22) starch (23) maltose (24) wave (25) muscular (26) peristalsis (27) sphincters (28) gastric (29) hydrochloric (30) protease (31) bile (32) droplets (33) emulsion (34) three (35) duodenum (36) villi (37) increase (38) absorption (39) water (40) salt (41) faeces

Chapter 5

(1) carbon dioxide (2) stomata (3) lower (4) water (5) roots (6) chlorophyll (7) light (8) carbon dioxide or $6CO_2$ (9) water or H_2O (10) light energy (11) chlorophyll (12) sugar or $C_6H_{12}O_6$ (13) oxygen or $6O_2$ (14) light (15) carbon dioxide (16) temperature (17) chloroplasts (18) palisade (19) mesophyll (20) xylem (21) water (22) phloem (23) sugar (24) growth (25) food (26) photosynthesis produces oxygen which is necessary

for the respiration of all living things (27) photosynthesis converts light energy into chemical energy, producing substances necessary for plant growth. These are used as food, directly or indirectly, by all heterotrophic organisms (28) oxygen (29) carbon dioxide

Chapter 6

(1) plasma (2) water (3) glucose, fatty acids, glycerol, amino acids, vitamins, minerals, albumin, fibrinogen, antibodies, hormones, urea, and carbon dioxide (4) capillary (5) tissue fluid (6) food (7) oxygen (8) waste (9) bi-concave (10) bone marrow (11) nucleus (12) haemoglobin (13) iron (14) oxygen (15) oxyhaemoglobin (16) irregular (17) lobed (18) granules (19) phagocytic (20) engulf disease-causing bacteria (21) large round (22) granules (23) antibodies (24) disease (25) clot (26) heart (27) arteries (28) thick (29) capillaries (30) permeable (31) veins (32) valves (33) heart (34) pulmonary circulation (35) systemic circulation (36) atrium (37) right (38) left (39) tricuspid (40) bicuspid (41) ventricles (42) right (43) pulmonary (44) left (45) aorta

Chapter 7

(1) vessels (2) water (3) minerals (4) leaves (5) phloem (6) glucose (7) leaves (8) storage areas (9) growing points (10) water (11) evaporation (12) spongy (13) stomata (14) under (15) guard (16) temperature, humidity, wind (17) brings water and minerals to leaves, essential for absorption of carbon dioxide, cools leaves (18) drought (19) xylem (20) transpiration (21) sugar (22) leaves (23) storage (24) growing

Chapter 8

(1) breakdown (2) energy (3) life (4) oxygen (5) carbon dioxide (6) water (7) energy (8) oxygen (9) energy (10) lactic acid (11) alcohol (12) yeast (13) bacteria (14) ethanol, citric acid, oxalic acid (15) the fermentation process of yeast produces carbon dioxide which forms bubbles as it escapes from the dough (16) alcohol (17) poisonous (18) lactic acid (19) contraction (20) debt.

Chapter 9

(1) oxygen (2) carbon dioxide (3) whole body surface (4) of the small body volume (5) respiratory (6) large area, moist, thin, in contact with capillaries, well-ventilated (7) diaphragm (8) flatter (9) intercostal muscles (10) backbone (11) sternum (12) cage (13) increase (14) thoracic (15) air pressure in the pleural cavity is always lower than in the lungs, and this pressure difference stretches the elastic lung tissue until it almost fills the thorax, as the volume of the thorax increases so does the volume of the lungs (16) pressure (17) germs (18) dust (19) mucus (20) goblet (21) cilia (22) mouth (23) swallowed (24) trachea (25) bronchi (26) bronchioles (27) alveoli (28) capillaries (29) gas (30) larger (31) as red cells squeeze through the capillaries they expose a large surface area to the capillary walls for oxygen absorption (32) friction slows down blood flow thus increasing the time available for oxygen absorption

Chapter 10

(1) maintenance (2) internal environment (3) feed (4) man (5) mammals (6) birds (7) blood (8) tissue (9) carbon dioxide, oxygen, temperature, food, urea, poisons, osmotic pressure (10) sugar (11) amino acids (12) protein (13) A, D, B_{12} (14) iron, copper, potassium (15) removing harmful substances produced by disease organisms, and those taken into the body such as certain drugs and alcohol (16) clotting (17) fibrinogen (18) heat (19) warms (20) bile (21) waste (22) metabolism (23) excess (24) urea (25) urine (26) urea contains nitrogen derived from the breakdown of excess amino acids in the body (27) glomeruli (28) Bowman's capsules (29) tubules (30) glucose, amino acids, vitamins, minerals (31) reabsorption (32) can maintain a constant body temperature (33) man, dog (34) poikilothermic (35) they have no temperature control mechanism (36) warm-blooded animals can be cooler than their surroundings, and cold-blooded animals can be warmer than their surroundings (37) sweat (38) evaporates (39) heat (40) superficial (41) vasodilation (42) over-heated (43) close (44) radiation (45) shivering (46) heat (47) erector (48) by stopping cold air from reaching the skin (49) by producing a layer of still air around the body which insulates it

Chapter 11

(1) growth movements (2) the movement occurs only at a plant's growing points (3) phototropism (4) A (5) A (6) B (7) C (8) green (9) the features of sets A and B were due to lack of light, and growth in light from only one direction respectively (10) positively (11) chlorophyll (12) geotropism (13) upwards (14) negative (15) downwards (16) positive (17) auxin (18) direction (19) tip (20) increases (21) elongation (22) towards

Chapter 12

(1) slugs, earthworms (2) water (3) hydrostatic (4) exoskeleton (5) lime (6) chitin (7) the periodic shedding of the cuticle and its replacement by a new, larger one (8) endoskeleton (9) bones (10) skull, backbone, rib-cage (11) shoulder blades, hips, limbs (12) to support soft tissues; to give shape to the body; to protect soft tissues; to produce blood cells; to form a system of rods and levers moved by muscles (13) calcium, phosphate (14) humerus, femur (15) red and white blood cells (16) synovial (17) elbow, knee, finger knuckles (18) they move in one plane only (19) shoulder, hip (20) they consist of a ball-shaped end of bone which fits into a hollow socket in another bone (21) ligaments (22) tendons (23) does not move when the muscle contracts (24) does move when the muscle contracts (25) antagonistic (26) flexors (27) extensors

Chapter 13

(1) the brain and spinal cord (2) fibres (3) insulated (4) eyes, ears, nose, taste buds, touch receptors (5) dendron (6) body (7) axon (8) central nervous system (9) cell (10) central nervous system (11) dendrites (12) axon (13) muscle or gland (14) threshold (15) all-or-none principle (16) the frequency at which they pass along a nerve (17) synapse (18)

behaviour in which a stimulus results in a response which is unlearned, and occurs quickly without conscious thought (19) eye blink, sweating, salivation, coughing (20) lowest (21) spinal cord (22) regulation of temperature, blood pressure, rate of heart-beat, and breathing (23) organs of balance, stretch receptors in muscles (24) tone (25) muscular (26) walking, cycling (27) grey (28) cerebral cortex (29) neurone cell bodies (30) conscious (31) vision, talking, emotion, memory (32) ducts (33) hormones (34) blood (35) brain (36) it controls other endocrine glands (37) to influence growth, cause vasoconstriction, and to contract the uterus during childbirth (38) neck (39) glucose (40) respiration (41) physical (42) mental (43) sudden, violent effort (44) escape from predators (45) stress, fear, anger (46) glucose (47) increasing the rate of conversion of glucose into glycogen in the liver (48) oestrogens (49) to control development of secondary sexual characteristics; prepare the uterus for pregnancy; and maintain the uterus during pregnancy (50) testes (51) testosterone (52) secondary sexual characteristics in males

Chapter 14

(1) light (2) touch (3) sound (4) taste (5) olfactory (6) stretch receptors and semi-circular canals (7) hard (8) soft (9) rough (10) smooth (11) temperature (12) compare temperatures (13) they give warning of damage to the body (14) taste-buds (15) sweet, sour, salt, bitter (16) olfactory organs (17) they give warning of poisons or other harmful substances (18) by the skull, the conjunctiva, tears, and the blink reflex (19) extrinsic (20) movement (21) hole (22) pupil (23) light (24) depth (25) radial muscles contract making the pupil larger, circular muscles make it smaller (26) circular (27) contract (28) releases (29) suspensory (30) more convex (31) increases (32) bend (33) radial (34) less convex (35) presbyopia (36) elasticity (37) near (38) hypermetropia (39) the distance between the lens and the retina is less than normal (40) converging (41) myopia (42) the eyeball is abnormally elongated (43) diverging (44) retina (45) rods and cones (46) cones are sensitive to colour and work in bright light, rods are insensitive to colour and work in dim light (47) forwards (48) owls, humans, apes (49) distances (50) orientation (51) movements towards food, movements away from danger (52) kinesis (53) response of woodlice to dry conditions (54) response to stimuli from one direction (55) movement of sperms to ova, and maggots away from light

Chapter 15

(1) asexual (2) mitosis (3) hereditary (4) copy (5) change (6) generation (7) sexual (8) gametes (9) meiosis (10) fertilization (11) zygot (12) it has inherited two different sets of hereditary information: one set from each parent (13) that there are separate male and female sexes (14) that one organism contains both male and female sex organs (15) external (16) internal (17) reproductive system (18) copulation (19) that fertilization is more certain because the eggs and sperms are in a confined space (20) binary (21) spores (22) the tip of a vertical hypha forms a sporangium which bursts, releasing spores (23) midges, groundsel, syca-

mores, houseflies (24) living (25) die (26) birds, mammals (27) parental (28) young have a good chance of survival

Chapter 16

(1) pedicel (2) receptacle (3) whorls (4) female (5) gynoecium (6) ovary (7) ovules (8) embryo sac (9) egg (10) gamete (11) ovule (12) ovary wall (13) male (14) androecium (15) filament (16) anther (17) pollen sacs (18) pollen (19) gametes (20) coralla (21) by their colour, scent, and the nectar produced by nectaries at their bases (22) pollinate (23) calyx (24) to protect the other floral parts during the bud stage of development (25) its petals are the same size and shape (26) radially (27) it can be cut in two along many vertical planes to produce identical halves (28) its petals are different shapes and sizes (29) bilaterally (30) it can be cut into two identical halves along only *one* vertical plane (31) transfer of pollen from anthers to stigmas in the same flower or between flowers on the same plant (32) transfer of pollen from one plant to another of the same species (33) cross (34) hereditary information of two different plants is intermixed (35) evolutionary (36) unisexual flowers, protandry, protogyny (37) petals (38) nectaries (39) abundant pollen, small light pollen grains, large anthers hanging outside flowers, 'feathery' stigmas, flowers above the leaves or opening before them (40) increase (41) pollen (42) stigmas (43) stinging nettle, hazel, willow (44) large, coloured and scented petals, often with honey guides; sticky or spiky pollen grains; anthers and stigmas situated where insects will brush against them (45) attract (46) they pollinate the flowers (47) white dead nettle, sweet pea, buttercup (48) tubular (49) pollen tube (50) ovule (51) micropyle (52) embryo (53) bursts open (54) male (55) female (56) radicle (57) plumule (58) seed (59) cotyledons (60) cotyledons (61) endosperm (62) testa (63) it avoids over crowding, and increases the chances of a species spreading into new territory (64) by wind, animals, and self-dispersal (65) the development of a seedling from a seed (66) water, air, warmth (67) dry (68) fresh (69) asexual (70) stem tubers, runners (71) cuttings and grafting (72) it produces exact copies of parent plants, and enables us to produce plants which cannot be grown from seeds

Chapter 17

(1) that the embryo develops inside the female's reproductive system (2) the developing embryos are kept warm, fed, protected, and the female is free to lead a normal life (3) milk (4) mammary (5) puberty (6) ovulation (7) month (8) Graafian (9) corups luteum (10) stimulate development of the uterus lining in preparation for a fertilized ovum (11) oviduct (12) uterus (13) thirty-six (14) dies (15) menstruation (16) fourteen (17) testes (18) epididymis (19) seminal vesicle (20) prostate (21) sperm ducts (22) penis (23) semen (24) stimulate swimming movements of sperm tails, and nourish the sperms (25) pregnancy (26) nine (27) embryo (28) uterus (29) it produces enzymes which digest the cells in the uterus wall (30) placenta (31) food (32) oxygen (33) blood (34) carbon dioxide (35) nitrogenous waste (36) it is suddenly cut off from its supply of food and oxygen (37) uterus (38) abdominal (39)

head (40) umbilical (41) bleeding (42) breech (43) Caesarian section (44) lactation (45) antibodies (46) it is pure, changes to meet the baby's needs, is easily digested (47) syphilis (48) gonorrhoea (49) German measles (50) placenta (51) deafness, blindness, and heart disease (52) follicle stimulating hormone – F.S.H. (53) the ovaries are sensitive to F.S.H. and produce several follicles

Chapter 18
(1) a group of organisms of the same kind living in a particular environment (2) food, space, and fertile mates are plentiful (3) food (4) predators (5) environmental (6) the maximum number of a species which it will support (7) sigmoid (8) flattened 'S' (9) disease, and poor nutrition (10) agricultural (11) food (12) water (13) disease (14) slowing down (15) fewer people are needed to produce food and other goods, so large families are not needed for survival (16) death (17) birth (18) rapid (19) population (birth) control (20) condom (21) penis (22) vagina (23) sperms (24) cervix (25) 6 to 8 hours (26) vasectomy (27) cutting, tying, or blocking the sperm tubes (28) oviducts

Chapter 19
(1) change (2) oldest (3) higher (4) acquired (5) young inherit characteristics acquired by parents in the course of their lives (6) natural selection (7) double (8) generation (9) constant (10) struggle (11) reproductive (12) survival (13) they help an organism to survive in the struggle for existence (14) inheritance (15) variation (16) natural selection (17) inheritance (18) features with survival value (19) species (20) they cause corresponding changes in the process of natural selection, and so different features of organisms may suddenly gain survival value (21) ice ages, mountain building, spread of deserts, continental drift (22) small (23) unchanged (24) intermediate

Chapter 20
(1) laws (2) hereditary (3) offspring (4) discontinuous (5) they were distinctly different and produced no intermediate forms when crossed (6) crossing organisms which differ in one or more characteristics (7) monohybrid (8) all tall (9) dominant (10) it had 'dominated' the dwarf characteristic (11) self-pollinated (12) tall (13) dwarf (14) 3:1 (15) recessive (16) it had 'receded' during the F_1 generation (17) characteristics (18) pairs (19) separate (20) one member of each pair (21) the factors come together again and the pairs are restored (22) chromosomes (23) meiosis (24) the factors are carried by the chromosomes (25) Both factors and chromosomes occur in pairs. Factors and chromosomes separate at meiosis and one from each pair goes into each gamete. Factor and chromosome pairs are restored at fertilization (26) locus (27) homologous (28) alleles (29) homozygous (30) heterozygous (31) genotype (32) its visible appearance (33) a change in a gene, or a chromosome, usually involving an alteration in a DNA molecule (34) heat, chemicals such as mustard gas, radiation (35) gametes (36) mutations are usually recessive (37) natural (38) harmful to the organism (39) unfavourable (40) favourable (41) evolutionary (42) proteins (43)

proteins are the building materials of life, and form enzymes and hormones which control the chemistry of living things (44) genotype (45) genes (46) DNA (47) insulin, hormones, vitamins, vaccines

Chapter 21
(1) food (2) living (3) host (4) pathogenic (5) micro-organisms (6) common cold virus, food poisoning bacteria, malarial protozoa, ringworm fungus (7) chicken pox, measles, polio, influenza (8) inside living organisms (9) DNA (10) virus factories (11) leprosy, cholera, VD, dental caries (12) reproduce by dividing in two (13) 20 to 30 (14) spores (15) boiling, disinfectant (16) malaria, amoebic dysentery (17) malaria (18) female *Anopheles* mosquitoes (19) removing breeding places, insecticide sprays (20) ringworm, athlete's foot (21) contact with infected people, spores (22) intestines (23) eaten (24) colds, pneumonia (25) droplets of moisture containing germs are breathed, coughed, or sneezed from infected people (26) water (27) contact with infected people, and insects (28) contact with infected people and objects (29) small pox, measles (30) vectors (31) houseflies, mosquitoes (32) cancer (33) heart disease, emphysema, bronchitis (34) cirrhosis of the liver, heart disease, cancer, stomach ulcers (35) twenty (36) thirteen (37) sebaceous (38) follicles (39) supple, water-proof, and partly germ-proof (40) mucus (41) cilia (42) oesophagus (43) swallowed and passed out of the body in the faeces (44) stomach acid (45) digestive enzymes (46) white (47) phagocytes (48) lymph nodes, liver, spleen, and bone marrow (49) opsonins make germs more likely to be eaten by phagocytes; lysins dissolve germs, agglutinins stick germs together, and antitoxins neutralize poisons formed by germs (50) a suspension of dead germs, inactivated germs, or harmless germs similar to pathogenic types (51) antibodies (52) immune (53) typhoid fever, poliomyelitis, cholera, bubonic plague (54) active (55) the body actively produces the antibodies (56) passive (57) the body gains immunity without doing any work

Chapter 22
(1) decay (2) saprotrophic (3) fungi (4) humus (5) the lignin and cellulose of plants, and the hard parts of animals (6) liquid (7) crumbs (8) nitrates, phosphates, ammonium salts (9) plant (10) they provide air spaces which increase drainage and oxygen levels in the soil; they help retain water and soil minerals; and help prevent wind erosion (11) clay, silt, sand, gravel (12) physical (13) chemical (14) water (15) expands (16) roots (17) being pounded against each other while rolling down a hill; being washed along a river bottom (18) carbon dioxide (19) carbonic (20) limestone (21) soluble (22) wash, or drain (23) soil (24) carbon dioxide produced by respiring soil organisms (25) organic (26) decomposing (dissolving) underlying rocks (27) most soil organisms and plant roots need oxygen for respiration (28) size of its particles; the larger the particles the greater the amount of air (29) minerals (30) adsorption (31) capillarity (32) stronger (33) clay (34) water (35) the maximum amount of water which it can retain (36) causing organisms to decay and producing humus and minerals (37) aerate (38) passing it

through their bodies and ejecting it at the soil surface as a worm cast (39) dragging leaves into their burrows where they are left to decay (40) on top of the rocks from which it was formed (41) alluvial (42) rivers (43) glaciers (44) valley bottoms and plains (45) topsoil is situated above subsoil, it is black, and contains more humus and organisms (46) sand (47) clay (48) humus (49) lime (50) crumbs (51) it is well aerated; drains freely yet retains plenty of moisture and dissolved minerals; it warms quickly in spring; it is easily cultivated (52) large (53) humus, clay (54) good drainage and aeration; germinating roots penetrate it easily; and it is easily cultivated (55) it loses water and minerals quickly (56) small (57) it is able to draw water from below by capillarity, and it retains its minerals (58) its heavy consistency; lack of air; tendency to dry out into hard lumps (59) their mineral supplies are continually replaced by animal droppings and decaying organisms (60) minerals (61) they receive no droppings and contain few decaying organisms (62) ploughing to expose lumps to frost; adding sand, humus, and lime (63) adding humus (peat and manure) (64) manure, dried blood, bone-meal (65) slowly (66) they must first decay (67) quickly (68) sulphate of ammonia, super-phosphate of lime

Chapter 23

(1) protein (2) atmosphere (3) nitrates (4) soil (5) fixing (6) atmospheric (7) carbohydrates (8) nitrates (9) bacteria and fungi (10) ammonia (11) nitrites (12) nitrates (13) nitrifying (14) they are involved in the continuous circulation of nitrogen between and within the physical and biological worlds (15) carbon dioxide (16) by respiration of animals and plants, and by decay bacteria, and by combustion, all of which release carbon dioxide gas (17) carbon (18) it involves the circulation of carbon between the atmosphere and living organisms (19) heat (20) warms (21) clouds (22) water (23) rain (24) water (25) green plants (26) primary (27) animals (28) plants (29) secondary (30) animals (31) trophic (32) earthworms, woodlice, millipedes (33) saprophytic bacteria, and fungi (34) humus (35)

minerals (36) healthy plant growth (37) four (38) energy (39) smaller (40) numbers (41) biomass (42) web (43) each food is eaten by more than one consumer

Chapter 24

(1) community (2) habitat (3) forest, sea shore, pond (4) population (5) producers, consumers and their parasites, scavengers, decomposers, their physical environment, and the circulation between the two of carbon, oxygen, water, etc. (6) biotic (7) abiotic (8) niche (9) lungs, gills, skeletons, digestive systems (10) niche (11) sucking mouth parts of insects, desert adaptations of plants (12) ecological succession (13) climax community (14) rainfall, humidity, temperature, wind, light (15) landscape features such as lakes, rivers, valleys, mountains (16) soil (17) organisms

Chapter 25

(1) harms living things (2) coal (3) oil (4) sulphur dioxide, carbon dioxide, carbon monoxide, oxides of nitrogen (5) cutting off light for photosynthesis, and blocking stomata (6) damaging leaf cuticle, and preventing chlorophyll formation (7) metal and stone (8) sunlight (9) nitrogen (10) exhaust (11) brains (12) young (13) power stations (14) vehicle exhausts (15) sulphuric (16) nitric (17) leaves (18) roots (19) acid (20) minerals (21) aluminium (22) mercury (23) kills fish and other aquatic life (24) crushing and washing coal, fitting cleaning systems to chimneys, burning fuel more efficiently, removing pollutants from engine exhausts, reducing speed limits (25) decomposed (26) oxygen (27) fish, crustaceans, aquatic insects, and worms (28) cyanide, lead, copper, mercury, zinc (29) crop (30) chemicals enter organisms such as fish, near the beginning of food chains, and spreads through other organisms often reaching humans (31) X-rays (32) leukaemia (33) radios, motor cycles, machinery (34) deafness, mental depression (35) destruction of habitats for our own use (36) pollution (37) killing for profit and fun (38) planting trees, making ponds where cultivation is difficult (39) pesticide (40) wildlife (41) hunting